The Physics of Rock Failure and Earthquakes

Physical modeling of earthquake generation processes is essential to further our understanding of seismic hazard. However, the scale-dependent nature of earthquake rupture processes is further complicated by the heterogeneous nature of the crust. Despite significant advances in the understanding of earthquake generation processes, and the derivation of underlying physical laws, controversy remains regarding what the constitutive law for earthquake ruptures ought to be, and how it should be formulated. It is extremely difficult to obtain field data to define physical properties along a fault during a rupture event, at sufficiently high spatial and temporal resolution to resolve the controversy. Instead, laboratory experiments offer a means of obtaining high-resolution measurements that allow the physical nature of shear rupture processes to be deduced.

This important new book is written using consistent notation, providing a deeper understanding of earthquake processes from nucleation to their dynamic propagation. Its key focus is a deductive approach based on laboratory-derived physical laws and formulae, such as a unifying constitutive law, a constitutive scaling law, and a physical model of shear rupture nucleation. Topics covered include: the fundamentals of rock failure physics, earthquake generation processes, physical scale-dependence, and large-earthquake generation cycles and their seismic activity.

Providing cutting-edge information on earthquake physics, this book is designed for researchers and professional practitioners in earthquake seismology and rock failure physics, and also in adjacent fields such as geology and earthquake engineering. It is also a valuable reference for graduate students in earthquake physics, rock physics, and earthquake seismology.

Mitiyasu Ohnaka has been a Professor Emeritus at the Earthquake Research Institute, the University of Tokyo, since his retirement in 2001. Previously, he worked at the ERI in the fields of rock physics, experimental seismology, and the physics of earthquakes, from 1970 onwards, as well as holding various positions such as Honorary Professor at University College London, and invited lecturer or visiting scholar at many worldwide institutions, including the Kavli Institute for Theoretical Physics at UC Santa Barbara. Professor Ohnaka has also worked widely in Japan, supervising researchers and students, and delivering undergraduate and post-graduate lectures, at institutions from the University of Tokyo to Yamagata University and more. He is the co-author of *The Physics of Earthquake Generation, Earthquakes and Faults*, and *The Role of Water in Earthquake Generation* (these three in Japanese), and *Theory of Earthquake Premonitory and Fracture Processes* (Polish Scientific Publishers, 1995). Professor Ohnaka was Executive Committee member of the International Association of Seismology and Physics of the Earth's Interior (IASPEI) from 1991 to 1995, and also Chair of the Sub-Commission on Modeling the Earthquake Source from 1991 to 2001 in IASPEI. He is a member of the Seismological Society of Japan, and the American Geophysical Union.

The Physics of Rock Failure and Earthquakes

MITIYASU OHNAKA

The University of Tokyo, Professor Emeritus

CAMBRIDGE
UNIVERSITY PRESS

CAMBRIDGE
UNIVERSITY PRESS

University Printing House, Cambridge CB2 8BS, United Kingdom

One Liberty Plaza, 20th Floor, New York, NY 10006, USA

477 Williamstown Road, Port Melbourne, VIC 3207, Australia

4843/24, 2nd Floor, Ansari Road, Daryaganj, Delhi - 110002, India

79 Anson Road, #06-04/06, Singapore 079906

Cambridge University Press is part of the University of Cambridge.

It furthers the University's mission by disseminating knowledge in the pursuit of education, learning and research at the highest international levels of excellence.

www.cambridge.org
Information on this title: www.cambridge.org/9781108445719

First published 2013
First paperback edition 2017

A catalogue record for this publication is available from the British Library

Library of Congress Cataloging in Publication data
Onaka, Michiyasu, 1940–
The physics of rock failure and earthquakes / Mitiyasu Ohnaka.
pages   cm
Includes bibliographical references and index.
ISBN 978-1-107-03006-0
1. Seismology.   2. Rock mechanics.   3. Earthquakes.   I. Title.
QE534.3O53   2013
551.22028′7 – dc23   2012035059

ISBN 978-1-107-03006-0 Hardback
ISBN 978-1-108-44571-9 Paperback

Contents

Preface

Over the past four decades, great progress has been made in scientifically understanding earthquake source processes; in particular, advances in the field of earthquake physics have contributed substantially to a profound understanding of earthquake generation processes in terms of the underlying physics. Yet, a fundamental problem has remained unresolved in this field. The constitutive law governing the behavior of earthquake ruptures provides the basis of earthquake physics, and the governing law plays a fundamental role in accounting quantitatively for the entire process of a scale-dependent earthquake rupture, from its nucleation to its dynamic propagation to its arrest, in a unified and consistent manner. Therefore, it is critically important to strictly formulate the constitutive law for earthquake ruptures, based on positive facts, from a comprehensive viewpoint.

Over the past two decades, however, there has been controversy regarding what the constitutive law for earthquake ruptures ought to be, and how it should be formulated. For the physics of earthquakes to be a quantitative science in the true sense, it is essential to resolve this controversy. Regrettably, the resolution of seismological data observed in the field is not high enough to end the controversy. In order to resolve this controversy, therefore, it is critically important to formulate the constitutive law based on positive facts elucidated by high-resolution laboratory experiments properly devised for the purpose intended, by correctly recognizing the real situation of seismogenic fault properties. Without a rational formulation of the law governing real earthquake ruptures, the physics of earthquakes cannot be a quantitative science in the true sense. Hence, there is an urgent need to rationally formulate the constitutive law for earthquake ruptures, based on positive facts, from a comprehensive viewpoint. Resolution of this controversy is a necessary step towards a more complete, unified theory of earthquake physics.

As described in Chapter 1, individual faults embedded in the Earth's crust are inherently inhomogeneous. Fault inhomogeneity has profound implications for the strict formulation of the constitutive law for earthquake ruptures. The process of an earthquake rupture at shallow crustal depths is not a simple process of frictional slip failure on a uniformly precut weak fault, but a more complex process, including the fracture of initially intact rock at some local strong areas on an inhomogeneous fault (see Chapter 1). Accordingly, the constitutive law for real earthquake ruptures must be formulated as a unifying law that governs not only frictional slip failure at precut-interface (or frictional contact) areas on faults but also shear fracture at some local strong areas on the faults. This requirement must be met when we strictly formulate the constitutive law for real earthquake ruptures (Chapters 3 and 4).

In addition, rupture phenomena, including earthquakes, are inherently scale-dependent; indeed, some of the physical quantities inherent in shear rupture exhibit scale-dependence (see Chapters 5 and 6). To quantitatively account, in a unified and consistent manner, for scale-dependent physical quantities inherent in the rupture over a broad scale range, it is critically important to formulate the governing law so as to incorporate the scaling property inherent in the rupture breakdown. This is another requirement that must be met for the constitutive formulation for scale-dependent earthquake ruptures (Chapters 3 and 4).

Accordingly, the properties of fault heterogeneity and of physical scaling are the keys to rational formulation of the constitutive law for earthquake ruptures. In light of these properties, it is possible to rationally formulate the governing law (or constitutive law) for earthquake ruptures, based on the basic research on the physics of rock fracture and friction.

The primary reason I had for writing this book was that there are no published books on earthquake physics written deductively in a consistent manner based on the basic research on the physics of rock fracture and friction, achieved by high-resolution laboratory experiments. The time is ripe to write such a book, because underlying physical laws, such as a unifying constitutive law and a constitutive scaling law, and a physical model of shear rupture nucleation have been derived from high-resolution laboratory experiments properly devised for the purpose intended.

Chapter 1 of this book is an introductory chapter mostly devoted to the description of seismogenic fault inhomogeneities. Fundamental items of rock fracture/friction mechanics are described in Chapter 2. The central theme of this book is described in the remaining chapters. A key characteristic of this book is that the constitutive law for shear rupture, including earthquake ruptures, is formulated as a unifying law which governs not only frictional slip failure on a precut rock interface but also the shear fracture of intact rock, and into which the scaling property inherent in shear-rupture breakdown is incorporated. This is the common thread that runs through the entire book. In terms of a single constitutive law, the process of a shear rupture generation – from its stable, quasi-static nucleation to its unstable, dynamic propagation – is accounted for quantitatively in a unified and consistent manner, and scale-dependent physical quantities inherent in the rupture over a broad range from laboratory-scale to field-scale are treated consistently and quantitatively in a unified manner.

High-resolution laboratory experiments on shear rupture on an inhomogeneous fault are best suited for fully elucidating the physical nature of a scale-dependent shear rupture generation process from its nucleation to the subsequent dynamic rupture. Based on these experiments, therefore, the shear rupture nucleation process is physically modeled (Section 5.1), and observed data on seismic nucleation are consistently accounted for in quantitative terms based on the physical model (Section 5.2). In addition, strong motion source parameters such as peak slip velocity and peak slip acceleration are theoretically derived from the laboratory-derived constitutive equation, and discussed in quantitative terms (Section 5.3). Chapter 6 focuses on the root cause of scale-dependent physical quantities, and it is shown that the scale-dependence of scale-dependent physical quantities, such as slip acceleration, nucleation zone size and the duration time of nucleation, is attributed to the scale-dependent breakdown displacement or the characteristic length representing the geometric irregularity of rupturing surfaces. In Chapter 7, the final chapter, it is shown that large-earthquake generation cycles and accompanying seismic activity can be accounted for consistently under the premise that the governing law for earthquake ruptures is a slip-dependent constitutive law, and that the seismogenic layer and individual faults therein are heterogeneous. In addition, the final section of this chapter focuses on the predictability of large earthquakes.

Since 1985, I have had many opportunities to present my research findings at international meetings, and to discuss outstanding issues regarding earthquake phenomena with leading researchers from around the world. In these international meetings, I also had opportunities to personally get to know leading senior colleagues in the field of earthquake seismology. From the early 2000s, some of those who expressed a positive interest in my leading-edge research findings and deductive approach to addressing outstanding issues regarding earthquake phenomena encouraged me to write a book about the physics of earthquakes on the basis of laboratory-derived physical laws or formulae and physical model of shear rupture nucleation. Since I myself had intended to write such a book, their encouragement gave me the inspiration and keen desire to do so. Hence, I am grateful to international meetings' organizers/conveners for inviting me to take part in their meetings, and to scientific participants for fruitful or critical discussions. In particular, I wish to thank the following: the late Leon Knopoff, the late Keiiti Aki, Massimo Cocco, Shamita Das, James H. Dieterich, Raul Madariaga, Mitsuhiro Matsu'ura, Takeshi Mikumo, Peter Mora, James R. Rice, Christopher H. Scholz, and Xiang-chu Yin.

In the course of my research I worked with my research associates and graduate students, whom I thank for their collaboration. I am grateful to Takeshi Mikumo and Piotr Senatorski for their comments on part of the original manuscript. I thank Philip Meredith for suggesting Cambridge University Press (CUP) as the publisher of this work, and anonymous reviewers for critical reading of the book proposal submitted to CUP and helpful suggestions. I also thank Barbara Wiggin and Mary Rose Reade for their help in enhancing the manuscript's readability. Last but not least, I would like to thank the editors at CUP, Susan Francis, Laura Clark, Tom O'Reilly, and other staff for their help in the production of this book.

Mitiyasu Ohnaka
The University of Tokyo

1   Introduction

Fracture (or failure) phenomena are observed over a very broad range of size scales, from atomistic-scale to microscopic-scale to macroscopic-scale fractures. A shear failure (or rupture) of laboratory-scale, whether the shear failure of intact rock or frictional slip failure on a precut rock interface, would be of the order of 10^{-3} to 1 m. In contrast, shear rupture phenomena occurring in the Earth's interior, including microearthquakes and huge earthquakes, encompass a much broader range of size scales from 10^{-1} to 10^{6} m. Rupture phenomena over such a broad scale range covering both laboratory-scale and field-scale are encompassed by continuum mechanics. This book deals with shear failures (or ruptures) of a scale range of laboratory-scale and field-scale within the framework of continuum mechanics.

It has been established that the source of shallow focus earthquakes at crustal depths is shear rupture instability along a fault embedded in the Earth's crust, which is composed of rock. At the same time, laboratory experiments have demonstrated that a rock specimen fails (or ruptures) in shear mode under combined compressive stress environments, and that the shear failure (or rupture) of rock is governed by constitutive law. These facts physically mean that earthquake rupture processes are governed by the constitutive law. This enables a deeper understanding of the process of earthquake generation in terms of the underlying physics, if the constitutive law for earthquake ruptures is properly formulated by taking into account the real situation of seismogenic fault zone properties such as fault heterogeneities.

In light of this, we have to correctly recognize the real situation of seismogenic environments and fault zone properties, in order to strictly formulate the constitutive law for earthquake ruptures, and to quantitatively account, in a unified and consistent manner, for the entire process of scale-dependent earthquake rupture in terms of the underlying physics. Since correct recognition of seismogenic fault zone properties is an indispensable prerequisite for strict formulation of the constitutive law for earthquake ruptures, this introductory chapter is mainly devoted to a description of seismogenic fault zone properties.

Since an unstable, dynamic rupture can occur only in the brittle through semi-brittle layer of the Earth's interior, the occurrence of earthquake ruptures is generally restricted to shallow crustal depths, called the *seismogenic layer*, except in plate-boundary subduction zones where deeper earthquakes can occur. The brittle through semi-brittle layer away from plate-boundary subduction zones is usually limited to crustal depths shallower than 20 km, and temperatures and lithostatic pressures in the layer range from the Earth's surface temperature to roughly 500–600 °C, and from atmospheric pressure to about 500 MPa, respectively. The seismogenic layer and individual faults embedded therein are inherently heterogeneous; in other words, heterogeneity is a key property of the seismogenic layer and preexisting faults therein. This fact cannot be denied, even if a part of the preexisting fault zone in a superficial layer of the Earth's crust is narrow and remarkably planar, as some geologists have noted. It is not possible to look at the whole structure of a fault zone from an isolated part of the fault zone viewed in the superficial layer. A view from an isolated part cannot be universalized.

Indeed, seismic observations have demonstrated that a major or great earthquake at shallow crustal depths never occurs alone, but is accompanied by aftershocks, and often preceded by seismic activity (small to moderate earthquakes) enhanced in a relatively wide region surrounding the source fault during the process leading up to the event (see Chapter 7). This is a reflection of the fact that the seismogenic layer and individual faults therein are heterogeneous. If the seismogenic layer and individual faults were homogeneous, then not only foreshocks but also aftershocks could never occur. Specifically, the occurrence of a mainshock rupture on a fault is accompanied by a rapid stress drop and slip displacement on the finite fault, and an adequate amount of the elastic strain energy accumulated in the medium surrounding the fault is dissipated during the rupture. Consequently, the mainshock rupture arrests within a finite region (the fault area), and the rupture arrest results in the re-distribution of local stresses on and around the fault area, leading to aftershock activities in and around the source region. Note that the occurrence of such a sequential process in a finite region is due to heterogeneities of the seismogenic layer and preexisting faults therein.

Seismological observations and analyses (e.g., Kanamori and Stewart, 1976; Aki, 1979, 1984; Beroza and Mikumo, 1996; Bouchon, 1997; Zhang *et al.*, 2003; Yamanaka and Kikuchi, 2004) have revealed that individual faults embedded in the seismogenic layer are heterogeneous, and contain what are called "asperities" (e.g., Lay *et al.*, 1982) or "barriers" (e.g., Aki, 1979) in the field of earthquake seismology. The presence of "asperities" or "barriers" on a fault is a clear manifestation of the fact that a real fault contains strong local areas highly resistant to rupture growth, with the rest of the fault having low (or little) resistance to rupture growth. These seismological observations are consistent with geological observations of structural heterogeneity and geometric irregularity for real faults, as described below.

In general, real faults embedded in the seismogenic crust are nonplanar, being segmented and bifurcated (e.g., Sibson *et al.*, 1986; Wesnousky, 1988). In addition, individual surfaces of fault segments exhibit geometric irregularity with band-limited self-similarity (e.g., Aviles *et al.*, 1987; Okubo and Aki, 1987), and there are gouge layers in between fault surfaces for mature faults (Sibson, 1977). These geometric irregularities and structural heterogeneities for real faults (or fault zones) play a prominent role in causing heterogeneous distributions of not only stresses acting on individual faults but also fault strength and resistance to rupture growth, because these physical quantities are structure-sensitive. The resistance to rupture growth has a specific physical meaning in the framework of fracture mechanics. For the definition of the resistance to rupture growth, see Section 2.2.

Let us consider a specific case where a fault consists of a number of discrete segments which form an echelon array with individual segments nearly parallel to the general trend of the fault (Segall and Pollard, 1980). The stepover zones in such an echelon array possibly have the highest strength, equal to the fracture strength of intact rock, and the resistance to rupture growth may be high enough to impede or arrest the propagation of ruptures at these sites. In addition, nonplanar fault segment surfaces exhibiting geometric irregularity with band-limited self-similarity contain various wavelength components. When such geometrically irregular fault segment surfaces, pressed under a compressive normal stress, are sheared in the brittle regime, the prime cause of frictional resistance is the shearing strength of interlocking asperities (Byerlee, 1967). In this case, the local strength at the sites of interlocking asperities is strong enough to equal the shear fracture strength of initially intact rock in the brittle regime.

Thus, the sites of the zones of segment stepover and/or interlocking asperities are potential candidates for strong areas of high resistance to rupture growth. These sites may act as "barriers" or "asperities." When such strong areas act as "barriers," stress will build up and elastic strain energy will accumulate in the elastic medium surrounding these sites until they break. When a large rupture breaks through these sites and links them together, they will be regarded as "asperities."

Local stress drops at "asperities" on real seismic faults obtain values as high as 50 to 100 MPa (e.g., Papageorgiou and Aki, 1983a and 1983b; Ellsworth and Beroza, 1995; Bouchon, 1997), which is high enough to equal the breakdown stress drop of intact rock tested under seismogenic environmental conditions simulated in the laboratory (see Section 3.4). This strongly suggests that the earthquake rupture process at shallow crustal depths is not a simple process of frictional slip failure on a uniformly precut weak fault, but a more complex process including the fracture of initially intact rock at some local areas on an inhomogeneous fault.

Such local strong areas on a fault, called "asperities" or "barriers," are required for an adequate amount of elastic strain energy to accumulate in the elastic medium surrounding the fault, owing to tectonic loading. Since the elastic strain energy accumulated provides the driving force to bring about an earthquake or to radiate seismic waves, local strong areas highly resistant to rupture growth on a fault play a much more important role in generating a large earthquake or in radiating strong motion seismic waves than does the rest of the fault having low (or little) resistance to rupture growth. Thus, there is no doubt that real faults in the Earth's crust are inherently heterogeneous.

In light of this, the constitutive law for earthquake ruptures must be formulated as a unifying law that governs not only frictional slip failure at precut weak interface (or frictional contact) areas on faults but also the shear fracture of intact rock at some local strong areas on the faults (Chapters 3 and 4).

In addition, rupture phenomena, including earthquakes, are inherently scale-dependent. Indeed, some of the physical quantities inherent in shear rupture exhibit scale-dependence (Chapters 5 and 6). It is therefore essential to formulate the governing (or constitutive) law in such a way that the scaling property inherent in the rupture breakdown is incorporated into the law; otherwise, scale-dependent physical quantities inherent in the rupture over a broad scale range cannot be treated consistently and quantitatively in a unified manner in terms of a single constitutive law.

As described above, the constitutive law governing the behavior of earthquake ruptures provides the basis of earthquake physics, and the governing law plays a fundamental role in accounting quantitatively, in a unified and consistent manner, for the entire process of a scale-dependent earthquake rupture, from its nucleation to its dynamic propagation to its arrest, on a heterogeneous fault. Therefore, it is critically important to strictly formulate the constitutive law for real earthquake ruptures, based on positive facts, from comprehensive viewpoints. Without a rational formulation of the law governing real earthquake ruptures, the physics of earthquakes cannot be a quantitative science in the true sense.

However, the resolution of seismological data observed in the field is not high enough not only to strictly formulate the constitutive law, but also to fully elucidate the physical nature of a scale-dependent earthquake rupture generation process from its nucleation to the subsequent dynamic propagation on a heterogeneous fault. This is because it is not possible to preliminarily deploy a series of high-resolution instruments for measuring local shear stresses (or strains) and local slip displacements along the fault on which a pending earthquake is expected to occur at a crustal depth.

In contrast, in laboratory experiments, the experimental method can be properly devised for the purpose intended, and high temporal and spatial resolution measurements can be made at a series of locations along the preexisting fault on which a shear rupture occurs. Therefore, high-resolution laboratory experiments properly devised for the purpose intended enable us not only to strictly formulate the constitutive law for shear rupture but also to fully elucidate the physical nature of a scale-dependent shear rupture process from its nucleation to the subsequent dynamic rupture on an inhomogeneous fault. In particular, in order to strictly formulate the constitutive law as a unifying law that governs not only frictional slip failure at precut interface areas on a fault but also the shear fracture of intact rock at some local strong areas on the fault, we need detailed data obtained in high-resolution laboratory experiments on the shear failure of intact rock and frictional slip rupture on a precut rock interface. I have devoted myself to conducting such leading-edge research through high-resolution laboratory experiments properly devised for the purpose intended, and contributed to the derivation of underlying physical laws, such as a unifying constitutive law and a constitutive scaling law, and to the elucidation of the physical nature of the scale-dependent shear rupture generation process, to achieve a deeper understanding of the physical process from earthquake nucleation to its dynamic propagation in terms of the underlying physical laws.

As mentioned above, real faults embedded in the Earth's crust are inherently inhomogeneous, and the earthquake rupture process at shallow crustal depths is not a simple process of frictional slip failure on a uniformly precut weak fault, but a more complex process, including the fracture of initially intact rock at some local strong areas on an inhomogeneous fault. In this book, therefore, the constitutive law is formulated as a unifying law which governs not only frictional slip failure at precut interface areas on a fault but also shear fracture at some local strong areas on the fault, and into which the scaling property inherent in shear-rupture breakdown is incorporated, based on positive facts elucidated in high-resolution laboratory experiments properly devised for the purpose intended (Chapters 3 and 4). This enables one to quantitatively account for the physical behavior of scale-dependent shear ruptures, including earthquake rupture nucleation and strong motion during the dynamic rupture propagation, in a unified and consistent manner, in terms of a single constitutive law (Chapters 5 and 6), and consequently to enhance the physics of earthquakes to a more complete, quantitative science in the true sense. This deductive approach based on the results of high-resolution laboratory experiments is the prominent feature of this book, and sets this book apart from others in the fields of rock failure physics and earthquake physics.

Fundamentals of rock failure physics

2.1 Mechanical properties and constitutive relations

2.1.1 Elastic deformation

When an external stress is applied to a real material, the material necessarily deforms, and this may eventually lead to fracture; in other words, fracture is commonly preceded by deformation. If there is a unique relation between the applied stress and the resultant deformation (or strain), the material is referred to as *perfectly elastic*. Perfect elasticity implies that there is no energy loss during the loading and unloading processes, which means that the energy stored in the specimen during the loading process is completely released during the unloading process. In other words, the elastic deformation is instantaneous and is completely recoverable when the applied stress is removed. When there is a linear relation between the stress applied to a material and the resultant deformation (or strain), the material is called *linearly elastic*, which is usually valid for small deformation.

Geomechanical phenomena in geological and tectonic settings in the Earth are described quantitatively in terms of basic equations, which include mechanical constitutive laws governing deformation or rupture. The relation between an externally applied force and the mechanical response of a material to the applied force is called a (*mechanical*) *constitutive relation*.

A typical, well-known example of such constitutive relations is Hooke's law; this law holds for a material having the property of the aforementioned linear elasticity. If it is assumed that the matter of a linearly elastic body is homogeneous and isotropic, Hooke's law is simply written as

$$\sigma_{ij} = \lambda \delta_{ij} \varepsilon_{kk} + 2\mu \varepsilon_{ij}, \tag{2.1}$$

where σ_{ij} is the stress tensor, ε_{ij} is the strain tensor, δ_{ij} is the unit tensor called Kronecker's delta ($\delta_{ij} = 0$ for $i \neq j$, and $\delta_{ij} = 1$ for $i = j$), and λ and μ are Lamé's constants. Lamé's constants prescribe the elastic property of a material.

There are two independent elastic constants for a homogeneous and isotropic body with the property of linear elasticity. In Eq. (2.1), Lamé's constants have been used as the two independent constants. However, there are other elastic constants defined as those best suited to specific uses. The other elastic constants are Young's modulus E, Poisson's ratio ν, the modulus of rigidity G, and the bulk modulus of elasticity K. These elastic constants are directly related to Lame's constants as follows:

$$E = \frac{\mu(3\lambda + 2\mu)}{\lambda + \mu}, \tag{2.2}$$

$$v = \frac{\lambda}{2(\lambda + \mu)}, \tag{2.3}$$

$$G = \mu, \tag{2.4}$$

$$K = \lambda + \frac{2}{3}\mu. \tag{2.5}$$

The elastic constants represent the elastic property of a material, and the material property depends on ambient conditions. Accordingly, the elastic constants also depend on ambient conditions.

2.1.2 Ductile deformation

As the stress (or load) applied to a real material increases, the material may fail (or fracture) within the elastic limit; otherwise, it continues to deform permanently beyond the elastic limit, and eventually it will fail (or fracture) due to the breakdown of inter-atomic bonds. The deformation beyond the elastic limit is not recoverable when the applied stress is removed, and the residual deformation at zero stress is referred to as *permanent deformation*.

The term *ductile deformation* is used when permanent deformation takes place without losing its ability to resist the applied load. The transition from elastic to ductile deformation occurs at the yield stress point. A small amount of ductile deformation is widely observed preceding the shear failure of rock tested under confining pressures, even in the brittle regime. This is because rocks are inherently inhomogeneous. Rocks are made up of various types of mineral grains, and contain a variety of mechanical flaws and stress concentration sources, such as microcracks, pores, and grain boundaries. In compressive loading of one such rock, elastic deformation is usually limited to the first 40–50% of the peak strength. Above the 40–50% level of the peak strength, the number of microcracks increases progressively as the rock is loaded. At higher loads, time-dependent crack growth and the interaction and coalescence of neighboring cracks become progressively more important, forming a thin, planer zone of higher crack density, which eventually results in the macroscopic fracture surfaces in the post-peak region, where the shear strength degrades with ongoing slip (slip-weakening). Thus, when rock in the brittle regime fails by shear mode under confining pressure, non-elastic deformation necessarily precedes the macroscopic shear fracture. The mechanism of this non-elastic (or ductile) deformation in the brittle regime is called *cataclasis*, which involves micro-fracturing and friction acting on microcrack surfaces and fragment particle interfaces developed in rock during loading.

In addition, individual crystalline solids of mineral grains in rocks contain various types of crystallographic defects such as point defects, dislocations, and twin boundaries in their crystal lattices. These crystallographic defects cause plastic deformation of the minerals

Table 2.1 Constitutive law parameters for high-temperature plastic creep flow

Rock name	A (MPa^{-n}s^{-1})	n	Q (kJ mol^{-1})	References
Heavitree quartzite (vacuum-heated, dry samples)	4×10^{-6}	4.0	300	Kronenberg and Tullis (1984)
Heavitree quartzite (α-quartz, 0.4% water added samples)	2.2×10^{-6}	2.7	120	Kronenberg and Tullis (1984)
Heavitree quartzite (α-quartz, dry samples)	3.2×10^{-5}	1.9	123	Hansen and Carter (1982)
Simpson quartzite (α-quartz, 0.4% water added samples)	2.0×10^{-2}	1.8	167	Hansen and Carter (1982)
Mt. Burnett dunite (dry samples)	1.3×10^{3}	3.3	465	Carter and Ave'Lallemant (1970)
Mt. Burnett dunite (wet samples)	0.1	2.1	226	Carter and Ave'Lallemant (1970)
Anita Bay dunite (dry samples)	3.2×10^{4}	3.6	535	Chopra and Paterson (1981)
Anita Bay dunite (wet samples)	1×10^{4}	3.4	444	Chopra and Paterson (1981)

contained in rocks at high temperatures. Dissolution creep (pressure solution) also occurs under stress (or pressure) in the presence of water. Accordingly, the mechanisms of ductile deformation (or flow) of rock are the cataclasis, crystal plasticity involving dislocation creep and diffusion creep, and dissolution creep.

Rocks exhibiting brittle behavior at room temperature deform plastically at high temperatures under confining pressures. Another typical example of mechanical constitutive relations may be the constitutive law for plastic flow, in particular, steady-state creep flow at high temperatures. In the case of plastic flow, the physical quantity that is uniquely related to stress is not strain but strain rate. It has been established by laboratory experiments that the steady-state creep flow of rock at a high temperature under a confining pressure obeys the following law (e.g., Carter, 1976; Kirby, 1983):

$$\dot{\varepsilon} = A f(\sigma_{\text{diff}}) \exp\left(-\frac{Q}{RT}\right), \tag{2.6}$$

where $\dot{\varepsilon}$ is the strain rate, $f(\sigma_{\text{diff}})$ is a function of the differential flow stress σ_{diff} defined as the difference between flow stress σ and confining pressure P_c, T is absolute temperature, A is a constant, Q is the activation energy, and R is the gas constant. The specific functional form of $f(\sigma_{\text{diff}})$ is commonly represented by a power law; that is,

$$f(\sigma_{\text{diff}}) = \sigma_{\text{diff}}^n, \tag{2.7}$$

where n is an exponent depending on the material and ambient conditions.

Equation (2.6) is a constitutive equation that prescribes the relation between flow stress σ_{diff} and strain rate $\dot{\varepsilon}$ at temperature T. The creep strain rate depends exponentially on the temperature, and hence the strain rate is very sensitive to temperature. For reference, the plastic flow law parameters A, n, and Q determined experimentally for representative rocks are listed in Table 2.1.

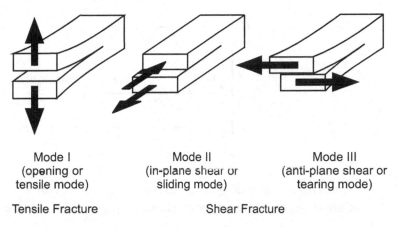

Mode I
(opening or
tensile mode)

Mode II
(in-plane shear or
sliding mode)

Mode III
(anti-plane shear or
tearing mode)

Tensile Fracture   Shear Fracture

Fig. 2.1   The three fundamental modes of fracture.

2.1.3 Fracture

If increasing stress is applied to a real material, the deformation eventually concentrates in a narrowly localized zone in the material, the fracture finally occurs, and new crack surfaces are created as a result of the breakdown of inter-atomic bonds in the localized zone. Therefore, fracture is characterized by the fact that the strength deteriorates with ongoing relative displacement between the fracturing surfaces during the breakdown process. When the relative displacement is perpendicular to the fracture plane to open the crack, it is called *tensile fracture* (mode I). When the relative displacement is parallel to the fracture plane, it is called *shear fracture*. Shear fracture consists of two fundamental modes: one of which is called *sliding mode* or *in-plane shear mode* (mode II), in which the relative displacement (or slip) is perpendicular to the crack edge, and the other is called *tearing mode* or *anti-plane shear mode* (mode III), in which the relative displacement (or slip) is parallel to the crack edge. These three different fracture modes are illustrated in Figure 2.1.

A fracture that occurs within the elastic limit is called a *brittle fracture*, though a fracture including a small amount of permanent deformation may also be categorized as a brittle fracture type for practical purposes. In contrast, a fracture that occurs after an extensive amount of permanent deformation is referred to as a *ductile fracture*. Once fracture occurs in a real material, the stability or instability of the fracture process is governed by how progressively the strength during the fracture process degrades with ongoing relative displacement. Thus, the constitutive relation for fracture is determined by the transient response of the traction on the fracturing surfaces to the relative displacement. The constitutive relation for rock fracture and for frictional slip failure is an important, fundamental theme of this book, and hence it is described in detail in Chapter 3. In this subsection, we evaluate the breaking strength of atomic bonds that hold atoms together, and the constitutive relation for the bond breaking process.

To evaluate the cohesive strength (or atomic bond strength) σ_c of a material with ideal crystal structure that does not contain any flaws or defects (Orowan, 1948; Gilman, 1959), let r_0 be the equilibrium spacing between atomic planes under no applied stress (see Figure 2.2).

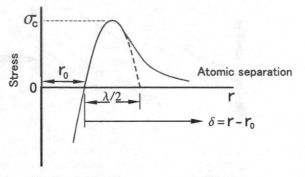

Fig. 2.2 The tensile stress σ required for separating atomic planes, as a function of distance r. The equilibrium spacing between atomic planes under no applied stress is denoted by r_0. Fracture occurs when $\sigma = \sigma_c$.

The tensile stress σ required for separating the atomic planes is a function of the distance r, and increases with an increase in r until the stress σ attains the cohesive strength σ_c, at which atomic bonds are broken, and after which the strength degrades with ongoing displacement between the atoms (Figure 2.2). In this case, the stress σ is a function of the displacement δ within the cohesive zone near the crack tip; that is,

$$\sigma = f(\delta), \tag{2.8}$$

where δ is the displacement from equilibrium, defined by $\delta = r - r_0$. If the specific functional form of $f(\delta)$ is known, relation (2.8) is a self-consistent constitutive relation governing the bond breaking process.

Let us assume that the stress versus displacement curve can be approximated by a sine curve with wavelength λ (Figure 2.2). Under this assumption, the stress σ is simply expressed as follows:

$$\sigma = \sigma_c \sin\left(\frac{2\pi\delta}{\lambda}\right). \tag{2.9}$$

At small displacements where $(2\pi\delta/\lambda) \ll 1$ holds, Eq. (2.9) is reduced to

$$\sigma \cong \sigma_c \left(\frac{2\pi\delta}{\lambda}\right). \tag{2.10}$$

If it is assumed that Hooke's law holds at these small displacements, then

$$\sigma = E\left(\frac{\delta}{r_0}\right), \tag{2.11}$$

where E is Young's modulus. From Eqs. (2.10) and (2.11), we have

$$\sigma_c = \frac{\lambda}{2\pi}\frac{E}{r_0}. \tag{2.12}$$

The energy G_c required for breaking atomic bonds is equal to the work done up to the breakdown of the bonds, so that

$$G_c = \int_0^{\lambda/2} \sigma_c \sin\left(\frac{2\pi\delta}{\lambda}\right)d\delta = \frac{\lambda\sigma_c}{\pi}. \tag{2.13}$$

Thus, we have from Eqs. (2.12) and (2.3)

$$\sigma_c = \sqrt{\frac{EG_c}{2r_0}} \qquad (2.14)$$

The cohesive strength of a homogeneous material that does not contain any flaws or defects can be theoretically estimated from Eq. (2.14). If we take typical values of $E = 10^{11}$ N/m^2, $r_0 = 3 \times 10^{-10}$ m, and $G_c = 1$ J/m^2 ($= 1$ N m/m^2), then we have

$$\sigma_c = 1.29 \times 10^{10}\,\text{N/m}^2 \approx E/7.8. \qquad (2.15)$$

The strength of this order of magnitude has been achieved in laboratory tests when conducted on thin specimens (fibers or rods) that were prepared and handled very carefully before testing (Gilman, 1962). It has been shown that $\sigma_c \cong E/10$ and $G_c \cong Er_0/10$ are appropriate measures of the theoretical strength and energy, respectively, required for breaking atomic bonds of a variety of inorganic materials (Gilman, 1962).

However, the fracture strength of a real material such as rock is extremely low when compared with the theoretical strength. This is because such materials as rock are inherently heterogeneous and contain a variety of mechanical flaws and stress concentration sources, such as microcracks, pores, and grain boundaries. This must be kept in mind when we deal with rock fracture and earthquake rupture.

2.1.4 Friction

The surface of a solid, even when best-prepared, is made up of asperities. When two surfaces of solid materials are placed in contact, they do not touch over the entire area (apparent area of contact) A_a, but they make contact only at a number of asperities. The sum of the areas of all the asperity contacts constitutes the real area of contact A_r ($< A_a$). The interaction between mating surfaces takes place at these contacting portions at which they are in atom-to-atom contact with each other. It is known that inter-atomic forces are powerful but of very short range, in the order of a few Angstroms. An adsorbed film is in general formed on the surface of a solid material. The main constituent of the adsorbed film may be molecules of water vapor and/or oxygen. The thickness of the film may be in the order of a few molecularly thick layers (Rabinowicz, 1965). The presence of adsorbed film on mating solid surfaces interrupts the atom-to-atom interaction at contacting portions on the surfaces, because the inter-atomic forces are of very short range. The film acts as a lubricating agent to separate contacting materials by interposing itself between the mating surfaces.

Nevertheless, adhesion occurs at the junctions where the film has broken up during the normal load application or during sliding. In this case, the shearing strength of such adhesive junctions is the prime cause of frictional resistance (Bowden and Tabor, 1954). For rock containing typical, hard silicate minerals such as quartz, feldspar, and pyroxene, the penetration of hard asperities into the film on the opposing surface easily occurs during the normal load application or during sliding, and asperities at contacting portions fail by brittle fracture, rather than by plastic shear (Byerlee, 1967). In

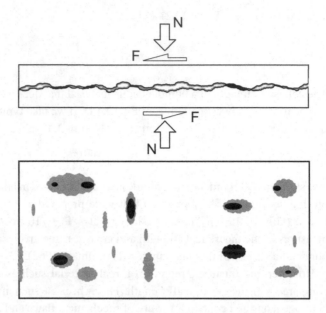

Gray portions: solid–film–solid contact areas at which solid surfaces are in
contact with each other through adsorbed films

Black portions: solid–solid contact (film–broken) areas at which solid surfaces
are in direct contact with each other

Fig. 2.3 Two surfaces of solid materials placed in contact with each other.

this case, frictional strength (or resistance) is primarily due to brittle fracture of these
asperities.

Let us denote the sum of the solid–solid contact (film–broken) areas by A_{r1}, which is a
fraction of the entire contact areas A_r, and the rest of the contact (solid–film–solid contact)
areas by $A_{r2} = A_r - A_{r1}$ (see Figure 2.3). When two solid surfaces are placed together by a
normal load N, total frictional force F is given by

$$F = \mu_{r1} A_{r1} \left(\frac{N}{A_r} \right) + \mu_{r2} A_{r2} \left(\frac{N}{A_r} \right), \tag{2.16}$$

where μ_{r1} and μ_{r2} are the frictional coefficients for the solid–solid contact areas and for the
solid–film–solid contact areas, respectively. The average frictional coefficient μ between
the surfaces is

$$\mu = \frac{F}{N} = \mu_{r1} \left(\frac{A_{r1}}{A_r} \right) + \mu_{r2} \left(\frac{A_{r2}}{A_r} \right). \tag{2.17}$$

The frictional coefficient μ_{r1} represents the strength of the solid material, and μ_{r2} represents
the shearing strength of the adsorbed films. It is thus obvious that μ is less than μ_{r1}, but
greater than μ_{r2}. It has been demonstrated in laboratory experiments that μ_{r1} is much
greater than unity, though μ_{r2} is less than unity; for example, $\mu_{r1} = 4$–18 and $\mu_{r2} = 0.3$
have been estimated for Solenhofen limestone (Ohnaka, 1975).

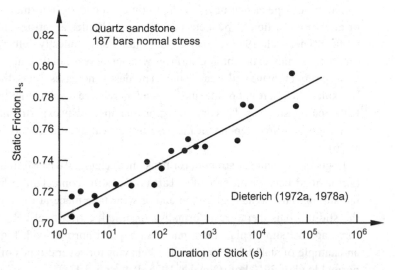

The time-dependence of static frictional coefficient μ_s for rock: an example of the experimental results. Reproduced from Dieterich (1972a).

At a higher normal load, the asperities that are in contact with the opposing surface deform more in the direction normal to the mating surfaces, causing the two mating surfaces to make closer contact at more asperities that have been out of touch in between the surfaces. This results in a larger increase in the sum of the real areas of contact A_r. As the normal load increases, A_{rl} may gradually approach A_r. At a sufficiently high normal load at elevated temperature, the entire area A_a of the mating rock surfaces may come into real contact $(A_{rl} = A_r = A_a)$, and in this case the shear resistance becomes equal to (but never exceeds) the shear strength of intact rock. In other words, the shear strength of intact rock is the upper limit of frictional strength. Indeed, frictional strength conforms to the shear fracture strength of intact rock at crustal depths where the lithostatic pressure and temperature are high enough, in the brittle–plastic transition regime (Stesky, 1978; see also Ohnaka, 1992, 1995b).

Static frictional strength modestly increases with an increase in stationary contact time t between two mating surfaces at a constant normal stress. The time-dependence of static frictional coefficient μ_s for rock obeys a logarithmic law (Dieterich, 1972a, 1978a):

$$\mu_s = \mu_0 + C \log(Dt + 1), \tag{2.18}$$

where μ_0 is the static frictional coefficient at $t = 0$, and C and D are constants. According to Dieterich (1978a), the constants μ_0 and C are insensitive to normal stress. Average values of μ_0 are 0.7–0.8, average values of C are 0.016, 0.022, 0.020, and 0.012 for sandstone, granite, quartzite, and greywacke, respectively, and D is approximately equal to 1.0 (Dieterich, 1978a). An example of the experimental results is shown in Figure 2.4. This time-dependence has been subsequently confirmed by other researchers (Beeler *et al.*, 1994; Marone, 1998a).

The time-dependence of μ_s is due to time-dependent deformation at mating asperities, or alternatively, time-dependent penetration of harder asperities into the opposing softer surface (Dieterich, 1978a). Such asperity junctions modestly deform non-elastically in the direction normal to the mating surfaces with an increase in time at a constant normal stress, and consequently the real area of contact modestly increases. Thus, the time-dependence can be explained as a result of the time-dependent relative displacement in the direction normal to the mating surfaces. This interpretation is supported by experiments on indentation creep, which gives a time-dependent increase in the real area of contact (Scholz and Engelder, 1976).

In general, frictional strength (or resistance) changes with sliding even under a constant normal load, because the sum of real areas of asperity contact on sliding surfaces varies with the sliding, and because adhesion and/or wear take place at the real contact areas during the sliding. It is therefore informative to describe frictional behavior as a plot of shear force against slip displacement under an applied normal load. Figure 2.5(a) shows such an example of stable frictional sliding behavior for several types of rock under a constant normal load of 2910 kgf (Ohnaka, 1975; Paterson, 1978).

Under certain conditions, spontaneous jerking motions, referred to as *stick-slip*, occur during frictional sliding. Such examples are shown in Figures 2.5(b) and (c). As exemplified in Figure 2.5(b), frictional sliding can be stable under low normal loads even for granite, but is accompanied by stick-slip oscillations under high normal loads, and the amplitude of stick-slip oscillation becomes greater at a higher normal load (Ohnaka, 1973, 1975). As exemplified in Figure 2.5(c), frictional sliding is affected by the roughness of the sliding surfaces, and stick-slip oscillations occur more readily on the smoother surfaces. In other words, frictional sliding on smooth surfaces is always accompanied by distinctly audible stick-slip instabilities ((I) in Figure 2.5(c)), whereas frictional sliding on very rough surfaces occurs stably or semi-stably without any distinct stick-slip instability ((III) in Figure 2.5(c)). Frictional sliding on intermediately rough surfaces (between (I) and (III)) takes place in a jerky manner (stick-slip) for hard rock such as granite and monzonite. Frictional behavior of soft rock such as marble and limestone is more strongly influenced by the roughness of the sliding surfaces (Ohnaka, 1973, 1975). The stick-slip behavior is in general affected not only by the normal force across sliding surfaces and surface conditions such as roughness, but also by other conditions such as the stiffness of the loading system, material properties and temperature (Ohnaka, 1973; Dieterich, 1978a; Paterson, 1978).

When shear force (or shear stress) is applied at a constant rate, frictional resistance between sliding surfaces increases linearly with time during the stage where the sliding surfaces stick to each other. After the frictional resistance attains a critical level, the shear force (or shear stress) drops, with ongoing slip displacement, to a residual friction level. Accordingly, if shear force (or shear stress) and slip displacement during stick-slip processes are respectively plotted against time, they are expressed as shown in Figure 2.6. In this manner, the "stick" and the subsequent "slip" occur cyclically on a precut rock interface. Such frictional stick-slip failure events occurring repeatedly on a precut rock interface have attracted particular attention as a possible source mechanism of shallow focus earthquakes which take place on preexisting faults embedded in the seismogenic crust (e.g., Brace and Byerlee, 1966, 1970; Brace, 1972; Byerlee and Brace, 1968; Dieterich, 1972b; Scholz

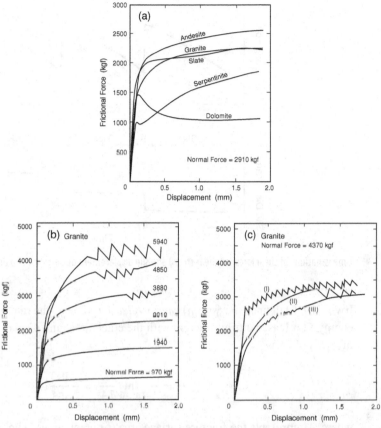

Fig. 2.5 (a) An example of stable frictional sliding behavior for several types of rock under a constant normal load of 2910 kgf. (b) Frictional sliding behavior for granite rock under a constant normal load of 970, 1940, 2910, 3880, 4850, or 5940 kgf. (c) The effect of surface roughness on frictional sliding behavior for granite rock under a constant normal load of 4370 kgf. (I) polished surfaces, (II) ground surfaces, and (III) rough surfaces. Reproduced from Ohnaka (1975).

et al., 1972). A law governing stick-slip cycle processes on a precut rock interface was first formulated by Dieterich (1978b, 1979, 1981, 1986) and Ruina (1983), and the formulation will be described in Chapter 4.

2.2 Basics of rock fracture mechanics

2.2.1 Energy release rate and resistance to rupture growth

Fracture occurs when the energy available for crack growth (i.e., fracture) is equal to or greater than the resistance to crack growth (or fracture). The energy approach for fracture criterion was first proposed by Griffith (1920), and greatly developed by Irwin (1956).

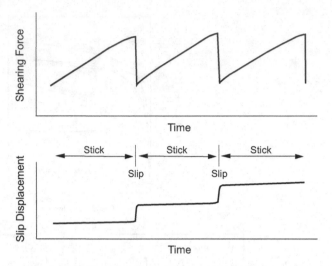

Fig. 2.6 Time variations of shear force (or shear stress) and slip displacement during stick-slip processes.

Irwin (1956) defined the (*strain*) *energy release rate* G as the rate of change in the potential energy Ω released during fracture with the crack area, and G is expressed mathematically in terms of Ω as

$$G = - \lim_{\delta A \to 0} \frac{\delta \Omega}{\delta A} = - \frac{\partial \Omega}{\partial A}, \qquad (2.19)$$

where A represents the fracture surface area (or crack area). The potential energy Ω of an elastic body is defined as the (elastic) strain energy $\overline{W}$ stored in the body minus the work $\overline{U}_{\mathrm{ex}}$ done by external forces; that is,

$$\Omega = \overline{W} - \overline{U}_{\mathrm{ex}}. \qquad (2.20)$$

Thus, it is clear that the dimensions of G are [energy/area] = [J/m^2]. The energy release rate G acts as the driving force for fracture, and hence, G is also called the *crack driving force* or the *crack extension force*.

The *resistance to rupture growth* (*or crack growth*) is defined as the critical energy per unit area, G_{c}, required for a rupture front to grow further, which is referred to as the *fracture energy*. In general, G_{c} may include not only the surface energy per unit area of newly created crack, but also the energy per unit area required for non-elastic deformation at the tip of an extending crack, or other type of energy dissipation associated with the crack tip. In addition, G_{c} can be affected by the geometric irregularity (or roughness) of a nonplanar fracture surface. If the fracture surface is completely flat, G_{c} is scale-independent because the dimensions of G_{c} are [energy/area] = [J/m^2]. However, the fracture surface of a heterogeneous body such as rock is not a flat plane but has geometric irregularity with a fractal nature. In this case, the real area of the fracture surface is larger than the

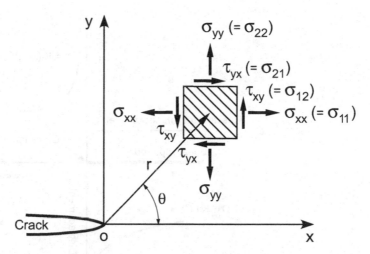

Stress components near the tip of a crack in an elastic material.

apparent area measured under the assumption that the fracture surface is a flat plane. The G_c defined by Eq. (2.44) in subsection 2.2.4 is calculated under the tacit assumption that the fracture surface is a flat plane, and hence G_c estimated from Eq. (2.44) for a heterogeneous body such as rock is scale-dependent. This is an important fact that must be taken into account when we estimate the fracture energy of inhomogeneous materials such as rock, and of earthquakes that take place on inhomogeneous faults, of which the surfaces are geometrically irregular (see Chapters 3 and 6).

For a crack tip to grow, G must be equal to or greater than G_c; that is,

$$G \geq G_c. \tag{2.21}$$

This is the fracture criterion expressed mathematically in terms of energy. In particular, $G = G_c$ for quasi-static crack growth on a flat plane in an elastic body, and in this case

$$G_c = 2\gamma_s, \tag{2.22}$$

where γ_s is the surface energy per unit area. (Note that two surfaces are created when a material fractures.)

2.2.2 Stress concentration and cohesive zone model

We will consider here elastic brittle fracture. By the term *elastic brittle fracture* we mean an idealized case where the behavior of material is purely elastic up to the fracture. Let us consider a homogeneous and isotropic material which includes a single crack, and assume the material to be linearly elastic outside the crack. Within the framework of linear elasticity theory, the stresses σ_{ij} outside the crack, near its tip (Figure 2.7) can be well approximated

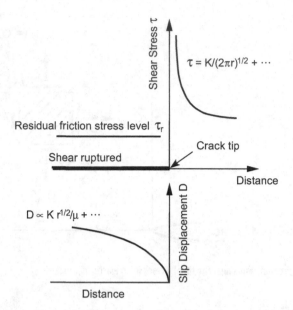

Fig. 2.8 Shear stress and slip displacement distributions near the tip of a shear crack in an elastic material.

by the following relation (Irwin, 1957; Williams, 1957):

$$\sigma_{ij} = \frac{K f_{ij}(\theta)}{\sqrt{2\pi r}}, \tag{2.23}$$

where r is the distance from the crack-tip, θ is the angle measured from the plane ahead of the crack, K is the stress intensity factor, and $f_{ij}(\theta)$ is a dimensionless function of θ. Specific functional forms of $f_{ij}(\theta)$ are known, but they are not listed here because they are unnecessary for the objectives of this book. Refer to typical textbooks of fracture mechanics for more information about K and $f_{ij}(\theta)$ (e.g., Lawn and Wilshaw, 1975; Anderson, 1995).

In this elastic crack model, σ_{ij} goes to infinity as r approaches zero outside the crack, while σ_{ij} remains zero or is equal to a finite level of residual friction stress inside the crack; in other words, the stresses drop instantaneously to zero or to a residual stress level at the crack-tip, and there is no zone where the stresses decrease transitionally to zero or to a residual stress level (see Figure 2.8, in which the case of shear fracture is illustrated). The singular stresses at the crack tip for any finite K, which is proportional to a finite stress applied remotely from the crack if the crack length is kept constant, inevitably lead to the conclusion that the elastic crack model is unrealistic and not physically reasonable, because no real materials can sustain infinitely high stress.

To avoid such an unrealistic prediction of singular stresses at a crack tip in a brittle material, Barenblatt (1959) postulated a zone of cohesive force near the crack tip, and this zone is referred to as the *cohesive zone* (Figure 2.9). An atomic bond separation in the zone

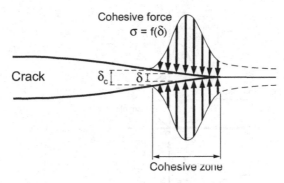

Fig. 2.9   Barenblatt's cohesive force model.

is allowed by an external stress applied remotely from the crack, while the separation is resisted by atomic cohesive force in the zone. The stress singularity caused by an externally applied stress at the outer edge of the cohesive zone is canceled by the singularity due to the cohesive force acting in the zone. As a result, stresses at the crack tip are held finite and the crack surfaces close smoothly (or the crack tip has a cusp shape).

In the cohesive zone model, the cohesive stress σ is a function of the relative displacement δ between fracture surfaces within the cohesive zone (ref. Eq. (2.8)), and the fracture energy is given by (Rice, 1968)

$$G_c = \int_0^{\delta_c} f(\delta)d\delta, \tag{2.24}$$

where δ_c is the critical displacement at which σ decreases to zero (Figure 2.9).

The concept of the cohesive zone for tensile fracture is applicable to shear fracture (or shear rupture) (Ida, 1972; Palmer and Rice, 1973); however, we have to keep in mind that the process zone behind the front of shear fracture is physically not identical to the cohesive zone for tensile fracture. In macroscopic tensile fracture, the relative displacement which opens the crack is perpendicular to the macroscopic fracture plane, and hence the fracturing surfaces do not interact with each other during the breakdown process. Therefore, the tensile traction is not affected by the geometric irregularities of the fracturing surfaces. Accordingly, the size of the breakdown zone is extremely small. By contrast, in shear fracture, the relative displacement (or slip) proceeds on the fracture plane, and the fracturing surfaces are in mutual contact and are interactive during the breakdown process. Hence, the shear traction and the size of the breakdown zone are both strongly influenced by the geometric irregularity of shear-fracturing surfaces (see Chapters 3, 5, and 6). Accordingly, the underlying physics of the constitutive relation (between the stress and the relative displacement) for shear fracture is different from that of the constitutive relation for tensile fracture.

The above facts are critically important, and therefore must be taken into consideration in order to establish the constitutive law for earthquake ruptures, because earthquake sources are shear rupture instabilities that take place on faults preexisting in the seismogenic layer in the Earth's interior, and because the seismogenic layer and individual faults

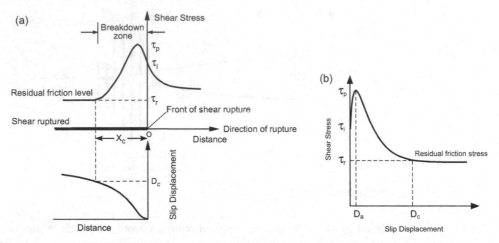

Fig. 2.10 (a) Breakdown zone behind the front of shear rupture. (b) Laboratory-derived constitutive relation for shear failure.

therein are inherently heterogeneous. Thus, the role played by the geometric irregularity of shear-rupturing surfaces in the constitutive law for shear rupture must be carefully examined and confirmed by appropriately designed laboratory experiments (see Chapter 3).

2.2.3 Breakdown zone model for shear failure

In the case of shear fracture, the zone over which the shear strength degrades transitionally to a residual friction stress level on the rupture surfaces behind the rupture front is called the *breakdown zone*. Figure 2.10(a) shows a physical model of the breakdown zone, which is theoretically derived based on laboratory-derived constitutive relations for shear failure.

Let us theoretically calculate spatial distributions of shear stress and slip displacement in the breakdown zone behind the tip of a dynamically propagating shear-crack. Spatial distributions of shear stress and slip displacement in the breakdown zone can be theoretically calculated, assuming the constitutive relation for shear failure (Ida, 1972). For this calculation, we adopt a laboratory-derived constitutive relation (see Chapter 3), which is illustrated in Figure 2.10(b). The laboratory-derived constitutive relation for shear failure can be formulated as the shear traction τ being expressed as a function of slip displacement D, as follows (Ohnaka and Yamashita, 1989; see also Section 4.3):

$$\tau(D) = (\tau_i - \tau_r)[1 + \alpha \log(1 + \beta D)] \exp(-\eta D) + \tau_r, \tag{2.25}$$

where τ_i denotes initial stress on the verge of slip ($D = 0$), τ_r denotes a residual friction stress level, and α, β, and η are constants.

Let us use a Cartesian coordinate system with the xz plane being the crack plane and the y-axis normal to the crack plane. For simplicity, we assume a two-dimensional shear crack in an unbounded elastic medium. To make the problem more tractable, we seek a solution for a semi-finite crack moving steadily in the positive x direction with a rupture speed V. The origin of the coordinate system is fixed at the moving crack tip, so that the crack is located on the negative x-axis.

The relation between shear stress and relative displacement on the crack plane plays a central role in the present analysis. According to Aki and Richards (1980, p. 853; 2002, p. 539) and Ohnaka and Yamashita (1989), we have

$$\tau_{yz}(x) = \frac{\mu}{2\pi} \left(1 - V^2/V_S^2\right)^{1/2} \mathrm{P} \int_{-\infty}^{0} \frac{\Delta w'(\xi)}{\xi - x} d\xi \qquad (2.26)$$

for an anti-plane shear crack (mode III), and

$$\tau_{xy}(x) = \frac{2\mu}{\pi} \frac{V_S^2}{V^2} \left[\left(1 - V^2/V_P^2\right)^{1/2} - \frac{\left(1 - V^2/2V_S^2\right)^2}{\left(1 - V^2/V_S^2\right)^{1/2}} \right] \mathrm{P} \int_{-\infty}^{0} \frac{\Delta u'(\xi)}{\xi - x} d\xi \quad (2.27)$$

for an in-plane shear crack (mode II), where μ is the rigidity, τ_{yz} and τ_{xy} are shear stress components, V_P and V_S are longitudinal (or primary) and shear (or secondary) wave velocities, $\Delta u(x)$ and $\Delta w(x)$ are relative displacements in the x and z directions, $\Delta u'(x)$ and $\Delta w'(x)$ denote the derivatives of Δu and Δw with respect to x, respectively. The letter P before the integral sign denotes the Cauchy principal value. If we denote the pre-integral factors associated with crack velocity V by $\mu C(V)$, then Eqs. (2.26) and (2.27) are reduced to a simpler form as

$$\tau(x) = \mu C(V)\mathrm{P} \int_{-\infty}^{0} \frac{D'(\xi)}{\xi - x} d\xi, \qquad (2.28)$$

where $\tau(x)$ is the shear stress and $D(\xi)$ is the slip displacement on the crack plane.

The inverse expression of Eq. (2.28) is written as (Ohnaka and Yamashita, 1989)

$$\sqrt{-x} D'(x) = -\frac{A}{2} - \frac{1}{\pi^2 \mu C(V)} \mathrm{P} \int_{-\infty}^{0} \frac{\sqrt{-\xi}}{\xi - x} \tau(\xi) d\xi, \qquad (2.29)$$

where asymptotic relation $D(x) \to A\sqrt{-x}$ with $x \to -\infty$ has been considered (A being constant). Since a semi-infinite crack has been assumed in the present analysis, the integral in Eq. (2.29) diverges if τ_r is non-zero at some distance from the crack tip. For simplicity, we assume for the present calculation that $\tau_r = 0$ on the entire crack plane (i.e., $\tau_p = \Delta\tau_b$). Under this assumption, the constitutive equation expressed as Eq. (2.25) is reduced to

$$\tau(D(x)) = \tau_i[1 + \alpha \log(1 + \beta D)] \exp(-\eta D) \quad (x < 0). \qquad (2.30)$$

The initial strength τ_i can be regarded as constant during the dynamic breakdown process.

For convenience sake, we introduce a parameter x_c defined as the distance from the crack tip to the point at which relation $\tau(-x_c) = \chi \tau_p$ holds, where χ is a fixed numerical value of a small fraction (for instance, $\chi = 0.1$ or 0.15). If we further introduce another parameter

x_{max}, which is greater than x_c, and beyond which relation $\tau \cong 0$ practically holds, then we have, by integrating both sides of Eq. (2.29),

$$D(x) = A\sqrt{-x} - \frac{1}{\pi^2 \mu C(V)} \int_{-x_{max}}^{0} \tau(\xi) \log \left| \frac{\sqrt{-\xi} + \sqrt{-x}}{\sqrt{-\xi} - \sqrt{-x}} \right| d\xi. \tag{2.31}$$

Considering the condition of finite stress at the crack tip, and introducing the dimensionless slip displacement ϕ defined by $\phi(X) = D(x)/D_c$, where D_c is the breakdown displacement, we finally have from Eq. (2.31) (Ohnaka and Yamashita, 1989)

$$\phi(X) = \frac{A\sqrt{x_{max}}}{D_c} \left[\sqrt{X} - \frac{1}{2\Gamma} \int_0^1 \sigma(Y) \log \left| \frac{\sqrt{Y} + \sqrt{X}}{\sqrt{Y} - \sqrt{X}} \right| dY \right]. \tag{2.32}$$

In this equation, X and Y are dimensionless distances defined as $X = -x/x_{max}$ and $Y = -\xi/x_{max}$, respectively, and Γ is a dimensionless parameter defined by

$$\Gamma = \int_0^1 \frac{\sigma(Y)}{\sqrt{Y}} dY, \tag{2.33}$$

where $\sigma(Y)$ is the dimensionless shear stress at a dimensionless distance Y in the breakdown zone, measured from the crack-tip, and given by

$$\sigma(Y) = \sigma(\phi(Y))$$
$$= \frac{\tau(\phi)}{\tau_p} = \frac{\tau_i}{\tau_p}[1 + \hat{\alpha} \log(1 + \hat{\beta}\phi)] \exp(-\hat{\eta}\phi). \tag{2.34}$$

In this equation, $\hat{\alpha}$, $\hat{\beta}$, and $\hat{\eta}$ are defined as $\hat{\alpha} = \alpha$, $\hat{\beta} = \beta D_c$, and $\hat{\eta} = \eta D_c$, respectively. Equation (2.32) has the following constraint:

$$\phi(1) = D(-x_{max})/D_c. \tag{2.35}$$

The spatial distribution of slip displacement as a function of the distance from the crack tip is obtained by solving Eq. (2.32). The integral equation (2.32) is nonlinear, and hence it has been solved numerically (Ohnaka and Yamashita, 1989).

Consequently, spatial distributions of shear stress and slip displacement in the breakdown zone can be derived theoretically as illustrated in Figure 2.10(a), if the laboratory-derived constitutive relation as shown in Figure 2.10(b) is adopted. From Figure 2.10(a), one can see that the shear stress attains its peak value not at the crack-tip but within the breakdown zone behind the crack-tip. This is a direct result of the laboratory-derived constitutive relation in which the shear strength has its peak value at a non-zero value of slip displacement.

2.2.4 J-integral and energy criterion for shear failure

Let us consider an arbitrary counter-clockwise path Σ around the tip of a crack in a brittle material (see Figure 2.11). The J integral is defined by Rice (1968)

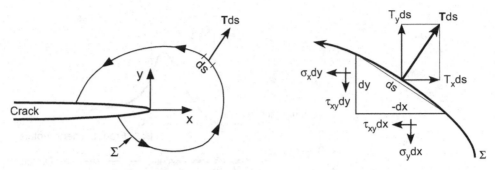

An arbitrary counter-clockwise path Σ around the tip of a crack, and T_i and σ_{ij} components at a location of length increment ds along the path Σ.

as

$$J = \int_{\Sigma} \left(W dy - T_i \frac{\partial u_i}{\partial x} ds \right),$$

(2.36)

where W is the strain energy density, $T_i (= \sigma_{ij} n_j)$ are components of the traction vector, u_i are components of the displacement vector, ds is a length increment along the contour Σ, and n_j are components of the unit vector normal to Σ. The strain energy density is defined as

$$W = \int_0^{\varepsilon_{ij}} \sigma_{ij} d\varepsilon_{ij}.$$

(2.37)

Rice (1968) demonstrated that the value of the J integral is independent of the path of integration around the crack, and showed that J is equal to the energy release rate G.

The J integral is rewritten as follows:

$$J = \int_{\Sigma} \left[\left\{ W - \left(\sigma_x \frac{\partial u}{\partial x} + \tau_{xy} \frac{\partial v}{\partial x} + \tau_{xz} \frac{\partial w}{\partial x} \right) \right\} dy + \left\{ \tau_{xy} \frac{\partial u}{\partial x} + \sigma_y \frac{\partial v}{\partial x} + \tau_{yz} \frac{\partial w}{\partial x} \right\} dx \right].$$

(2.38)

given that $T_i ds$ is expressed as (see Figure 2.11)

$$T_x ds = \sigma_x dy - \tau_{xy} dx$$
$$T_y ds = \tau_{xy} dy - \sigma_y dx$$
$$T_z ds = \tau_{xz} dy - \tau_{yz} dx,$$

(2.39)

and that $u_x = u$, $u_y = v$, and $u_z = w$.

If we evaluate the J integral by employing path independence to choose Σ as the path from P^- to O along the lower fracture surface, and from O to P^+ along the upper fracture surface (see Figure 2.12), then we have from Eq. (2.38)

$$J_P = \int_{\Sigma} \tau_{xy} \frac{\partial u}{\partial x} dx = \int_P^O \tau(D) \left(\frac{\partial u^-}{\partial x} - \frac{\partial u^+}{\partial x} \right) dx = -\int_P^O \tau(D) \frac{\partial D}{\partial x} dx = \int_0^{D_P} \tau(D) dD$$

(2.40)

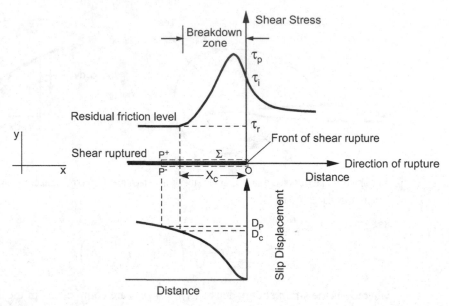

Fig. 2.12   Breakdown zone model, and a counter-clockwise path $\Sigma(P^- \to 0 \to P^+)$ surrounding the front edge of the breakdown zone.

for the breakdown zone model shown in Figure 2.12, given that $dy = 0$ on Σ. In the above equation, u^+ and u^- denote the displacements on the upper and lower fracture surfaces, respectively, in the x direction; D denotes the relative displacement given by $D = u^+ - u^-$; τ_{xy} denotes the shear traction, which is a single-valued function of D ($\tau_{xy} = \tau(D)$); and D_P denotes the relative displacement D at P on the fracture surfaces.

Figure 2.13 shows a specific functional form of $\tau(D)$ derived from laboratory experiments on shear failure of rock (see Chapter 3). In Figure 2.13, τ_i denotes the initial stress on the verge of slip displacement D, τ_p denotes the peak shear strength, τ_r denotes the residual friction stress level, D_a denotes the displacement at which τ has its peak value τ_p, and D_c denotes the *breakdown displacement*, defined as the critical slip displacement required for τ to degrade to τ_r. As demonstrated in the previous subsection, the breakdown zone model shown in Figure 2.12 can be derived theoretically from the laboratory-derived constitutive relation$\tau(D)$ shown in Figure 2.13.

When P lies outside the breakdown zone, as shown in Figure 2.12, we have

$$J_P - \tau_r D_P = \int_0^{D_P} [\tau(D) - \tau_r]dD = \int_0^{D_c} [\tau(D) - \tau_r]dD + \int_{D_c}^{D_P} [\tau(D) - \tau_r]dD.$$

$$(2.41)$$

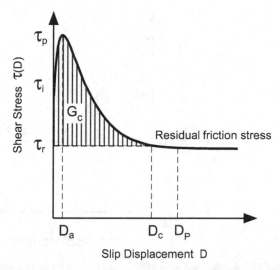

Fig. 2.13 An illustration of the laboratory-derived slip-dependent constitutive relation $\tau(D)$ for the shear failure of rock. The vertical-lined area gives the shear fracture energy, or the resistance to shear rupture growth, G_c expressed in Eq. (2.44).

Since

$$\int_{D_c}^{D_F} [\tau(D) - \tau_r]dD = 0, \tag{2.42}$$

Eq. (2.41) is reduced to

$$J_P - \tau_r D_P = \int_0^{D_c} [\tau(D) - \tau_r]dD, \tag{2.43}$$

from which we find that $J_P - \tau_r D_P$ is independent of P. Equation (2.43) was first derived by Palmer and Rice (1973).

In shear rupture, the rupture surfaces rub against one another under a compressive normal load. Accordingly, the energy is dissipated during frictional sliding on the rupture surfaces. The terms $\tau_r D_P$ and $\tau_r D_c$ in Eq. (2.43) denote such energy dissipation. Hence, the energy release rate for shear rupture is given by the left-hand side of Eq. (2.43), and the integral of the right-hand side of Eq. (2.43), which is equal to the vertical-lined area in Figure 2.13, gives the fracture energy for shear rupture. In other words, the shear fracture energy, or the resistance to shear rupture growth, G_c is expressed as

$$G_c = \int_0^{D_c} [\tau(D) - \tau_r]dD. \tag{2.44}$$

Thus, Eq. (2.43) is equivalent to the Griffith criterion $G = G_c$ for tensile fracture.

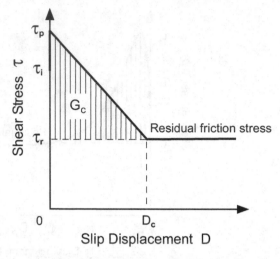

Fig. 2.14 A simplified, linear slip-weakening relation. The vertical-lined area gives the shear fracture energy, or the resistance to shear rupture growth, G_c expressed in Eq. (2.44).

2.2.5 Relation between resistance to rupture growth and constitutive relation parameters

The resistance to rupture growth is a basic concept and a distinct physical quantity in the framework of fracture mechanics, and is defined as the energy G_c required for the rupture front to grow further. Once the functional form of constitutive relation $\tau(D)$ is specified as shown in Figure 2.13, the resistance to rupture growth can be calculated from Eq. (2.44), and expressed in terms of constitutive relation parameters $\Delta\tau_b$ and D_c as (Ohnaka and Yamashita, 1989)

$$G_c = \frac{1}{2}\Gamma\Delta\tau_b D_c. \tag{2.45}$$

In this equation, Γ is the dimensionless parameter defined by Eq. (2.33). The constitutive relation parameter $\Delta\tau_b$ is defined as $\Delta\tau_b = \tau_p - \tau_r$, and referred to as the *breakdown stress drop*.

If a simplified, linear slip-weakening relation is assumed (Figure 2.14), then $\Gamma = 1$ can be derived from Eq. (2.44). However, laboratory-derived slip-dependent constitutive relations are found to be nonlinear. Ohnaka and Yamashita (1989) numerically calculated Γ from Eq. (2.33), using nonlinear constitutive relations observed during dynamic stick-slip failure on a precut fault in the laboratory, and obtained $\Gamma \cong 0.5$ when τ_i/τ_p has a typical value ranging from 0.5 to 0.7. On the other hand, $\Gamma = 0.8 - 1$ was obtained from experimental data on nonlinear constitutive relations for the shear fracture of intact rock (Tsukuba granite) samples tested in the laboratory (see Figure 2.15. Plotted data are from Ohnaka *et al.* (1997)). Thus, Γ can be regarded as virtually constant, and it is confirmed that the assumption that Γ is of the order of unity is valid even for nonlinear constitutive relations. Hence, $\Gamma \cong 1$ will be used later in the analysis to be done in Section 5.3.

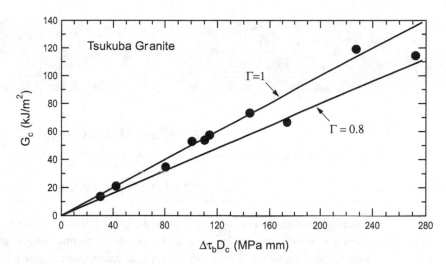

Fig. 2.15   Relation between G_c and $\Delta\tau_b D_c$ for laboratory data on the shear failure of intact Tsukuba granite. Data from Ohnaka *et al.* (1997).

Equation (2.45) indicates that the resistance to rupture growth, or G_c, is directly related to breakdown stress drop $\Delta\tau_b$ and breakdown displacement D_c. As will be shown in Chapter 3, D_c is scale-dependent. Hence, the resistance to rupture growth G_c is necessarily scale dependent. The scale-dependence of G_c is fully discussed later in Chapter 6.

Laboratory-derived constitutive relations for shear failure

3.1 Shear failure of intact rock

3.1.1 Method and apparatus used

In order to establish the constitutive law for the shear failure (or fracture) of intact rock in the brittle to semi-brittle (or brittle–plastic transition) regimes under seismogenic crustal conditions, it is necessary to carefully observe the entire shear failure process during which the post-failure (or slip-weakening) behavior is stabilized. A sophisticated high-pressure testing apparatus having a stiff loading frame and electronic servo-controls with a 16-bit resolution digital system was constructed for this purpose in our laboratory at the Earthquake Research Institute, the University of Tokyo (Ohnaka *et al.*, 1997). The post-failure behavior can be stabilized even in the brittle regime by enhancing the stiffness of the loading system and properly choosing servo-control variables.

The entire system includes the load frame assembly with which servo-hydraulic actuators are integrated, a triaxial pressure cell, servo-hydraulic intensifiers (for confining pressure and pore pressure), a hydraulic power supply system, a heater and cascade temperature controller, a water cooling system, and a computer-automated system controller with software designed for use with multiple channel servo-hydraulic valves and a heater (for details, see Ohnaka *et al.*, 1997).

The spherical triaxial pressure cell with an outer diameter of 500 mm and an inner diameter of 200 mm was specially designed for this specific apparatus to minimize unnecessary stress concentrations (Figure 3.1). This triaxial pressure cell was made by machining forged special steel (SUS630), to ensure high mechanical strength and toughness under confining pressures of up to 500 MPa, and temperatures of up to 500 °C. The maximum allowable temperature is limited to 500 °C by the use of a liquid medium to supply the confining pressure. A commercially available liquid (SYLTHERM800) composed of Si-compound ($n(SiO(CH_3)_2)Si_2O(CH_3)_6$), developed by DOW CORNING as a heat transfer fluid, was used as the confining pressure medium. The outside of the spherical pressure cell was cooled by running water during the experiments.

The apparatus was designed to utilize two different shapes of test specimen: circular cylinder and rectangular solid. Three principal stresses σ_1, σ_2, and σ_3 ($0 < \sigma_3 < \sigma_2 < \sigma_1$; compressive stress being defined as positive) are applied independently to the specimen by actuators with servo-controls. In the triaxial test under the condition that $\sigma_3 < \sigma_2 < \sigma_1$, the rectangular solid specimen is used. In this test, the minimum principal stress σ_3 is first applied by confining fluid pressure, then the intermediate principal stress σ_2 is

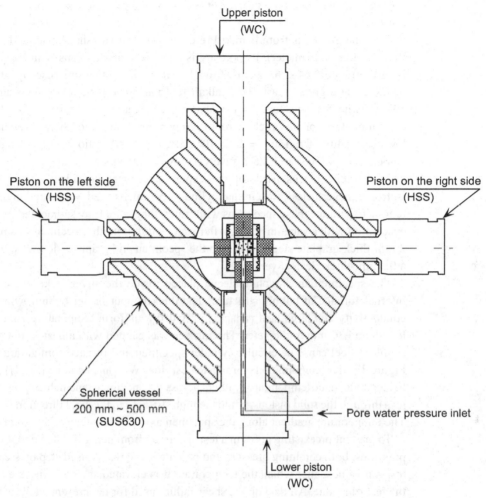

Fig. 3.1   Cross-section of the spherical triaxial pressure vessel used (front view) (Ohnaka *et al.*, 1997).

applied by synchronized operation of right and left lateral axial pistons, and finally the maximum principal stress σ_1 is applied by synchronized operation of upper and lower vertical axial pistons. In the triaxial test under the condition that $\sigma_3 = \sigma_2 < \sigma_1$, the circular cylinder specimen is used. In this test, the minimum principal stress σ_3 ($= \sigma_2$) is first applied by confining fluid pressure, and then the maximum principal stress σ_1 is applied by synchronized operation of the vertical axial pistons, while the lateral pistons are servo-controlled so as not to make contact with the specimen.

Using this apparatus and circular cylinder test specimens (40 mm in length and 16 mm in diameter), many experiments were conducted to reveal the constitutive properties of the shear failure of intact Tsukuba granite in the brittle to semi-brittle regimes under laboratory-simulated seismogenic crustal conditions (Ohnaka *et al.*, 1997; Odedra, 1998; Odedra *et al.*, 2001; Ohnaka, 2003; Kato *et al.*, 2003a, 2003b, 2004). Granite rock was chosen as a representative rock type because it is one of the major constituents of the continental crust.

Tsukuba granite is from Ibaraki Prefecture, central Honshu, Japan, and is fine-grained (0.2–1.3 mm in diameter). Its porosity is 0.6–0.9%, and its density in a dry environment is in the range 2.64–2.65 g/cm^3. Modal analysis by optical microscopy of thin-sections of the granite gives: quartz 30%, alkali feldspar 30%, plagioclase feldspar 30%, biotite 5%, chlorite 5%, muscovite <1%, magnetite and accessory minerals <1%. The physical parameters for this rock at room temperature and atmospheric pressure are as follows: modulus of rigidity $= 2 \times 10^4$ MPa, Poisson's ratio $= 0.12$, longitudinal wave velocity $= 4.4$ km/s, and shear wave velocity $= 2.9$ km/s.

The granite blocks used for the experiments were considerably homogeneous and isotropic; nevertheless, the circular cylinder specimens used were cored in the same orientation. To saturate these specimens with water, they were fully submerged in water after air evacuation with a vacuum pump. After water saturation, the specimen was placed inside a 2 mm-thick graphite sleeve, and then the specimen and graphite sleeve were placed inside a 0.3 mm-thick clean silver jacket.

The specimen and graphite sleeve placed inside the silver jacket were heated by an internal furnace of bobbin structure. The thick copper heater bobbin, with its high heat conductivity and large heat capacity, provided a uniform temperature zone of 52 mm in length and 30 mm in diameter. The furnace was capped with a heat insulator housed in a stainless steel can. The circular cylinder specimen and furnace configuration is shown in Figure 3.2. The combination of Tungsten carbide (WC) and ceramic (Si_3N_4) piston spacers helped reduce the temperature gradient across the specimen, and helped prevent excess heat loss through the solid pistons, while the high stiffness in the axial direction was maintained. The temperature gradient along the specimen axis was less than 2 °C /cm at 480 °C.

To prevent preexisting pores in a test specimen from being isolated under high confining pressures, both confining pressure and pore pressure were raised step by step to individual test run values, after which the temperature was elevated at a constant rate of 3 °C/min to the test run value. All data on the shear failure used for the present analysis were obtained by deforming Tsukuba granite specimens at a strain rate of 10^{-5}/s (corresponding to an axial displacement rate of 0.4 μm/s) during which confining pressure, pore pressure, and temperature were held constant independently by servo-controls.

In particular, pore pressure in the specimen was servo-controlled through the pressure of water injected into an axial hole in the lower loading piston connected to the specimen. The graphite sleeve has a much higher porosity than the granite specimen, so it served to minimize a possible pore pressure gradient in both axial and circumferential directions of the specimen. In addition, the slow, stable deformation process including dilatancy, servo-controlled at the rate of 0.4 μm/s, facilitated water fluid transfer from the specimen surface to its inner portions. Thus, constant pore water pressure in the developing fracture network was successfully maintained during deformation (see Kato *et al.*, 2003a). The shear failure of intact Tsukuba granite proceeded stably even in the purely brittle regime under the crustal conditions simulated in the laboratory. It has been demonstrated in the present experimental configuration that the effective stress law holds for Tsukuba granite at a strain rate of 10^{-5}/s during stable shear failure (Odedra *et al.*, 2001; Kato *et al.*, 2003a).

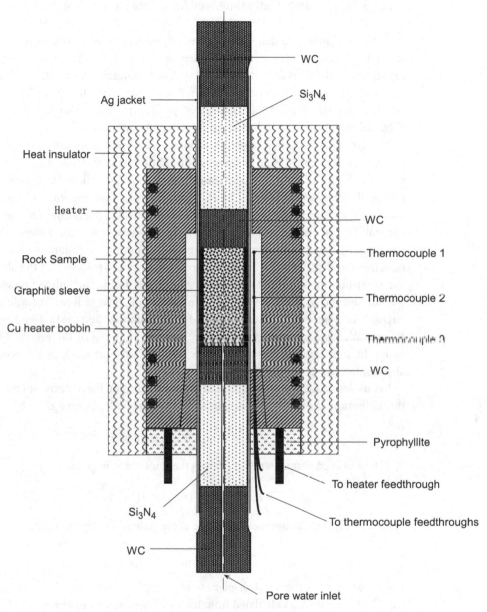

Ag jacket

Heat insulator

Heater

Rock Sample

Graphite sleeve

Cu heater bobbin

WC

Si₃N₄

Si₃N₄

WC

WC

Thermocouple 1

Thermocouple 2

Thermocouple 3

WC

Pyrophyllite

To heater feedthrough

To thermocouple feedthroughs

Pore water inlet

Fig. 3.2 Circular cylinder specimen and furnace configuration (Ohnaka *et al.*, 1997).

The analog output signals of axial load, axial displacement, confining pressure, pore pressure, and temperature, directly measured during the tests, were digitized at a sampling frequency of 1 Hz using a multi-channel analog-digital converter of 16-bit resolution, and the digitized data immediately transferred to a permanent disk for storage.

3.1.2 Constitutive relations derived from data on the shear failure of intact rock

In order to demonstrate that there is a unifying constitutive law that governs both frictional stick-slip failure on a precut rock interface and the shear failure (or fracture) of intact rock, experimental data sets of both frictional stick-slip failure on a precut rock interface and the shear failure of intact rock are needed. In this section, we will deal with laboratory data on the shear failure of intact rock, and show specifically how the constitutive relation for the shear failure of intact rock is derived from the laboratory data.

The compressive failure strength of intact rock under confining pressure P_c has been conventionally represented in terms of the differential strength σ_{diff}^f defined by $\sigma_{diff}^f = \sigma_1^f - \sigma_3$, where σ_1^f is the maximum principal stress at failure, and σ_3 is the minimum principal stress, which is equal to P_c. However, rock failure commonly occurs by shear-faulting (or shear mode) under combined compressive stress environments. In addition, the stability or instability of the shear failure process, once the failure has occurred, is governed by how progressively the shear strength during the failure process degrades with ongoing slip displacement. In other words, the transient response of the shear traction on the shear-faulting surfaces to the slip displacement during the failure process is the key to quantitative analysis of the stability or instability of shear failure. Accordingly, the failure strength defined as σ_{diff}^f is not suitable for formulating the constitutive law for the shear failure. We need to represent the failure strength in terms of the resolved shear strength along the fault surfaces (or macroscopic shear failure surfaces), as a function of ongoing slip displacement.

Let us define the fracture angle θ as the angle between the σ_1-axis and the fault plane. In this definition, the resolved shear stress τ along the fault plane is given by

$$\tau = \frac{1}{2}(\sigma_1 - \sigma_3)\sin 2\theta, \tag{3.1}$$

and the resolved normal stress σ_n across the fault plane is given by

$$\sigma_n = \sigma_3 + \frac{1}{2}(\sigma_1 - \sigma_3)(1 - \cos 2\theta). \tag{3.2}$$

Apparent slip D_{app} along the fault plane is evaluated from the relation:

$$D_{app} = \frac{\Delta l}{\cos \theta}, \tag{3.3}$$

where Δl denotes the axial displacement of the specimen. A plot of τ calculated from Eq. (3.1) against D_{app} calculated from Eq. (3.3) is shown as an example in Figure 3.3. D_{app} includes elastic deformation of the specimen, and this elastic deformation is represented by line A in Figure 3.3. The elastic deformation must be subtracted for evaluating net slip along the fault plane. Thus, the amount of net slip D along the fault plane is given by

$$D = \frac{\Delta l}{\cos \theta} - D_{el}, \tag{3.4}$$

where D_{el} denotes the elastic deformation of the specimen. Note that the amount of net slip on the fault plane is the relative displacement between both walls of the fault zone thickness, which is described below.

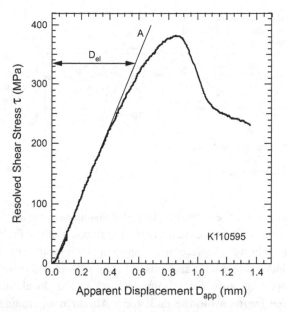

Fig. 3.3 A plot of the resolved shear stress τ against the apparent slip displacement D_{app} (Ohnaka *et al.*, 1997). The straight line A in the figure represents the elastic deformation D_{el} of the sample.

The shear failure (or fracture) is an inhomogeneous and nonlinear process during which inelastic shear deformation concentrates in a highly localized zone, resulting in bond-shearing and the release of the shear stress along the shear-fracturing surfaces with ongoing slip displacement. The fault zone (or shear zone) may thus be defined as a thin zone in which concentration of shear deformation is highly localized, and in which the macroscopic fracture surfaces are eventually formed. The fault zone may contain asperities on the fracture surfaces, gouge fragments, and/or highly damaged (or high crack density) thin layers consisting of subsidiary, minute cracks developed in the vicinity of the macroscopic fracture surfaces. Note, therefore, that no matter how thin it may be, the actual fault zone formed in a heterogeneous material such as rock necessarily has its own thickness. The effective fault zone thickness may be defined as the thickness of a highly damaged zone characterized by inelastic deformation, and the outside of the fault zone is primarily characterized by elastic deformation.

It should be noted that the shear traction used for the constitutive formulation is neither the shear stress acting on individual asperities, gouge fragments, and/or cracks contained in the fault zone, nor the shear stress acting on the real, macroscopic fracture surfaces eventually formed in the fault zone, because these stresses are not observable. The constitutive law for shear fracture (or rupture) needs to be formulated in terms of observable quantities, in such a way that scale-dependent physical quantities inherent in the rupture are scaled consistently in terms of the law. Since the shear stress acting on both walls of the fault zone thickness (see Figure 3.4) is observable, this has commonly been used as the shear traction for the constitutive formulation (e.g., Ruina, 1985; Ohnaka, 2003).

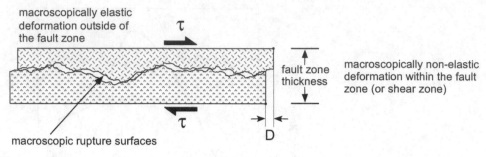

macroscopically elastic
deformation outside of
the fault zone

τ

fault zone
thickness

macroscopically non-elastic
deformation within the fault
zone (or shear zone)

τ

D

macroscopic rupture surfaces

Fig. 3.4   Fault zone model.

Likewise, the corresponding slip displacement used for the constitutive formulation is the relative displacement between both walls of the fault zone thickness (see Figure 3.4). The relative displacement between both walls of the fault zone thickness may include integrated amounts of slip associated with individual asperity fractures, growth of subsidiary, minute cracks, and local displacement between contacting gouge fragments in the fault zone. Although an amount of slip associated with an individual asperity fracture, growth of an individual minute crack, or local displacement between contacting gouge fragments in the fault zone is not observable, the overall, relative displacement between both walls of the fault zone thickness is observable. Note, therefore, that both shear traction and slip displacement used for the constitutive formulation are defined in a macroscopic sense. The relation between the shear traction and the corresponding slip displacement or the slip velocity in the breakdown zone behind a rupture front is particularly important in formulating the constitutive law for the shear rupture.

A great number of laboratory-derived relations between the shear stress τ calculated from Eq. (3.1) and the slip displacement D calculated from Eq. (3.4) were obtained for the shear failure of Tsukuba granite (e.g., Ohnaka *et al.*, 1997; Odedra, 1998; Ohnaka, 2003; Kato *et al.*, 2003a). Figure 3.5 shows a typical example of the laboratory-derived relations between τ and D, and this figure exemplifies a slip-dependent constitutive relation for the shear failure of intact granite. We find from Figure 3.5 that τ initially increases with an increase in D, and that after τ has attained its peak value τ_p at $D = D_a$, τ degrades to a residual friction stress level τ_r with ongoing slip. Note that the slip-strengthening (or displacement hardening) phase precedes the slip-weakening phase, and that the rate of slip-strengthening decreases as τ approaches its peak value τ_p. This is because:

(1) Elastic deformation of rock is usually limited to the first 40–50% (stress level τ_{i0} in Figure 3.5) of the peak shear strength (τ_p in Figure 3.5).
(2) Non-elastic (or ductile) deformation takes place as a result of the occurrence of microcracking above the stress level τ_{i0} under loading.
(3) At higher loads, crack interaction and coalescence become progressively more important, forming a thin, planar zone of higher crack density, which eventually results in the macroscopic shear fracture surfaces.

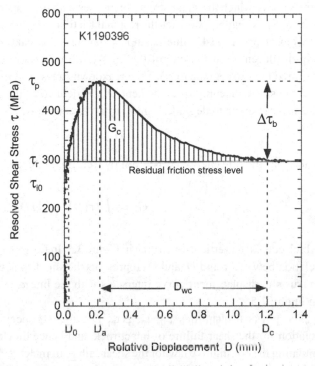

Fig. 3.5 A typical example of the laboratory-derived slip-dependent constitutive relations for the shear failure of intact Tsukuba granite (Ohnaka, 2003). In the figure, τ_{i0} is the initial stress beyond which the shear strength increases with ongoing displacement, τ_p is the peak shear strength, τ_r is the residual friction stress, $\Delta\tau_b$ is the breakdown stress drop defined as $\Delta\tau_b = \tau_p - \tau_r$, D_a is the relative displacement required for the shear strength to attain its peak value, D_c is the breakdown displacement defined as the critical slip displacement required for the shear strength to degrade to a residual friction stress level, D_0 is the relative displacement at which the curve showing the relation between the shear stress and the relative displacement intersects with the residual friction stress level, and D_{wc} is the slip-weakening displacement defined as $D_{wc} = D_c - D_a$. The vertical-lined area is equal to the shear fracture energy G_c defined by Eq. (3.6).

Thus, when an intact rock specimen in the brittle regime fails in shear mode, non-elastic deformation necessarily concentrates in the zone where the macroscopic shear rupture eventually occurs, and the rate of slip-strengthening decreases as the peak strength approaches. The relative displacement D up to D_a, at which the peak strength τ_p is attained, consists of integrated amounts of local slip caused by microcracking that necessarily occurs as a preparatory phase of the imminent macroscopic rupture. The preparatory phase of non-elastic deformation during which microcracking, crack-interaction and coalescence take place is therefore an integral part of the eventual macroscopic shear fracture. Without this preparatory phase, intact rock cannot fail macroscopically, so this phase is a crucial constitutive property of the shear fracture of intact rock, and therefore must be incorporated into the constitutive relation.

Note that the relation between τ and D shown in Figure 3.5 is self-consistent as the constitutive relation for the shear failure of intact rock. In Figure 3.5, τ_{i0} denotes the critical stress above which the shear stress τ increases with ongoing displacement D, τ_p denotes the peak shear strength, τ_r denotes the residual friction stress level, D_a denotes the displacement at which τ has its peak value τ_p, and D_c denotes the breakdown displacement, which is the slip displacement at the end of the breakdown process (that is, the critical displacement required for τ to degrade to τ_r). The breakdown stress drop $\Delta\tau_b$ is defined as $\Delta\tau_b = \tau_p - \tau_r$, and the slip-weakening displacement D_{wc} is defined as the slip displacement required for the shear strength to degrade from τ_p to τ_r. Accordingly,

$$D_{wc} = D_c - D_a. \tag{3.5}$$

The shear fracture energy G_c is given by (ref. Eq. (2.44))

$$G_c = \int_{D_0}^{D_c} [\tau(D) - \tau_r] dD, \tag{3.6}$$

which equals the vertical-lined area in Figure 3.5. In Eq. (3.6), $\tau(D)$ represents a constitutive relation between τ and D, and D_0 represents the slip displacement at which the shear stress versus slip displacement curve intersects with the line $\tau = \tau_r$ parallel to the abscissa axis in Figure 3.5.

The physical quantities τ_{i0}, τ_p, $\Delta\tau_b$, D_a, and D_c specifically prescribe a constitutive relation for the shear failure of intact rock, and hence they are crucial parameters for formulating the constitutive law for the shear failure of intact rock. In particular, the parameters τ_p, $\Delta\tau_b$, and D_c are critically important, because they substantially determine the fundamental property of the constitutive relation at a given ambient condition. These constitutive relation parameters were evaluated directly from the constitutive relation determined by each laboratory experiment at a given ambient test condition.

3.1.3 Geometric irregularity of shear-fractured surfaces and characteristic length

Actual rupture surfaces of heterogeneous materials such as rock are not flat planes, but inherently have geometric irregularity. It has been demonstrated that the breakdown process of shear rupture is greatly affected by the geometric irregularity of the rupturing surfaces, and that the characteristic length scale representing the geometric irregularity plays a key role in scaling scale-dependent physical quantities inherent in the rupture (Ohnaka and Shen, 1999; Ohnaka, 2003). It is therefore important to understand how the characteristic length representing the geometric irregularity of shear-fractured surfaces of intact rock can be quantified.

In the past, there was a persistent idea that rock fracture surfaces exhibit self-similarity at all scales; however, this does not hold true for shear-fractured surfaces of intact rock. Elaborate investigations lead us to conclude that the shear-fractured surfaces of intact granite commonly exhibit band-limited self-similarity. This is because the slipping process during the breakdown is the process that smoothes away any geometric irregularities of the shear-rupturing surfaces. Figure 3.6(a) shows an example of the shear-fractured surface profiles for the intact granite specimens tested (Ohnaka, 2003). The profiles of these shear-fractured surfaces were measured with a laser-beam profilometer, and the power spectral

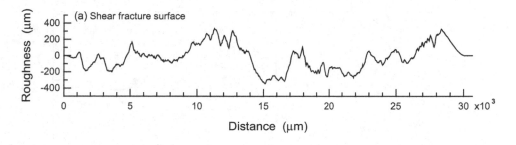

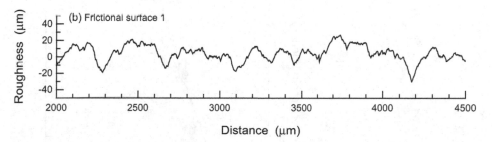

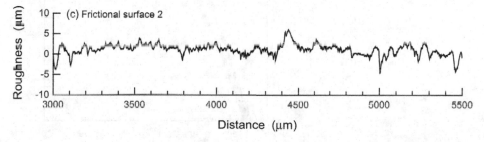

Fig. 3.6   (a) Shear fracture surface profile of an intact Tsukuba granite sample. (b) Profile of a precut Tsukuba granite interface with geometric irregularity (roughness) characterized by the characteristic length $\lambda_c = 200 \, \mu$m. (c) Profile of a precut Tsukuba granite interface with geometric irregularity (roughness) characterized by the characteristic length $\lambda_c = 100 \, \mu$m (Ohnaka, 2003).

density was calculated for the topographic profiles measured. Figure 3.7 shows a plot of the logarithm of the power spectral density against the logarithm of the wavelength for the profile shown in Figure 3.6(a) (the curve labeled "Shear fracture surface").

Since shear fracture surfaces are self-similar within a finite scale range, the characteristic length λ_c for a shear fracture surface can be defined as the critical wavelength beyond which the geometric irregularity of the shear fracture surface no longer exhibits self-similarity. The characteristic length λ_c thus defined represents a predominant wavelength component of the geometric irregularity of a shear-fractured surface. The amplitude and wavelength of a predominant component which departs from the self-similarity are not only limited by the specimen length, but also prescribed by the structural heterogeneity of the rock fabric which, in turn, will be prescribed by the spatial distribution, size, and shape of preexisting microcracks and rock-forming mineral grains contained in the rock specimen.

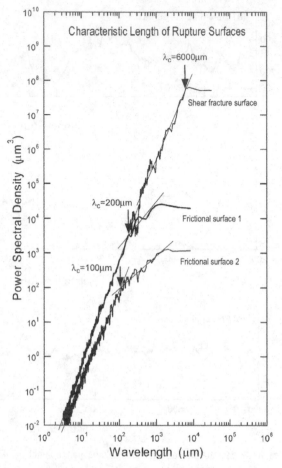

Fig. 3.7 Log–log plots of power spectral density against wavelength for the shear fracture surface profile of an intact Tsukuba granite sample, and profiles of precut Tsukuba granite interfaces with two different geometric irregularities (roughnesses), shown in Figure 3.6 (Ohnaka, 2003).

When the specimen size is fixed, the geometric irregularity and the resulting predominant wavelength component λ_c of the shear-fractured surfaces of the intact rock specimen are exclusively prescribed by the (structural) heterogeneity of the rock fabric. From Figure 3.7, we have $\lambda_c = 6$ mm for the fracture surface profile shown in Figure 3.6(a). Note that the characteristic length λ_c of 6 mm is indeed much shorter than the entire length ($\cong 40$ mm in this case) and width (the maximum size of 16 mm) of the fault formed obliquely across the circular cylinder specimen tested.

When a rupture surface has multiple band-limited self-similarities, a different fractal dimension can be calculated for each band, and in this case, a characteristic length λ_c can be defined as the corner wavelength that separates the neighboring two bands with different fractal dimensions. The characteristic length thus defined also represents a predominant wavelength component of the geometric irregularity of the rupture surface. The characteristic length defined as the wavelength at the upper fractal limit may be regarded as a special case.

3.2 Frictional slip failure on precut rock interface

3.2.1 Method and apparatus used

In the shear failure (or fracture) test of intact rock under confining pressure, described in the previous section, the breakdown process occurs almost simultaneously everywhere on the shear rupture surfaces, because the size of the test specimen is very small in relation to the breakdown zone length. The term *breakdown zone* is defined as the zone in which the shear strength degrades transitionally to a residual friction stress level with ongoing slip behind the rupture front (cf. Figure 2.10). In the test of frictional slip failure on a precut interface, a precut fault sufficiently long in relation to the breakdown zone length was used to observe not only the constitutive relation for frictional stick-slip failure but also how the slip failure nucleates and subsequently propagates on the fault plane. It is essential that the entire fault be sufficiently long compared to the length of the breakdown zone, in order to observe the breakdown process of a propagating frictional stick-slip failure, and the transition process from the formation of frictional slip failure nucleus to its unstable, high-speed propagation on the fault.

A biaxial testing apparatus with vertical and lateral actuators (or rams), independently servo-controlled, was utilized for the experiments on frictional stick-slip failure on a precut rock interface. Tsukuba granite was also chosen as the test sample. The experimental configuration was a double-direct-shear failure test of mode II (or sliding mode) type along two parallel preexisting faults (Ohnaka and Shen, 1999). In this configuration, three blocks of the granite rock with planar and parallel faces were set up in a biaxial testing apparatus, as shown in Figure 3.8. The stiffness of this loading system was 5×10^6 N/cm. One of the blocks was $250 \times 290 \times 50$ mm (labeled A in Figure 3.8), and the other two blocks were each $100 \times 290 \times 50$ mm (labeled B and C in Figure 3.8). Thus, the 290 mm long and 50 mm wide mating surfaces of both sides of the inner block B acted as two parallel faults (weak junctions) under an applied compressive normal load. The compressive normal load across the fault plane was first applied by a lateral ram, which was servo-controlled for the load to be held constant at a test run value. The tangential force (or shear load) on the fault plane was then applied, independent of the normal load across the fault plane, by a vertical ram, to push the inner block B between the two other blocks A and C (see Figure 3.8). The shear load application was servo-controlled such that the sample was deformed at a constant rate of 6.7×10^{-4} mm/s throughout the run, unless otherwise stated. The normal load and the shear load were measured directly with force transducers.

Whether or not an unstable frictional slip failure (or stick-slip failure) occurs on a precut fault depends on the stiffness of the loading system, and on the roughness of the fault

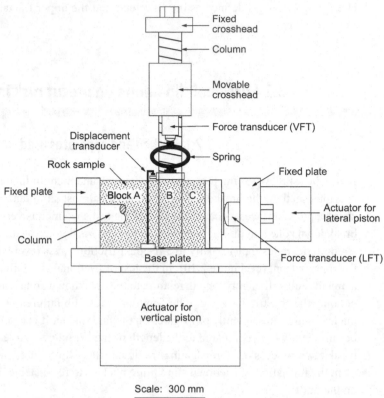

Scale: 300 mm

Fig. 3.8   A biaxial testing apparatus with vertical and lateral actuators (or rams), utilized for the experiments on frictional stick-slip rupture on a precut rock interface (Ohnaka and Shen, 1999).

surfaces. In the present experimental configuration, the loading system with a stiffness of 5×10^6 N/cm worked effectively to cause stick-slip failure instabilities on a smooth fault; however, it worked to cause stable frictional sliding on a rough fault. In order to cause stick-slip failure instabilities on the rough fault, the loading system stiffness was lowered to 6×10^5 N/cm by using an oval spring (see Figure 3.8). The loading system with this lowered stiffness worked effectively to cause stick-slip failure instabilities on the rough fault.

In the double-direct-shear failure test, the shear load application leads to the accumulation of shear stresses along the two parallel faults, and this eventually results in the onset of sliding mode shear rupture nucleation on either of the two faults. In these tests, blocks B and C were simply utilized as auxiliaries to bring about shear stress along the interface of blocks A and B, so that we analyzed the events of frictional stick-slip failure that began to nucleate first only on the fault between blocks A and B. This is because the primary interest is in investigating the physical process of shear rupture nucleation not contaminated by the effect of (parallel) fault–fault interaction.

To get constitutive information on how the local strength in the breakdown zone behind a rupturing front progressively degrades with ongoing local slip, a series of semiconductor

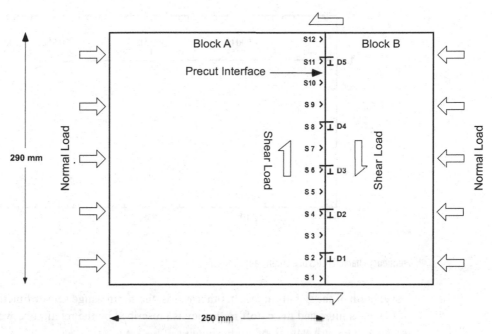

A precut rock interface (or precut fault) between blocks A and B, with two types of sensors in position (Ohnaka and Shen, 1999). S1 to S12 denote the positions at which 12 pairs of strain gauges were mounted to measure local shear strains in the direction parallel to the fault, and D1 to D5 denote the positions at which displacement sensors were mounted to measure local slips on the fault.

strain gauges, with an active gauge length of 2 mm, were mounted on block A at 25 mm intervals along the fault between blocks A and B, and at positions 5 mm from the fault; 12 pairs of strain gauges with mutually perpendicular orientations were used to measure local shear strains in the direction parallel to the fault (S1 to S12 in Figure 3.9). The local shear stresses along the fault were evaluated by multiplying these local shear strains by the modulus of rigidity ($\mu = 2 \times 10^4$ MPa) of the Tsukuba granite sample. The deployment of the series of 12 shear strain sensors at 25 mm intervals along the fault enabled observation of frictional stick-slip failure nucleation and its subsequent dynamic propagation on the fault.

Frequency characteristics of a semiconductor strain gauge sensor mounted on the sample block surface are determined by: (1) the frequency response of the semiconductor crystal used, (2) the cementing technique, and (3) the relation between the gauge length and the wavelength of strain signals. The theoretical limitation of the frequency response of the semiconductor crystal ($> 10^5$ MHz) is much higher than the frequency range of signals of present concern (< 1 MHz). Hence, the limitation of the frequency response of the crystal does not need to be considered. In the present experiments, strain gauges were carefully cemented with alpha-cyano-acrylate cement. Creep of this cement is negligible for the timescale of the present experiments. Accordingly, frequency characteristics of the strain gauge sensor are practically determined by the relation between the gauge length and the

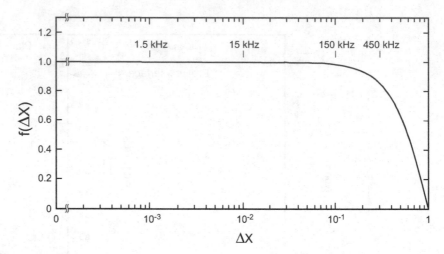

Fig. 3.10 Frequency characteristics of a strain gauge sensor.

wavelength of the signals alone. In other words, the strain gauge sensor functions as a kind of low pass filter, and the cutoff frequency is prescribed by the relation between the gauge length and the wavelength of strain signals.

The transfer function $f(\Delta X)$ of a strain gauge sensor with effective length L is given by (e.g., Ohnaka et al., 1983, 1986; Ohnaka and Kuwahara, 1990; Ohnaka and Shen, 1999)

$$f(\Delta X) = \frac{\sin(\pi \Delta X)}{\pi \Delta X}, \tag{3.7}$$

where $\Delta X = (fL\cos\theta)/c$. Here, θ is the angle between the direction of the gauge length and the direction of the signal propagation, c is the propagation velocity of the signal, f is the signal frequency, and the signal is assumed to be plane waves. When the propagation direction of a signal is parallel to the direction of the gauge length, the frequency response of a strain gauge sensor with a gauge length of 2 mm is flat from DC to 450 kHz (−1.4 dB), if the propagation velocity of the signal is assumed to be 3 km/s (Figure 3.10). This shows that the use of a strain gauge with an effective length of 2 mm as a sensor to measure local shear strain is sufficient for the present purpose.

Local slip displacements between the two sides of the fault are also required to investigate constitutive properties in the breakdown zone behind the propagating rupture front. To measure local slip at a point on the fault, two types of sensors were used: a displacement transducer of eddy current loss type (DTEC), and a metallic foil strain gauge (MFSG). Local slip displacements, unless otherwise stated, were measured at five points along the fault (D1 to D5 in Figure 3.9).

The displacement transducer of eddy current loss type utilizes the principle that a change in the distance between the sensor and the corresponding metal target is uniquely converted to a change in the inductance of the sensor coil to which high-frequency electric current has been supplied. The inductance change was converted to a change in the output voltage, which was recorded. The output frequency response of this displacement measuring system including the sensor, amplifier and linearizer was flat from DC to 20 kHz (−3dB), and the

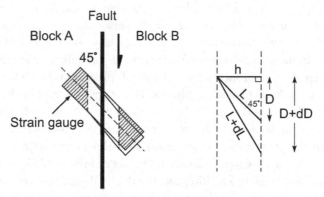

Fig. 3.11 Illustration of how to measure a local slip on a precut fault between two blocks A and B with a strain gauge. The strain gauge was cemented with α-cyano-acrylate cement (hatched portions) such that the strain gauge makes an angle of 45° with respect to the fault length direction. L and $L + dL$ indicate active gauge lengths before and after slip, respectively, and D and $D + dD$ indicate their components in the slip direction before and after slip, respectively (Ohnaka and Shen, 1999).

accuracy and resolution were within 2.9×10^{-4} mm. Displacement sensors of this type were used in most of the experiments on frictional stick-slip failure generated on faults with rough surfaces.

Frictional stick-slip failure on a precut fault with smoother surfaces occurs more violently than frictional stick-slip failure on a precut fault with rough surfaces, when other conditions are equal. This suggests that local slip displacements during slip failure on a smoother fault move more quickly, and may contain frequency components significantly higher than 20 kHz. For this reason, the displacement sensor of eddy current loss type was not used for measuring local slip displacements during frictional stick-slip failure on smooth faults. Instead, metallic foil strain gauges with an active gauge length of 2 mm were used as the displacement transducer in a series of experiments on frictional stick-slip failures on smooth faults.

Figure 3.11 illustrates how to measure a local slip between the two blocks with a metallic foil strain gauge. The deformation (dL/L) of a strain gauge sensor with active gauge length L is converted to the amount of slip (dD) on the fault by

$$dD = \sqrt{(L + dL)^2 - h^2} - D$$
$$= h\left(\sqrt{2(1 + dL/L)^2 - 1} - 1\right)$$
$$\approx h\left(2\frac{dL}{L}\right) = \sqrt{2}L\frac{dL}{L}. \tag{3.8}$$

Equation (3.8) shows that the maximum amount of measurable slip is limited by the maximum durable deformation of the strain gauge used; for example, when $L = 2$ mm, $dD = 2.8\,\mu$m for $dL/L = 10^{-3}$ and $dD = 28\,\mu$m for $dL/L = 10^{-2}$. The disadvantage of this technique is that strain gauge sensors must be renewed after several consecutive stick-slip failures. Nevertheless, this technique was used in a series of laboratory experiments on frictional stick-slip failure (Ohnaka et al., 1986, 1987a, 1987b; Ohnaka and Yamashita,

1989; Ohnaka and Kuwahara, 1990; Ohnaka and Shen, 1999), to measure dynamic local slips on the fault, because MFSG sensors have a high sensitivity (10^{-2} μm), and flat frequency response from DC to a frequency higher than 100 kHz.

In the experiments on frictional stick-slip failure generated on a rough fault, amplified signals of local shear strains and slip displacements along the fault, signals of the shear and normal loads remotely applied, and the overall deformation between both ends of the sample blocks A and B along the fault, were sampled at a preset frequency of 500 Hz synchronized by a single clock, with a multi-channel analog-to-digital converter with 14-bit resolution. In the experiments on frictional stick-slip failure on a smooth fault, all the amplified analog signals were bifurcated. One set of bifurcated signals was sampled at a frequency of 500 Hz using a multi-channel analog-to-digital converter with 14-bit resolution, and the other set of the signals was sampled at a frequency of 1 MHz using another multi-channel analog-to-digital converter with 12-bit resolution. In the experiments on frictional stick-slip failure on an extremely smooth fault, the amplified signals were sampled at a frequency of 1 MHz using a multi-channel analog-to-digital converter with 12-bit resolution. The overall response of the entire measuring and recording system was flat from DC to 200 kHz for the local shear strain signals, DC to 20 kHz (when DTEC sensors were used) or DC to 100 kHz (when MFSG sensors were used) for the local slip displacement signals, and DC to 2 kHz for the remotely applied shear and normal load signals (for details, see Ohnaka and Shen, 1999).

3.2.2 Geometric irregularity of precut fault surfaces and characteristic length

As mentioned in subsection 3.1.3, the breakdown process of shear rupture is affected by the geometric irregularity of the rupturing surfaces. In order to substantiate this, we need to prepare precut faults with different surface topographies (or roughnesses). To prepare faults with different surface roughnesses, precut fault surfaces were ground flat with a reciprocating surface grinder to an accuracy higher than 0.02 mm per 1 cm length, and then these flat surfaces were ground with carborundum grit of different grain sizes, ranging from coarse grit (#60, representative particle sizes of 210–250 μm) to fine grit (#2000, representative particle sizes of 7–9 μm). The profiles of these fault surfaces were measured with a diamond stylus profilometer with a tip radius of 2 μm. The resolution of the elevation data collected was 0.0084 μm (Ohnaka and Shen, 1999).

Figures 3.6(b) and 3.6(c) in the previous section show two examples of the topographic surface profiles of precut faults with two different roughnesses, which were used for the tests on frictional stick-slip failure. The power spectral densities for these fault surface profiles were calculated, and plotted in Figure 3.7 for comparison. From Figure 3.7, the corner wavelength of 200 μm is found for the rough fault surface shown in Figure 3.6(b), and the corner wavelength of 100 μm is found for the relatively smooth fault surface shown in Figure 3.6(c). These corner wavelengths represent the characteristic lengths for individual fault surfaces, and are much shorter than the entire fault length (290 mm) and width (50 mm). Note that the amplitude and wavelength of the predominant wave component of the geometric irregularity of the precut fault surfaces are neither relevant to the specimen

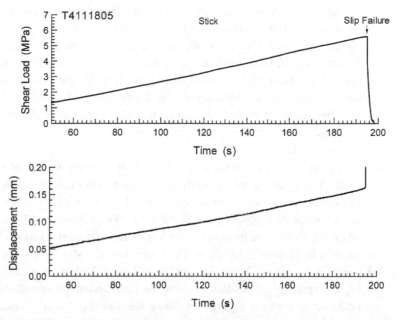

Fig. 3.12 Time records of remotely applied shear load and the relative displacement between both ends of the sample blocks A and B for a stick-slip rupture event (T4111805) on the rough fault. Reproduced from Ohnaka and Shen (1999).

size nor to the rock fabric, but related to the grain size of the carborundum grit used. Note also that these characteristic lengths are much shorter than the characteristic length of 6000 μm obtained for the shear-fractured surfaces of intact granite.

It will be shown later (see Section 3.3, and Chapters 5 and 6) that the characteristic length λ_c defined here plays a key role not only in scaling scale-dependent physical quantities inherent in the rupture, but also in unifying laboratory-derived constitutive relations for the shear fracture of intact rock and for frictional stick-slip failure on a precut rock interface.

3.2.3 Constitutive relations derived from data on frictional stick-slip failure

When shear load is applied at a constant rate with the vertical ram along a fault while an applied load normal to the fault is held constant, the shear stress along the fault builds up as shown in Figure 3.12. Figure 3.12 shows an example of recorded data on frictional stick-slip failure (stick-slip event T4111805) generated on a precut fault with rough surfaces (Ohnaka and Shen, 1999). In this figure, the shear load, which is the output signal from force transducer VFT in Figure 3.8, and the relative displacement measured between both ends of the sample blocks A and B along the fault in Figure 3.8 are plotted against time. In this experiment, the shear load was remotely applied at a constant deformation rate of 6.7×10^{-4} mm/s under the applied normal load of 6.2 MPa. The origin of time t was set such that $t = 0$ when the shear load was applied.

Figure 3.12 indicates that the remotely applied shear load increased with time at a constant rate until frictional stick-slip failure (event T4111805) occurred at $t = 195.08$ s. No strength degradation or slip acceleration preceding the event is discernible from Figure 3.12. If, however, local shear stresses are monitored at a series of different positions along the fault, it is clearly observable that the distribution of local shear stresses is non-uniformly induced and accumulated along the fault during the shear load application, and that the fault strength gradually degrades in a localized zone, prior to an eventual, dynamic slip failure that rapidly propagates over the entire fault at $t = 195.08$ s, as shown in Figure 3.13.

Figures 3.13 and 3.14 show examples of the time series plots of local shear stresses and local slip displacements, respectively, along a precut fault, which were observed during the nucleation process of frictional stick-slip failure event T4111805 (for the details of the slip failure nucleation process, see Section 5.1). From Figure 3.13, we can identify the time when the slip failure began to occur locally at individual positions S1 to S12 on the fault. Arrows in Figure 3.13 indicate the time when the slip failure began to occur locally at individual positions along the fault. From Figure 3.13, we can confirm that the slip failure began to nucleate at position S4 along the fault, and that it extended bi-directionally toward both ends of the fault. Figure 3.14 shows that slow slips began to proceed locally at positions D1, D2, and D3, prior to the imminent overall stick-slip failure, which actually occurred at $t = 195.08$ s. These local, slow slips correspond to local shear stress degradations at the same positions in the nucleation zone (note from Figure 3.9 that local slips at D1, D2, and D3 correspond to local shear stress degradations at S2, S4, and S6, respectively). In contrast, however, slip occurred abruptly without any slow preslip in the zone of dynamic rupture propagation (e.g., position D5 in Figure 3.14).

Figure 3.15 shows a plot of the local shear stress against the local slip displacement observed at one location during the nucleation of a frictional stick-slip failure event that proceeded quasi-statically on a precut fault with rough surfaces ($\lambda_c = 200$ μm). Figure 3.16 shows a plot of the local shear stress against the local slip displacement observed near the propagating front of a frictional stick-slip failure generated on a smoother fault ($\lambda_c = 100$ μm). These are typical examples of laboratory-derived constitutive relations that govern the instability or stability of frictional slip failure in the breakdown zone behind the propagating rupture front. The numerical values of the constitutive law parameters τ_i, τ_p, $\Delta\tau_b$, D_a, and D_c (or D_{wc}) for frictional stick-slip failure were determined from laboratory-derived constitutive relations as exemplified in Figures 3.15 and 3.16.

From Figures 3.15 and 3.16, one can see that the slip-strengthening phase precedes the slip-weakening process even for frictional slip failure as well as for the shear failure of intact rock. The slip-weakening mechanism for frictional slip failure is the microfracturing process at asperity junction areas (such as asperity interlocking and asperity ploughing) on the fault, and hence its microphysical mechanism is basically the same as that for the shear failure of intact rock. Note, however, that the mechanism of the slip-strengthening phase for frictional stick-slip failure is not the same as that for the shear failure of intact rock. The slip-strengthening on a precut fault with geometrically irregular surfaces under an applied normal load is primarily attained by a slip-induced increase in frictional resistance, due to an increase in the sum of asperity junction areas (such as asperity interlocking and asperity

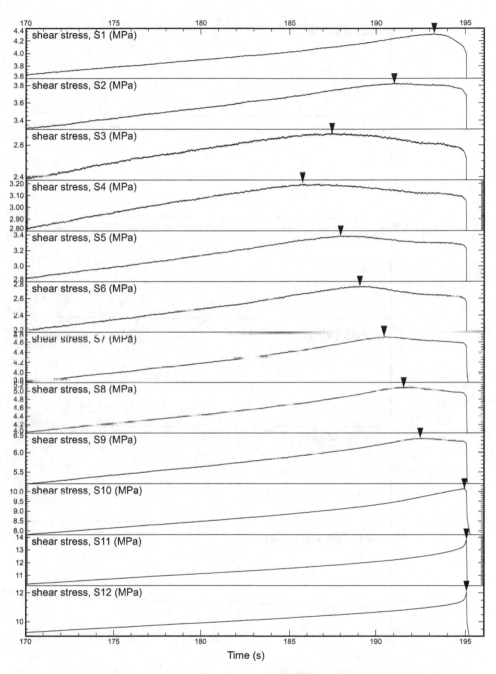

Fig. 3.13 Local shear stress versus time records at 12 positions (S1 to S12) along the fault for a stick-slip rupture event (T4111805) that occurred on the rough fault (Ohnaka and Shen, 1999). Arrows denote the time when slip failure began to occur locally.

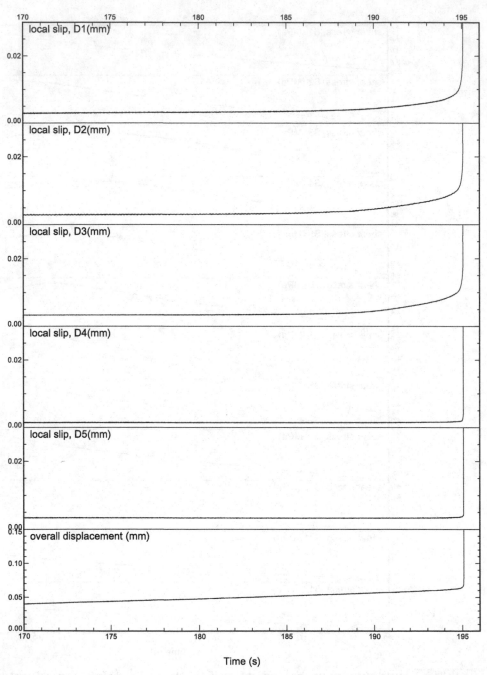

Fig. 3.14 Local slip displacement versus time records at five positions (D1 to D5) along the fault, and a time record of the relative displacement between both ends of sample blocks A and B (bottom frame) for the event T4111805 shown in Figures 3.12 and 3.13 (Ohnaka and Shen, 1999).

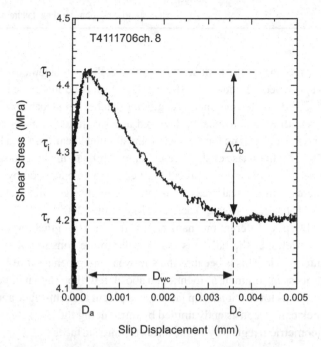

A typical example of the slip-dependent constitutive relations observed during the nucleation process leading to a stick-slip dynamic rupture on a precut rock interface with geometric irregularity (roughness) characterized by $\lambda_c = 200\,\mu m$ (Ohnaka, 2003)

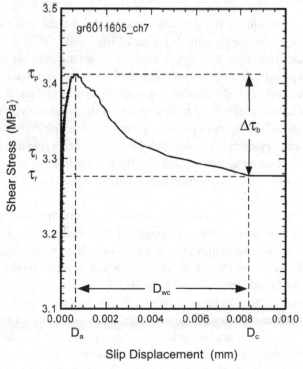

A typical example of the slip-dependent constitutive relations observed near the propagating front of a stick-slip rupture on a precut rock interface with geometric irregularity (roughness) characterized by $\lambda_c = 100\,\mu m$ (Ohnaka, 2003).

ploughing) on the fault with progressive slip displacement. The rate of slip-strengthening $\partial\tau/\partial D$ becomes lower, as slip displacement D comes close to the critical slip displacement D_a, and as the frictional strength approaches its peak value τ_p. At $D = D_a$ the frictional strength attains its peak value, and asperity junction areas resistant to breakdown on the precut fault can no longer sustain the applied shear load and begin to break down. Thus, at $D = D_a$ the phase of slip-strengthening shifts to the breakdown (or slip-weakening) phase. The strengthening of a precut fault can also be enhanced by a time- or slip-rate-dependent increase in the real areas of asperity contact (e.g., Dieterich, 1972a, 1978a; Beeler *et al.*, 1994; Marone, 1998a; Aochi and Matsu'ura, 2002).

Despite different mechanisms of the slip-strengthening phase, the constitutive relation for frictional slip failure is very similar to the constitutive relation for the shear failure of intact rock. This is because the slip-weakening behavior of both failures has the fracturing process in common, as mentioned above. It will be shown in Section 3.3 that the constitutive relations for frictional slip failure on a precut rock interface and for the shear failure of intact rock can be consistently unified by introducing the characteristic length λ_c representing the geometric irregularity of the shear rupture surfaces.

3.2.4 Laboratory-derived relationships between physical quantities observed during dynamic slip rupture propagation

The constitutive law properly formulated must account quantitatively and in a consistent manner for the dynamic behavior of frictional stick-slip rupture propagating on a precut fault. Therefore, in order to examine to what extent the constitutive formulations so far proposed can account quantitatively in a consistent manner for the dynamic behavior of frictional stick-slip rupture, we need experimental data on physical quantities such as slip velocity and slip acceleration observed during dynamic frictional slip ruptures. If some relationships between physical quantities are empirically derived from laboratory data, the laboratory-derived relationships have a significant role in determining what the constitutive formulation ought to be, and how the constitutive law should be formulated. It is therefore important to present experimental data on stick-slip rupture dynamically propagating on a precut fault.

Laboratory experiments on stick-slip rupture dynamically propagating on a precut fault have revealed a number of empirical relationships. It is better in the end to discuss such empirical relationships, in comparison to the corresponding relationships derived theoretically in the framework of fracture mechanics based on the constitutive law formulated for shear rupture (or earthquake rupture). Therefore, we present most of the laboratory-derived empirical relationships in Section 5.3, in which dynamic rupture propagation and generation of strong motion seismic waves are discussed in the framework of fracture mechanics based on the constitutive law for shear rupture. In this subsection, we will present only a few basic examples of the laboratory-derived empirical relationships.

Figure 3.17 shows an example of a time series of local shear stresses recorded at six positions (CH. 1 to CH. 6) and local slip displacements recorded at four positions (CH. 1, CH. 3, CH. 5, and CH. 6) along a fault for a dynamically propagating stick-slip rupture

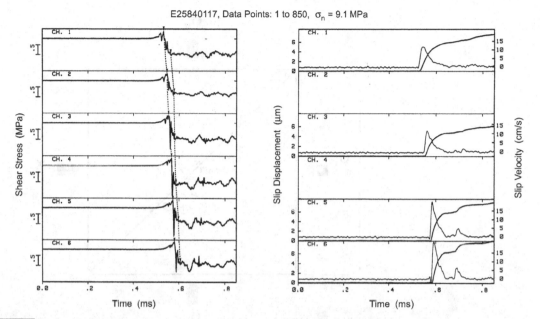

Fig. 3.17   Time series variations of local shear stresses recorded at six positions (CH.1 to CH. 6) and local slip displacements recorded at four positions (CH. 1, CH. 3, CH. 5, and CH. 6) along a precut fault for a stick-slip rupture event that propagated dynamically on the fault under the average normal stress $\sigma_n = 9.1$ MPa. The slip velocity, the time derivative of the slip displacement record smoothed by the moving average of nine points, is also plotted. The arrows indicate the time when the shear stress has its peak value on the verge of local slip failure at individual positions, and the time when the shear stress has just dropped to a residual friction stress level. Reproduced from Ohnaka and Kuwahara (1990).

(mode II type) that occurred on the fault under the normal stress $\sigma_n = 9.1$ MPa (Ohnaka and Kuwahara, 1990). The time series variation of slip velocities at the four positions shown in Figure 3.17 is the time derivative of individual slip displacement records smoothed by a moving average of nine points. The fault configuration used for these experiments (Ohnaka *et al.*, 1986, 1987a, 1987b; Ohnaka and Kuwahara, 1990) was a precut fault along the diagonal of a 5 cm thick square slab (28 cm × 28 cm) of Tsukuba granite (Figure 3.18). Semiconductor strain gauges with an active gauge length of 2 mm were used to measure local shear strains at the six positions along the fault, and the local shear stresses at these locations were evaluated by multiplying the local shear strains by the modulus of rigidity ($\mu = 20\,000$ MPa) of the granite sample used. Metallic foil strain gauge sensors with an active gauge length of 2 mm were used to measure local slips at the four positions on the fault. For the frequency characteristics of the strain gauge sensors and recording system used, refer to subsection 3.2.1.

From such laboratory data as shown in Figure 3.17, we find significant interrelationships between shear stress, slip displacement, and slip velocity observed at one location along a precut fault for a dynamically propagating stick-slip rupture (Ohnaka *et al.*, 1986; 1987a, 1987b; Ohnaka and Yamashita, 1989; Ohnaka and Kuwahara, 1990). Figure 3.19 shows a

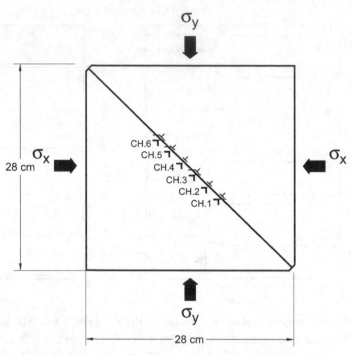

Fig. 3.18   Tsukuba granite rock sample (28 cm × 28 cm × 5 cm) with a precut interface along the diagonal to simulate a preexisting fault (Ohnaka and Kuwahara, 1990). Local dynamic shear strains were measured at six positions (CH.1 to CH.6), and local slip displacements were measured at four positions (CH.1, CH.3, CH.5, and CH.6) along the fault. The local shear stress at a position was obtained from the local shear strain measured at the position, multiplied by the rigidity of the granite sample.

typical example of (a) the observed relationship between local shear stress and local slip displacement, (b) the observed relationship between local shear stress and local slip velocity, and (c) the observed relationship between local slip velocity and local slip displacement, at one location along a precut fault for a stick-slip rupture that dynamically propagated on the fault under the condition of normal stress $\sigma_n = 4.01$ MPa (Ohnaka *et al.*, 1987b; Ohnaka and Yamashita, 1989). The observed relationship between local slip acceleration and local slip displacement is also shown in Figure 3.19(d), where the slip acceleration is the time derivative of the slip velocity.

From Figures 3.19(a)–(d) we can read the breakdown stress drop $\Delta\tau_b$, defined as the peak shear strength τ_p minus the residual friction stress τ_r, the breakdown displacement D_c, defined as the critical slip displacement required for the shear stress to degrade to the residual friction stress level, the peak slip velocity $\dot{D}_{max}$, and the peak slip acceleration $\ddot{D}_{max}$ ($\dot{D}$ and $\ddot{D}$ denote $\partial D/\partial t$ and $\partial^2 D/\partial t^2$, respectively). The slip displacement at which the slip velocity has a peak value is usually smaller than D_c, as exemplified in Figure 3.19(c). This means that the slip velocity usually has a peak value within the breakdown zone. The slip displacement at which the slip acceleration has a peak value is very small, as exemplified in Figure 3.19(d). This means that the slip acceleration has a peak

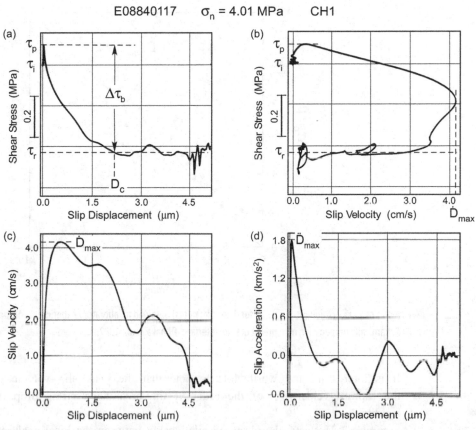

Fig. 3.19   A typical example of (a) the observed relationship between local shear stress and local slip displacement, (b) the observed relationship between local shear stress and local slip velocity, (c) the observed relationship between local slip velocity and local slip displacement, and (d) the observed relationship between local slip acceleration and local slip displacement, at one location along a precut fault for a stick-slip rupture that dynamically propagated along the fault under the condition of normal stress $\sigma_n = 4.01$ MPa. Data from Ohnaka *et al.* (1987b) and Ohnaka and Yamashita (1989).

value in close proximity to the front of a propagating slip rupture within the breakdown zone.

Figure 3.20 shows the observed relationship between the local breakdown stress drop $\Delta\tau_b$ divided by the rigidity μ (or local breakdown strain drop) and the local peak slip velocity $\dot{D}_{max}$ divided by the rupture velocity V (Ohnaka *et al.*, 1987a, 1987b). The black dots plotted in Figure 3.20 denote data points obtained in laboratory experiments on dynamically propagating stick-slip ruptures. The broken straight line in the figure indicates that the local breakdown strain drop is equal to the local peak slip velocity normalized to the rupture velocity; that is,

$$\frac{\Delta\tau_b}{\mu} = \frac{\dot{D}_{max}}{V}. \tag{3.9}$$

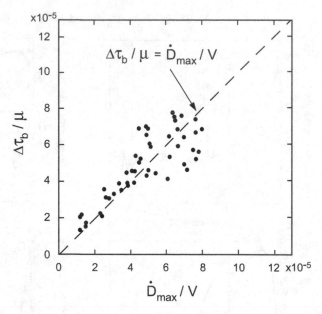

Fig. 3.20 A plot of $\Delta\tau_b/\mu$ against $\dot{D}_{max}/V$ for data (black dots) obtained in laboratory experiments on stick-slip ruptures dynamically propagating on a precut granite interface (Ohnaka *et al.*, 1987b).

It is interesting and worthwhile to note that the peak slip velocity normalized to the rupture velocity is of the order of the breakdown stress drop divided by the rigidity.

Figure 3.21 shows the observed relationship between the local peak slip acceleration and the local peak slip velocity. The black dots in the figure indicate data points obtained in the laboratory experiments on dynamically propagating stick-slip ruptures on a precut fault shown in Figure 3.18 (Ohnaka *et al.*, 1987a, 1987b; Ohnaka, 2003). The data points plotted in Figure 3.21 fall on and around the broken straight line denoting the following relationship (Ohnaka *et al.*, 1987a, 1987b):

$$\ddot{D}_{max} = 10^6(\dot{D}_{max})^2, \tag{3.10}$$

where $\dot{D}_{max}$ and $\ddot{D}_{max}$ are measured in m/s and m/s^2, respectively.

The experimental data presented above are only a few of the many data obtained in laboratory experiments on stick-slip ruptures dynamically propagating on a precut fault. As noted above, other data or laboratory-derived empirical relationships will be presented in Section 5.3, to account in quantitative terms in a consistent manner for the experimental data or laboratory-derived empirical relationships, in comparison to the corresponding relationships or formulae theoretically derived in the framework of fracture mechanics based on the constitutive law formulated for shear rupture (or earthquake rupture). Indeed, the laboratory-derived empirical formulae expressed as Eqs. (3.9) and (3.10) are theoretically derived in Section 5.3, together with other formulae.

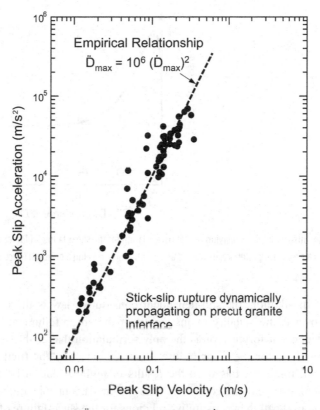

Fig. 3.21 A log–log plot of peak slip acceleration $\ddot{D}_{max}$ against peak slip velocity $\dot{D}_{max}$ for data (black dots) obtained in laboratory experiments on stick-slip ruptures dynamically propagating on a precut granite interface. Data from Ohnaka *et al.* (1987a, 1987b) with additional unpublished data.

3.3 Unifying constitutive formulation and a constitutive scaling law

3.3.1 Unification of constitutive relations for shear fracture and for frictional slip failure

As presented in Sections 3.1 and 3.2, high-resolution laboratory experiments have shown that constitutive relations for the shear fracture (or failure) of intact rock and for the frictional slip failure on a precut rock interface are commonly expressed as the shear traction τ being a function of the slip displacement D, as illustrated in Figure 3.22. This type of constitutive law is referred to as the *slip-dependent constitutive law*, since the shear traction τ during the breakdown process is a function of the slip displacement D. The slip-dependent constitutive law is self-consistent as the governing law for shear failure (or rupture), as pointed out in earlier papers (Rice, 1980, 1983, 1984; Rudnicki, 1980, 1988).

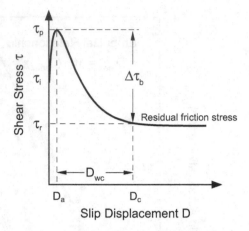

An illustration of the unifying constitutive relation for the shear failure of intact rock and for frictional stick-slip failure on a precut rock interface, expressed as shear traction τ being a principal function of slip displacement D.

In addition, the slip-dependent constitutive law is the only law that self-consistently governs the stability or instability of the shear failure of intact rock. Hence, the slip-dependent formulation is the only formulation that enables the unification of constitutive relations for the shear failure of intact rock and for frictional slip failure on a precut rock interface. Based on the results of high-resolution laboratory experiments, we will demonstrate in this section that the slip-dependent constitutive formulation indeed enables the unification of constitutive relations for the shear failure of intact rock and for frictional slip failure on a precut rock interface.

The specific functional form of the slip-dependent constitutive law is dictated by the following constitutive law parameters: τ_p (peak shear strength), τ_r (residual friction stress), D_a (critical slip displacement at which the peak shear strength is attained), and D_c (breakdown displacement defined as the critical slip displacement required for the shear traction to degrade to τ_r); or equivalently τ_p, $\Delta\tau_b$ (breakdown stress drop defined as $\Delta\tau_b = \tau_p - \tau_r$), D_a, and D_c or D_{wc} (slip-weakening displacement defined as $D_{wc} = D_c - D_a$). We will show below that these slip-dependent constitutive law parameters derived from laboratory data on the shear failure of intact rock and on frictional stick-slip failure on a precut rock interface can be unified by single formulae.

Figure 3.23 shows a plot of the logarithm of the breakdown stress drop $\Delta\tau_b$ against the logarithm of the breakdown displacement D_c for laboratory data on the shear failure of intact granite (Section 3.1), and on frictional stick-slip failure on precut faults with two different surface roughnesses (Section 3.2). In this figure, the black squares denote data on the shear failure of intact granite specimens, and the white and black triangles denote data on the frictional stick-slip failure on precut faults having surface roughnesses characterized by $\lambda_c = 100$ µm and $\lambda_c = 200$ µm, respectively. The $\Delta\tau_b$ versus D_c relation for actual earthquakes are also over-plotted in Figure 3.23 for later discussion in Chapter 6. In this section, we will concentrate on laboratory data alone. Laboratory data on

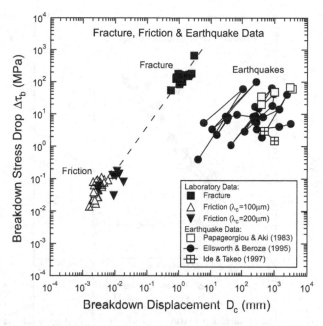

Fig. 3.23 Plots of the logarithm of the breakdown stress drop $\Delta \tau_b$ against the logarithm of the breakdown displacement D_c for laboratory data on the shear failure of intact Tsukuba granite and frictional stick-slip ruptures on precut Tsukuba granite interfaces having two different geometric irregularities (roughnesses) characterized by $\lambda_c -$ 100 and 200 μm. Field data on earthquakes are also over-plotted for comparison. Reproduced from Ohnaka (2003).

shear fracture and frictional stick-slip failure will be compared with field data on earthquakes, leaving scale-independent or scale-dependent physical quantities to be discussed in Chapter 6.

It can be seen in Figure 3.23 that $\Delta \tau_b$ for the shear failure of intact granite is highest among these data sets, while $\Delta \tau_b$ for the frictional stick-slip failure is lowest. More specifically, $\Delta \tau_b$ is in the range 10^2 to 10^3 MPa for the shear failure of intact rock, and in the range 1×10^{-2} to 2×10^{-1} MPa for frictional stick-slip failure. Thus, $\Delta \tau_b$ for the shear failure of intact rock is roughly four orders of magnitude greater than that for the frictional stick-slip failure. The large difference in $\Delta \tau_b$ between shear fracture and frictional stick-slip failure is partly attributed to a substantial difference between the cohesive strength of intact rock and the frictional strength on a precut rock interface. The difference is also partly attributed to a difference in the magnitude of the normal load (or confining pressure) applied during the experiments on the shear failure of intact rock and the frictional stick-slip failure on a precut fault. To cancel out these differences in strength and applied normal load, $\Delta \tau_b$ needs to be normalized to τ_p, given that τ_p is an increasing function of applied normal load (or confining pressure).

The breakdown displacement D_c for the shear failure of intact granite specimens ranged from 0.5 to 3 mm, whereas D_c for the frictional stick-slip failure occurring on precut rock interfaces with different surface roughnesses ranged from 1 to 20 μm (Ohnaka, 2003). Thus, D_c for the shear failure of intact rock specimens is two to three orders of slip amount

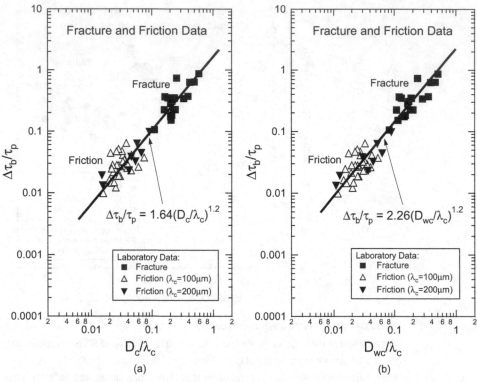

A plot of (a) the logarithm of $\Delta\tau_b/\tau_p$ against the logarithm of D_c/λ_c, and (b) the logarithm of $\Delta\tau_b/\tau_p$ against the logarithm of D_{wc}/λ_c, for laboratory data on the shear failure of intact Tsukuba granite and frictional stick-slip ruptures on precut Tsukuba granite interfaces having two different geometric irregularities (roughnesses) characterized by $\lambda_c = 100$ and 200 μm (Ohnaka, 2003). The formulae shown in the figures were derived from the double-error regression analysis of all data points plotted in individual figures (see text).

greater than D_c for the frictional stick-slip failure. As noted previously, the breakdown process is strongly affected by the geometric irregularity of the fault surfaces, and the geometric irregularity is represented by the characteristic length λ_c. Therefore, D_c needs to be properly normalized to λ_c in order to gain a unified comprehension for the present sets of data on fracture and on friction. Indeed, it will be confirmed later in subsection 3.3.2 that D_c scales with λ_c.

Figure 3.24(a) shows a plot of the logarithm of $\Delta\tau_b/\tau_p$ against the logarithm of D_c/λ_c for the same sets of data on shear fracture and on friction, shown in Figure 3.23 (Ohnaka, 2003). The black squares in Figures 3.24(a) and (b) denote data on the shear fracture and the white and black triangles denote data on the frictional stick-slip failure of precut faults with $\lambda_c = 100$ μm and $\lambda_c = 200$ μm, respectively. From Figure 3.24(a) we find that these different sets of data on shear fracture and on frictional stick-slip failure are unified by a single relationship in the $\Delta\tau_b/\tau_p - D_c/\lambda_c$ domain, and that the parameters $\Delta\tau_b/\tau_p$ and D_c/λ_c are interdependent. Note also from Figure 3.24(a) that the two subsets of data on the frictional stick-slip failure occurring on precut faults with different surface roughnesses characterized by $\lambda_c = 100$ μm and $\lambda_c = 200$ μm are also unified within experimental error.

Figure 3.24(a) indicates that the relationship between $\Delta\tau_b/\tau_p$ and D_c/λ_c is well represented by a power law of the form (Ohnaka, 2003)

$$\frac{\Delta\tau_b}{\tau_p} = \beta\left(\frac{D_c}{\lambda_c}\right)^M,$$ (3.11)

where β and M are numerical constants. A double-error regression analysis (York, 1966) of the data sets plotted in Figure 3.24(a) leads to the following values for β and M with their standard deviations: $\beta = 1.64 \pm 0.29$ and $M = 1.20 \pm 0.06$. The correlation coefficient for these data points is 0.933. In this analysis, equal weights of the values have been assumed for both $\Delta\tau_b/\tau_p$ and D_c/λ_c. Equation (3.11) represents a constraint to be imposed on the constitutive law parameters τ_p, $\Delta\tau_b$, and D_c.

The breakdown displacement D_c is the sum of the displacement D_a (required for the shear stress to increase to the peak strength τ_p) and the displacement D_{wc} (required for the shear strength to degrade from τ_p to the residual friction stress level τ_r). By definition, D_a is the displacement involved in the displacement-hardening (or slip-strengthening), and D_{wc} is the displacement involved in the slip-weakening. One may therefore expect that the relation between $\Delta\tau_b/\tau_p$ and D_{wc}/λ_c will also be represented by a power law of form similar to Eq. (3.11). Figure 3.24(b) shows a plot of the logarithm of $\Delta\tau_b/\tau_p$ against the logarithm of D_{wc}/λ_c for the fracture data and the frictional stick-slip failure data. We find from Figure 3.24(b) that the relation between $\Delta\tau_b/\tau_p$ and D_{wc}/λ_c obeys a power law of the form (Ohnaka, 2003)

$$\frac{\Delta\tau_b}{\tau_p} = \beta'\left(\frac{D_{wc}}{\lambda_c}\right)^M,$$ (3.12)

where β' and M are numerical constants. The double-error regression analysis for these data points gives the following values for β' and M with their standard deviations: $\beta' = 2.26 \pm 0.38$ and $M = 1.20 \pm 0.055$, and 0.954 for the correlation coefficient.

A unified comprehension of shear fracture and frictional stick-slip failure can also be gained from laboratory data on D_c (or D_{wc}) and D_a. Relational expression (3.12) is equivalent to relational expression (3.11), and the exponent of D_{wc}/λ_c in Eq. (3.12) coincides with the exponent of D_c/λ_c in Eq. (3.11). This leads to the conclusion that both D_{wc} and D_a are directly proportional to D_c. From Eqs. (3.11) and (3.12), we indeed have

$$D_{wc} = \left(\frac{\beta}{\beta'}\right)^{1/M} D_c = 0.766 D_c,$$ (3.13)

and from Eqs. (3.5) and (3.13), we have

$$D_a = \left[1 - \left(\frac{\beta}{\beta'}\right)^{1/M}\right] D_c = 0.234 D_c.$$ (3.14)

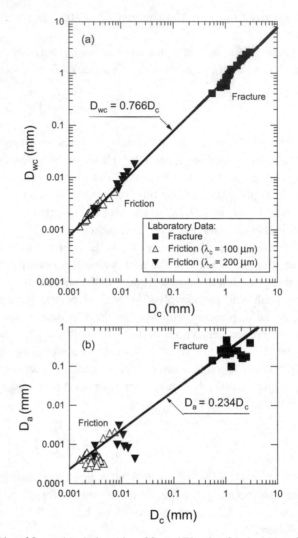

Fig. 3.25 (a) A plot of the logarithm of D_{wc} against the logarithm of D_c, and (b) a plot of the logarithm of D_a against the logarithm of D_c, for laboratory data on the shear failure of intact Tsukuba granite, and frictional stick-slip ruptures on precut Tsukuba granite interfaces having two different geometric irregularities (roughnesses) characterized by $\lambda_c = 100$ and $200\ \mu m$ (Ohnaka, 2003).

The proportional relationships between D_{wc} and D_c, and between D_a and D_c, expressed in Eqs. (3.13) and (3.14), respectively, can be confirmed directly from the present sets of data on D_{wc} and D_c for shear fracture and for frictional stick-slip failure. Figure 3.25(a) shows a plot of D_{wc} against D_c for the sets of data on shear fracture and on frictional stick-slip failure, and Figure 3.25(b) shows a plot of D_a against D_c for the same data sets. One can see from Figure 3.25 that the plot of D_{wc} against D_c is well represented by relational expression (3.13), and that the plot of D_a against D_c is represented by relational expression (3.14).

Figure 3.25 also indicates that experimental data on both the shear failure of intact rock and the frictional stick-slip failure on a precut rock interface are consistently unified by a single constitutive law.

The direct proportional relationship between D_c and D_{wc} indicates that the parameters D_c and D_{wc} are mutually equivalent, as noted above. In addition, Eqs. (3.13) and (3.14) show that the critical weakening displacement D_{wc} is predictable from the critical hardening displacement D_a. This implies that the preceding displacement-hardening process can substantially prescribe the displacement-weakening process. This suggests that the constitutive law parameters D_a, D_{wc}, and D_c may be described in terms of the common underlying physics.

It has been demonstrated that laboratory data on the shear failure of intact rock and the frictional stick-slip failure on a precut rock interface are both unified consistently by a single formula (3.11), or equivalently by formula (3.12). It has also been demonstrated that constitutive law parameters such as τ_p, $\Delta\tau_b$, D_a, and D_c are interdependent, and mutually constrained by Eqs. (3.5) and (3.11) to (3.14). Of these equations, only three equations are independent. Equations (3.5), (3.11), and (3.13) will be regarded as independent hereafter, and the following three constitutive parameters – τ_p, $\Delta\tau_b$, and D_c – will be regarded as fundamental.

3.3.2 A constitutive scaling law

As noted previously, the slip-dependent constitutive law for shear rupture is specifically prescribed by the following five parameters: τ_i, τ_p, $\Delta\tau_b$, D_a, and D_c (or D_{wc}). Of these, the displacement parameters D_a and D_c (or D_{wc}) are scale-dependent. Rewriting Eq. (3.11), we have

$$D_c = m(\Delta\tau_b/\tau_p)\lambda_c, \tag{3.15}$$

where $m(\Delta\tau_b/\tau_p)$ is a dimensionless parameter which is a function of $\Delta\tau_b/\tau_p$. The dimensionless parameter m is expressed as

$$m(\Delta\tau_b/\tau_p) = \left(\frac{1}{\beta}\right)^{1/M}\left(\frac{\Delta\tau_b}{\tau_p}\right)^{1/M} = 0.662\left(\frac{\Delta\tau_b}{\tau_p}\right)^{0.833}. \tag{3.16}$$

From Eqs. (3.13), (3.14), and (3.15), we have

$$D_{wc} = \left(\frac{\beta}{\beta'}\right)^{1/M} m(\Delta\tau_b/\tau_p)\lambda_c = 0.766 \times m(\Delta\tau_b/\tau_p)\lambda_c \tag{3.17}$$

and

$$D_a = \left[1 - \left(\frac{\beta}{\beta'}\right)^{1/M}\right] m(\Delta\tau_b/\tau_p)\lambda_c = 0.234 \times m(\Delta\tau_b/\tau_p)\lambda_c. \tag{3.18}$$

It is confirmed from Eqs. (3.15), (3.17), and (3.18) that all the displacement parameters D_c, D_{wc}, and D_a scale with λ_c, since $\Delta\tau_b/\tau_p$ is scale-independent. In particular, they are directly proportional to λ_c if $\Delta\tau_b/\tau_p$ is constant. Yet, the proportional relationship

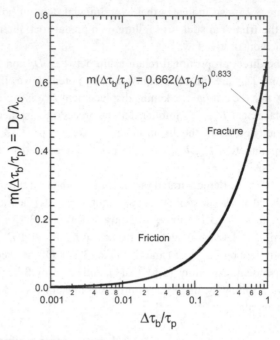

Fig. 3.26 Graphic representation of the dimensionless parameter $m = D_c/\lambda_c$ as a function of $\Delta\tau_b/\tau_p$ (Ohnaka, 2003).

between λ_c and either D_c, D_{wc}, or D_a is violated by the fluctuation of $\Delta\tau_b/\tau_p$, because $\Delta\tau_b/\tau_p$ in general can have different values according to the test conditions. In this sense, a specific scaling relation between D_c (or D_{wc} or D_a) and λ_c is entirely dependent on $\Delta\tau_b/\tau_p$. Figure 3.26 shows how the parameter m ($= D_c/\lambda_c$) depends on $\Delta\tau_b/\tau_p$. The parameter m is a monotonically increasing function of $\Delta\tau_b/\tau_p$, and has a maximum value of 0.66 at $\Delta\tau_b/\tau_p = 1$. For laboratory data on the shear failure of intact Tsukuba granite, $\Delta\tau_b/\tau_p$ had values ranging from 0.1 to 1 (Ohnaka, 2003), and m ranged from 0.1 to 0.66 (see Figure 3.26). By contrast, for laboratory data on the frictional stick-slip failure, $\Delta\tau_b/\tau_p$ had values ranging from 0.01 to 0.1 (Ohnaka, 2003), and m ranged from 0.015 to 0.1 (see Figure 3.26).

It has been demonstrated that the displacement-related constitutive law parameters D_c, D_{wc}, and D_a scale specifically with λ_c. The characteristic length λ_c has been defined as the critical wavelength beyond which the geometric irregularity of the rupture surfaces no longer exhibits self-similarity, or the corner wavelength that separates the neighboring two self-similar bands with different fractal dimensions (see Section 3.1). Therefore, the scale of D_c, D_{wc}, or D_a is directly related to the geometric property of the macroscopic rupture surfaces. The shear rupture that proceeds on such irregular rupturing surfaces is governed not only by the nonlinear physics of the constitutive law but also by the geometric property of the rupture-surface irregularity. This is because the shear-rupture surfaces are in mutual contact and interacting during slip (or breakdown process), in contrast with the tensile fracture that occurs when the material is pulled apart, as noted previously in Chapter 2.

The scale-dependence of scale-dependent physical quantities inherent in shear rupture is attributed to the scale-dependence of D_c, as theoretically demonstrated in Chapters 5 and 6. It thus follows that the constitutive scaling law (3.15) plays a fundamental role in not only unifying laboratory data on the shear fracture of intact rock and frictional stick-slip failure, but also scaling the scale-dependent physical quantities inherent in the rupture. This will be described in detail in Chapter 6.

3.3.3 Critical energy required for shear fracture and for frictional stick-slip failure

The laboratory-derived constitutive relation for the shear failure of intact rock and for frictional stick-slip failure on a precut rock interface has been shown in Sections 3.1 and 3.2, respectively. Once such a constitutive relation $\tau(D)$ is specifically given, the critical energy required for the front of a shear rupture to further grow (in other words, its resistance to rupture growth) G_c can be evaluated by integrating Eq. (2.44). Indeed, G_c for the shear failure of intact rock samples and for frictional stick-slip failure on a precut rock interface tested in the laboratory has been calculated directly from Eq. (2.44) (Ohnaka et al., 1997; Ohnaka, 2003). On the other hand, G_c can be expressed explicitly in terms of the constitutive law parameters $\Delta\tau_b$ and D_c as the form of Eq. (2.45). As noted in subsection 2.2.5, the dimensionless parameter Γ in Eq. (2.45) is of the order of unity. Thus, G_c can be estimated from the product of the constitutive law parameters $\Delta\tau_b$ and D_c, using Eq. (2.45).

The G_c defined by Eq. (2.44) can be rewritten as

$$G_c = G_{c1} + G_{c2},\qquad(3.19)$$

where

$$G_{c1} = \int_0^{D_a} [\tau(D) - \tau_r]dD\qquad(3.20)$$

and

$$G_{c2} = \int_{D_a}^{D_c} [\tau(D) - \tau_r]dD.\qquad(3.21)$$

In the case of the shear failure of intact rock under compressive loading in the brittle regime, numerous microcrackings occur as a preparatory phase of imminent macroscopic faulting, and inelastic deformation progressively concentrates in the zone where a macroscopic shear fault is eventually formed as a result of the crack coalescence (for details, see subsection 3.1.2). As the relative displacement D comes close to the critical displacement D_a, the shear strength approaches its peak value τ_p, and at $D = D_a$, the preparatory phase shifts to the breakdown (or slip-weakening) phase. The integrated energy expended for the numerous microcrackings and crack-coalescence in the preparatory phase is represented by G_{c1}, and G_{c2} represents the energy expended during the macroscopic shear faulting (or slip-weakening) process from the peak shear strength τ_p to the residual friction stress τ_r.

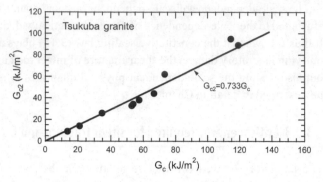

Relation between G_{c2} and G_c obtained from laboratory data on the shear failure of intact Tsukuba granite. Data from Ohnaka *et al.* (1997).

In the case of frictional slip failure on a precut rock interface under compressive loading, frictional resistance initially increases with ongoing slip displacement, due to an increase in the sum of asperity junction areas (such as asperity interlocking and asperity ploughing) with progressive slip displacement. As the slip displacement D gets close to the critical displacement D_a, the frictional strength approaches its peak value τ_p. At $D = D_a$, the phase of slip-strengthening shifts to the breakdown (or slip-weakening) phase. This is because asperity junction areas resistant to breakdown on the precut fault can no longer sustain the applied load, and begin to break down at $D = D_a$ (see subsection 3.2.3). Thus, G_{c1} represents the energy expended for an increase in the slip-induced frictional resistance to breakdown in the phase of slip-strengthening, and G_{c2} represents the energy expended during the breakdown (or slip-weakening) process from the peak shear strength τ_p to the residual friction stress τ_r.

Although the underlying micro-mechanism in the preparatory phase for frictional stick-slip failure on a precut rock interface is not the same as that in the preparatory phase for the shear failure of intact rock, the underlying physics of the macroscopic breakdown (or slip-weakening) process for frictional stick-slip failure on a precut rock interface is the same as the underlying physics for the shear failure of intact rock.

Equation (3.21) was employed by some scientists (e.g., Rice, 1980; Wong, 1982a, 1986) to estimate the shear fracture energy of intact rock. As noted above, however, the preparatory phase of inelastic deformation during which microcrackings and crack-coalescence develop in the brittle regime precedes the macroscopic shear faulting, and this preparatory phase is an integral part of the eventual macroscopic shear faulting. Without this phase, intact rock cannot fail macroscopically. Therefore, it is not only G_{c2} but also G_{c1} that is required for the shear failure of an intact material.

Figure 3.27 shows the relationship between G_c and G_{c2} obtained from experimental data on the shear failure of intact Tsukuba granite (data from Ohnaka *et al.* (1997)). From this figure, one can see that G_{c2} is directly proportional to G_c. Let G_{c1} be expressed in terms of G_c as $G_{c1} = \alpha_1 G_c$, and G_{c2} as $G_{c2} = \alpha_2 G_c$, where α_1 and α_2 are numerical constants, and $\alpha_1 + \alpha_2 = 1$. The double-error regression analysis gives: $\alpha_2 = 0.733$, $\alpha_1 (= 1 - \alpha_2) = 0.267$, and the correlation coefficient $r = 0.986$ (Ohnaka *et al.*, 1997). Thus, direct proportional relationships hold between G_{c1}, G_{c2}, and G_c. From this, it can be concluded that approximately 27% of the entire energy G_c is expended for the preparatory phase, and the rest (73%) of G_c is expended for the macroscopic slip-weakening phase.

Although the underlying micro-mechanism in the preparatory phase for frictional slip failure on a precut rock interface is not the same as that for the shear failure of intact rock, two different sets of experimental data on G_c for the shear failure of intact rock and for frictional slip failure on a precut rock interface can be unified, if the slip-dependent constitutive law on which the constraint (3.15) is imposed is assumed as the governing law for the shear rupture. This will be shown below.

Combining Eqs. (3.15) and (3.16) with Eq. (2.35) leads to the following expression for G_c (Ohnaka, 2003):

$$G_c = \frac{\Gamma}{2} \left(\frac{1}{\beta} \right)^{1/M} \Delta\tau_b \left(\frac{\Delta\tau_b}{\tau_p} \right)^{1/M} \lambda_c. \tag{3.22}$$

Equation (3.22) predicts that G_c scales with λ_c, indicating that G_c is scale-dependent (the scale-dependence of G_c will be discussed in detail in Chapter 6). Equation (3.22) can be rewritten as

$$\frac{G_c}{\tau_p \lambda_c} = \frac{\Gamma}{2} \left(\frac{1}{\beta} \right)^{1/M} \left(\frac{\Delta\tau_b}{\tau_p} \right)^{(M+1)/M}, \tag{3.23}$$

or equivalently

$$\frac{G_c}{\tau_p \lambda_c} = \frac{\Gamma\beta}{2} \left(\frac{D_c}{\lambda_c} \right)^{M+1}. \tag{3.24}$$

Equations (3.23) and (3.24) suggest that a unified comprehension can be gained of the energy required for the shear failure of intact rock and for frictional slip failure on a precut rock interface, if $G_c/(\tau_p\lambda_c)$ is plotted against $\Delta\tau_b/\tau_p$ or D_c/λ_c for the experimental data.

Figure 3.28(a) shows a plot of the logarithm of $G_c/(\tau_p\lambda_c)$ against the logarithm of $\Delta\tau_b/\tau_p$ for a set of experimental data on the shear failure of intact rock, and for a set of experimental data on frictional slip failure on a precut rock interface, both tested in the laboratory (Ohnaka, 2003). Figure 3.28(b) shows a plot of the logarithm of $G_c/(\tau_p\lambda_c)$ against the logarithm of D_c/λ_c for the same data sets. In Figures 3.28(a) and 3.28(b), black squares denote the points of data on the shear failure of intact granite, black triangles denote the points of data on frictional slip failure on a precut granite interface with the roughness characterized as $\lambda_c = 200$ μm, and white triangles denote the points of data on frictional slip failure on a precut granite interface with the smoother roughness characterized as $\lambda_c = 100$ μm.

From Figures 3.28(a) and 3.28(b), one can see that the shear fracture energy of intact rock samples and the frictional slip failure energy on a precut rock interface are both unified by a single law of the form (3.23) or (3.24). Since $M = 1.20$ has been obtained (see subsection 3.3.1), the exponent of $(\Delta\tau_b/\tau_p)$ is fixed to be $(M+1)/M = (1.20+1)/1.20 = 1.83$.

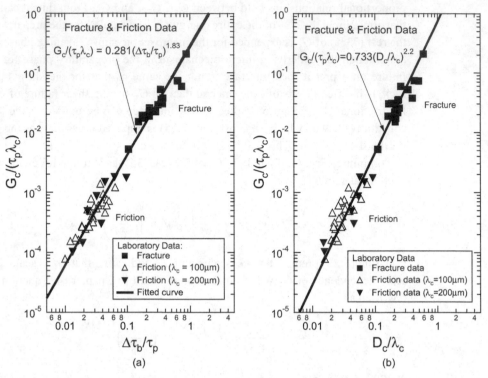

A plot of (a) the logarithm of $G_c/(\tau_p\lambda_c)$ against the logarithm of $\Delta\tau_b/\tau_p$, and (b) the logarithm of $G_c/(\tau_p\lambda_c)$ against the logarithm of D_c/λ_c, for laboratory data on the shear failure of intact Tsukuba granite and frictional stick-slip ruptures on precut Tsukuba granite interfaces having two different geometric irregularities (roughnesses) characterized by $\lambda_c = 100$ and 200 μm (Ohnaka, 2003). The formulae shown in the figures were derived from the double-error regression analysis of all data points plotted in individual figures (see text).

Under this condition, the double-error regression analysis of these data points gives (Ohnaka, 2003)

$$\frac{G_c}{\tau_p\lambda_c} = (0.281 \pm 0.038)\left(\frac{\Delta\tau_b}{\tau_p}\right)^{1.83}, \quad r = 0.984, \qquad (3.25)$$

where r is the correlation coefficient. Likewise, when the exponent of D_c/λ_c is fixed to be $M + 1 = 2.20$, the double-error regression analysis of the same data sets gives

$$\frac{G_c}{\tau_p\lambda_c} = (0.733 \pm 0.182)\left(\frac{D_c}{\lambda_c}\right)^{2.2}, \quad r = 0.961. \qquad (3.26)$$

From Eqs. (3.23) and (3.25), we know that $(\Gamma/2)(1/\beta)^{1/M} = 0.281$, from which $\Gamma = 0.85$ is obtained by substituting 1.64 for β and 1.20 for M. Likewise, we know from Eqs. (3.24) and (3.26) that $\Gamma\beta/2 = 0.733$, from which $\Gamma = 0.89$ is obtained by substituting 1.64 for β. The common result that $\Gamma = 0.85$–0.89 is reasonable, because Γ is of the order of unity.

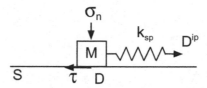

Fig. 3.29   A single-degree-of-freedom rigid-block spring model.

In Figures 3.28(a) and 3.28(b), G_c has been normalized to the product of τ_p and λ_c. Nevertheless, the normalized G_c is an increasing function of $\Delta\tau_b/\tau_p$ or D_c/λ_c. This may be attributed to a difference in experimental conditions (such as confining pressure, pore water pressure, and temperature) set up in the individual tests. From Figures 3.28(a) and 3.28(b), one can also see that the normalized G_c for the shear failure of intact rock is consistently greater than that for the frictional slip failure on a precut rock interface. This difference is attributed to a substantial difference between the cohesive bond strength for intact rock and frictional strength on a precut rock interface. The difference is also partly attributed to a difference in the magnitude of normal load (or confining pressure) applied during the experiments on the shear failure of intact rock and the frictional slip failure on a precut interface. Note that the interdependence of the parameters $\Delta\tau_b/\tau_p$ and D_c/λ_c shown in Figure 3.24(a) is reflected in Figures 3.28(a) and 3.28(b).

3.3.4 Stability/instability of the breakdown process

Whether shear failure occurs stably or unstably can never be accounted for in terms of the shear strength alone, but is accounted for in terms of the constitutive relation (between the shear strength and the slip displacement) and the elastic stiffness of the loading system. Let us confirm this fundamental fact specifically in this subsection.

For easier comprehension, we assume a single-degree-of-freedom rigid-block spring model as shown in Figure 3.29. In Figure 3.29, a rigid-block slider M on a surface S is compressed by applied normal stress σ_n, and the loading of the block slider M is caused by an imposed motion at the spring end. One may regard the resistance acting on the interface between block M and surface S as frictional resistance, or alternatively as cohesive strength greater than the frictional resistance. The resistance acting on the interface (or interface strength) is a function of normal stress σ_n.

Although this model is simple, it can be applied to the shear failure process of a small rock specimen in laboratory tests. This is because the distributions of shear stress and slip displacement on the rupturing surfaces can be regarded as spatially uniform in cases where the specimen size is small in relation to the breakdown zone length. To apply this simple model to an earthquake source, the interface between M and S represents the earthquake fault plane, the elasticity of the spring represents the elastic property of the rock body surrounding the fault, and the imposed loading rate at the spring end represents the tectonic loading rate.

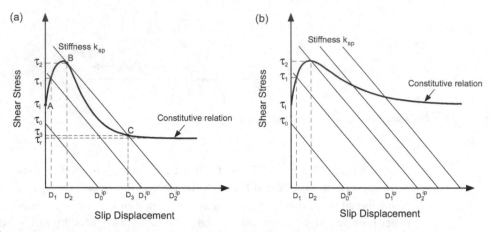

Fig. 3.30 Two case examples of the constitutive relation governing the interface between M and S in Figure 3.29.

First, let us consider the case where the strength on the interface between M and S is governed by the constitutive relation shown by a thick curve in Figure 3.30(a). The shear stress τ acting on the interface is caused by the imposed displacement D^{ip} at the spring end ($\tau = k_{sp}D^{ip}$, where k_{sp} is the spring stiffness). While the shear stress τ is lower than the interface strength, block M remains at rest. Such an example is illustrated in Figure 3.30(a) as a case in which the shear stress $\tau = \tau_0$ is caused by the imposed displacement D_0^{ip} at the spring end ($\tau_0 = k_{sp}D_0^{ip}$). When the shear stress caused by the imposed motion at the spring end becomes greater than the initial strength τ_i on the verge of the slip, block M slides, maintaining shear stress equal to interface strength, along the constitutive relation curve in Figure 3.30(a). Such an example is illustrated in Figure 3.30(a) as a case in which the shear stress $\tau = \tau_1$ is caused by the imposed displacement D_1^{ip} at the spring end ($\tau_1 = k_{sp}D_1^{ip}$). In this case, block M slides by D_1 on the interface (see Figure 3.30(a)). A further imposed motion at the spring end causes an increase in both the shear stress and the sliding displacement on the interface, along the constitutive relation curve in Figure 3.30(a). During the sliding process from the initial stage of point A $(0, \tau_i)$ to the stage of point B (D_2, τ_2) in the $D - \tau$ domain (Figure 3.30(a)), the slip on the interface between M and S can occur only by an increase in the imposed displacement at the spring end; otherwise, the slip does not occur. Hence, this sliding process is quasi-static and stable.

At the stage of point B (D_2, τ_2) and thereafter, however, a dynamic instability takes place, and the shear stress degrades dynamically with ongoing slip to a residual friction stress level τ_r on the interface between M and S, along the constitutive relation curve in Figure 3.30(a). This is because the release of the elastic strain energy stored in the spring is greater than the work done on the interface during the slip-weakening process from point B to point C along the constitutive relation curve shown in Figure 3.30(a); that is,

$$\frac{1}{2}(\tau_2 + \tau_3)(D_3 - D_2) > \int_{D_2}^{D_3} \tau(D)dD, \qquad (3.27)$$

where $\tau(D)$ denotes the constitutive relation between τ and D, and τ_3 and D_3 denote the shear stress and the slip displacement, respectively, at point C on the constitutive relation curve in Figure 3.30(a). Accordingly, the slip failure occurs spontaneously with the release of the elastic strain energy stored, even if there is no further increment of the imposed displacement at the spring end. Therefore, this slip failure process is dynamic and unstable, with the extra amount of released energy being consumed as kinetic energy.

Next, let us consider the case where the strength on the interface between M and S is governed by the constitutive relation shown by a thick curve in Figure 3.30(b). In this case, however large the amount of the imposed displacement at the spring end is, the release of the elastic strain energy stored in the spring never exceeds the work done on the interface during the slip-weakening process (see Figure 3.30(b)). An additional increment of the imposed displacement at the spring end is required for further progression of the slip-weakening. Accordingly, the slip failure is stable throughout the entire process in the case shown in Figure 3.30(b).

From these considerations, we can confirm that the stability/instability of slip failure (or shear rupture) is accounted for in terms of both the constitutive relation and the elastic stiffness of the loading system. We must keep in mind that the stability/instability of slip failure can never be discussed in terms of the strength alone. We should also keep in mind that slip failure (or shear rupture) can proceed quasi-statically and stably even in the brittle regime.

3.3.5 Breakdown zone size

When the length of a preexisting fault is sufficiently large in relation to the breakdown zone length, there is a transitional zone of strength degradation behind the propagating rupture front between the yet unbroken zone and the already broken-down zone along the fault; hence, the stress distribution in the vicinity of a propagating rupture front along the fault is spatially nonuniform. The transitional zone is what is referred to as the *breakdown zone* (see Figure 2.10). The constitutive relation between shear traction τ and slip displacement D holds in the breakdown zone. The breakdown zone ends at the distance X_c from the rupture front, and at the end of the breakdown zone, slip displacement D reaches the breakdown displacement D_c defined as the critical amount of slip required for the shear traction to degrade to a residual friction stress level. Accordingly, the breakdown zone size X_c is directly related to D_c. We specifically estimate below how large the breakdown zone size is.

For a model in which the front of the shear rupture governed by the laboratory-derived slip-dependent constitutive relation propagates dynamically at a steady speed V_c along a preexisting fault, it has been derived theoretically that X_c is directly related to D_c by (Ohnaka and Yamashita, 1989; see also Ohnaka, 2000)

$$X_c = \frac{1}{k}\frac{\mu}{\Delta\tau_b}D_c, \tag{3.28}$$

where μ is the modulus of rigidity of the elastic medium surrounding the fault, and k is a dimensionless quantity defined by

$$k = \frac{\Gamma}{\pi^2 \chi_c C(V_c)}. \tag{3.29}$$

In Eq. (3.29) χ_c represents a numerical parameter, $C(V_c)$ represents a function of the rupture velocity V_c with a functional form that depends on the longitudinal wave velocity V_P and the shear wave velocity V_S (see Eqs. (5.37) and (5.38)), and Γ represents a dimensionless parameter defined by Eq. (2.33); that is,

$$\Gamma = \int_0^1 \frac{\sigma(Y)}{\sqrt{Y}} dY, \tag{3.30}$$

where $\sigma(Y)$ denotes the dimensionless shear strength at a dimensionless distance Y measured from the rupture front in the breakdown zone. $\Gamma/\chi_c = 2.4\text{–}4.1$ has been evaluated for the model (Ohnaka and Yamashita, 1989), and we assume here that $\Gamma/\chi_c = 3.3$.

Let us estimate X_c specifically for shear ruptures. In the case of stick-slip failure on a precut Tsukuba granite interface in the laboratory, given that $V_P = 4.4$ km/s, $V_S = 2.9$ km/s, and $\mu = 20\,000$ MPa for the granite, we have from Eq. (3.29) that $k = 2.8$ for in-plane shear mode (mode II), and that $k = 2.9$ for anti-plane shear mode (mode III), if $V_c = 2$ km/s is assumed. Accordingly, we have from Eq. (3.28) that $X_c \cong 14$ cm under the assumption that $\Delta\tau_b = 0.1$ MPa and $D_c = 2$ μm for frictional stick-slip failure on a precut rock interface, and that $X_c \cong 7$ cm under the assumption that $\Delta\tau_b = 100$ MPa and $D_c = 1$ mm for the shear failure of intact granite. By contrast, if we assume that $\mu = 30\,000$ MPa, $V_P = 6$ km/s, $V_S = V_P/1.73$, and $V_c = 0.8V_S$ for typical earthquakes that take place in the seismogenic crust, we have from Eq. (3.29) that $k = 2.9$ for in-plane shear mode (mode II), and that $k = 3.5$ for anti-plane shear mode (mode III). Accordingly, we have from Eq. (3.28) that $X_c \cong 1$ km under the assumption that $\Delta\tau_b = 10$ MPa and $D_c = 1$ m for typical, large earthquakes. Thus, we notice that the breakdown zone length X_c for typical large earthquakes is about 10^4 times greater than that for shear failures observed in the laboratory. This suggests that X_c is scale-dependent. The scale-dependence of X_c will be discussed in more detail, together with other scale-dependent physical quantities inherent in the rupture, in later chapters (in particular, Chapter 6).

3.4 Dependence of constitutive law parameters on environmental factors

3.4.1 Introduction

The seismogenic layer is restricted to within the brittle regime through the semi-brittle regime of the Earth's interior. Temperatures in the layer range from the Earth's surface temperature to roughly 500–600 °C, and lithostatic pressures in the layer range from atmospheric pressure to about 500 MPa. These ranges in temperature and pressure are generally limited to depths shallower than 20 km in the Earth's crust. At a plate-boundary subduction zone along an oceanic trench, however, a cooling (oceanic) plate subducts underneath a continental plate at a rate of a few cm/year, and therefore there is a relatively low temperature zone in the subducting plate at depths much deeper than 20 km. Shear faulting resulting in an earthquake can therefore occur much deeper in such a subducting plate. Away from such plate-boundary subduction zones, however, earthquake generation is limited to within the aforementioned shallow layer, called the seismogenic layer. This is because an unstable, dynamic shear rupture can occur only in a fault zone embedded in a brittle layer or, at the most, in a semi-brittle layer.

The seismogenic environment and fault zone property exert an influence on the functional form of the constitutive law for earthquakes. The specific functional form of the slip-dependent constitutive law is dictated by the following constitutive law parameters: τ_i, τ_p, $\Delta\tau_b$, D_a, and D_c, as noted in Section 3.3; the influence of the seismogenic environment and fault zone property are implicitly exerted on the slip-dependent constitutive law through these constitutive law parameters. Of these parameters, τ_p, $\Delta\tau_b$, and D_c are fundamentally important, and therefore we focus our attention on how and to what extent these three constitutive law parameters are affected by the seismogenic environment and fault zone property. In elucidating how the constitutive law parameters for earthquakes are affected by seismogenic environments and fault zone properties, it is helpful to know how the constitutive law parameters for the shear failure of rock in the brittle regime through the semi-brittle regime are affected by environmental factors, such as temperature, confining pressure, and interstitial pore water, and by fault surface geometric irregularity. In this section, therefore, we will display the main results of those experiments obtained in the laboratory.

In the case of mature faults in the field, dynamic shear ruptures (earthquakes) have occurred repeatedly in fault zones; consequently, rock fragments or gouge particles are produced by shear pulverization of fault rock materials in the fault zones. These fragments or particles have been progressively pulverized during earthquake shear rupture processes that have recurred in the fault zones. If interstitial pores present in the fault zones are filled with liquid water, the pore water in the fault zones promotes the chemical alteration of gouge particles, leading to the argillation of fault gouge particles. Based on geological investigation, it has been reported that clay-rich fault gouges exist in mature fault zones.

If, for example, the entire fault zone of a particular fault consists of clay-matrix gouges alone, then the entire fault will be weak and non-elastically deformable. Therefore, an adequate amount of elastic strain energy cannot accumulate in the elastic medium (rock body) surrounding such a fault to produce a large earthquake or to radiate strong motion seismic waves. Aseismic creep movement can only occur along such a fault with tectonic loading.

However, the primary role in generating a large earthquake, or in radiating strong motion seismic waves, is played not by weak, non-elastically deformable areas on a fault but by strong areas highly resistant to rupture growth on the fault. Sizable strong areas highly resistant to rupture growth on a fault are necessary for a large amount of elastic strain energy to accumulate in the elastic medium surrounding the fault, under tectonic loading.

The release of the large amount of the accumulated elastic strain energy provides the driving force to bring about a large earthquake or to radiate strong motion seismic waves. In general, the size (or magnitude) of an earthquake and strong motion seismic wave radiation are determined by the amount of the released elastic strain energy, and the release of a large amount of the elastic strain energy is caused by the breakdown of sizable local strong areas highly resistant to rupture growth on the fault. It is therefore obvious that local strong areas highly resistant to rupture growth on a fault (called "asperities") play a much more important role in generating an earthquake than does the rest of the fault with low (or little) resistance to rupture growth.

In general, the rupture process of a real earthquake is not a simple frictional slip failure on a uniformly precut weak fault, but a more complex process including the fracture of initially intact rock on a heterogeneous fault. Some sites of high resistance to rupture growth (called "asperities" or "barriers") on a fault are strong enough to equal the shear failure strength of initially intact rock (see Chapter 1). Furthermore, the shear failure strength of intact rock under ambient crustal conditions provides the upper limit to frictional strength at precut rock interface areas on a fault. If, therefore, we are concerned with the generation process of typical earthquakes that radiate strong motion seismic waves, it is critically important to understand the constitutive relation for the shear failure of intact rock under seismogenic crustal conditions. On the other hand, if we are concerned with the aseismic creep behavior of the fault zone, it would be important to study the constitutive relation for non-elastically deformable, weak materials such as clay-matrix fault gouges.

As noted by earlier authors (e.g., Brace and Kohlstedt, 1980; Kirby, 1980), stresses in the crust cannot exceed the strength of intact rock under crustal ambient conditions. This is a basic idea in the estimation of the upper bound of tectonic stresses in the lithosphere. In addition, frictional strength on a precut rock interface conforms to the shear failure strength of intact rock at high pressures and temperatures close to the base of the seismogenic layer in the brittle–plastic transition regime. These also indicate the importance of understanding the constitutive properties of the shear failure of intact rock under crustal conditions. In this section, therefore, we focus on the constitutive properties of the shear failure of intact rock under crustal conditions.

3.4.2 Dependence of shear failure strength on environmental factors

In the brittle regime, the mechanical properties of rock are sensitive to stress (or pressure), but insensitive to temperature. Indeed, the effect of temperature on the shear failure strength of intact rock is negligible in the purely brittle regime. In the high-temperature plastic regime, the mechanical properties of rock are sensitive to temperature, but insensitive to stress (or pressure). Accordingly, the effect of stress (or pressure) is negligible in the plastic regime. In the intervening brittle–plastic transition regime, however, a thermally activated plastic flow property coexists with the brittle property, and therefore the mechanical properties of rock inherently depend on both stress (or pressure) and temperature. The effects of both stress (or pressure) and temperature on the shear failure strength are not negligible in the intervening brittle–plastic transition regime.

Interstitial pore water exerts an influence on the shear failure strength of rock through the mechanical effect of water pressure (according to the law of effective stress), and through the chemical effect of water molecules on some rock-forming minerals at stress concentrations (i.e., stress-aided corrosion). The mechanical effect of pore water pressure on the shear failure strength is exerted through the effective normal stress σ_n^{eff} defined by

$$\sigma_n^{eff} = \sigma_n - P, \tag{3.31}$$

where σ_n is the normal stress across shear failure fault surfaces, and P is interstitial pore water pressure.

The concept of effective stress was originally introduced in the field of soil mechanics. The use of effective stress is justified, and the failure criterion in terms of effective stress holds, if networks of pores in a material are interconnected, if the pore space is filled with liquid water, and if a change in pore water pressure is transmitted to the whole pore space in the material. Soils consist of discrete particles, and therefore it is easily understandable that a system consisting of soils and pore water meets the aforementioned conditions.

By contrast, the voids in crystalline rocks, such as granite, mostly consist of grain-boundary cracks, and hence the porosity of crystalline rock is very low. For example, the porosity of granite is 0.6–0.9%. Nevertheless, the law of effective stress has been validated based on laboratory experiments on the shear failure of granite tested at a moderate strain rate under confining pressures and pore pressures. In general, a rock sample in compression becomes dilatant prior to fracture, even under high confining pressures. This brings about an increase in porosity within the sample, prior to fracture.

Brace and Martin (1968) first corroborated in their laboratory experiments that the effective stress law holds for Westerly granite, when tested at a critical strain rate of 10^{-7}/s or slower in their experimental configuration. In general, the critical strain rate, at or below which the law of effective stress holds, depends on the configuration, permeability and geometry of a rock sample, and the viscosity of pore fluid. This suggests that the law of effective stress holds for granite, even when tested at a strain rate faster than 10^{-7}/s, if more suitable sample size and configuration are adopted. In fact, it has been demonstrated in our laboratory experiments (Kato et al., 2003a; Odedra et al., 2001) that the law of effective stress holds for Tsukuba granite tested at a strain rate of 10^{-5}/s (corresponding to an axial displacement rate of 0.4 μm/s) in the experimental configuration adopted by our rock physics laboratory (Ohnaka et al., 1997).

The shear failure strength (or peak shear strength) τ_p is a function of the effective normal stress σ_n^{eff}, ambient temperature T, and displacement rate $\dot{D}$ (or equivalently, strain rate $\dot{\varepsilon}$). If the effects of σ_n^{eff}, T, and $\dot{D}$ on τ_p are mutually independent and separable, τ_p can be expressed as

$$\tau_p(\sigma_n^{eff}, T, \dot{D}) = f(\sigma_n^{eff})g(T)h(\dot{D}), \tag{3.32}$$

where $f(\sigma_n^{eff})$, $g(T)$, and $h(\dot{D})$ are functions of σ_n^{eff}, T, and $\dot{D}$, respectively. In the case of the shear failure of intact rock under combined compressive stress environments in a triaxial test, the normal stress across a fault plane is not held constant, but changes with slip displacement D, during the breakdown process. In this case, τ_p is attained at $D = D_a$, and hence it is customary to express τ_p as a function of the normal stress to which the normal stress value at $D = D_a$ is assigned (see Figure 3.38(a) in the next subsection).

Since the effect of the deformation rate on the shear failure strength of intact rock has been conventionally expressed in terms of strain rate $\dot{\varepsilon}$, which is directly proportional to $\dot{D}$, Eq. (3.32) can be equivalently written as

$$\tau_p\big(\sigma_n^{\text{eff}}, T, \dot{\varepsilon}\big) = f\big(\sigma_n^{\text{eff}}\big)g(T)h'(\dot{\varepsilon}), \qquad (3.33)$$

where $h'(\dot{\varepsilon})$ is a function of $\dot{\varepsilon}$.

When T and $\dot{\varepsilon}$ are held constant so as to be $T = T_0$ (room temperature) and $\dot{\varepsilon} = \dot{\varepsilon}_0(10^{-5}/\text{s})$, τ_p is a function of σ_n^{eff} alone. If $f(\sigma_n^{\text{eff}})$ can be expanded in a power series of σ_n^{eff}, Eq. (3.33) can be rewritten as

$$\tau_p\big(\sigma_n^{\text{eff}}, T_0, \dot{\varepsilon}_0\big) = f\big(\sigma_n^{\text{eff}}\big) = c_0 + c_1\sigma_n^{\text{eff}} + c_2\big(\sigma_n^{\text{eff}}\big)^2 + \cdots, \qquad (3.34)$$

where $c_i(i = 0, 1, 2, \ldots)$ are constants, and specific functional forms of $g(T)$ and $h'(\dot{\varepsilon})$ are chosen such that $g(T_0) = 1$ and $h'(\dot{\varepsilon}_0) = 1$, respectively. If second- and higher-order terms are negligible, Eq. (3.34) reduces to the linear Coulomb criterion

$$\tau_p\big(\sigma_n^{\text{eff}}, T_0, \dot{\varepsilon}_0\big) = f\big(\sigma_n^{\text{eff}}\big) = c_0 + c_1\sigma_n^{\text{eff}}, \qquad (3.35)$$

where c_0 is the cohesive strength (that is, the inherent shear strength of the material), and c_1 is the coefficient of the internal friction of the material. The shear failure strength relative to that at $T = T_0$ is a function of T alone, since

$$\frac{\tau_p\big(\sigma_n^{\text{eff}}, T, \dot{\varepsilon}_0\big)}{\tau_p\big(\sigma_n^{\text{eff}}, T_0, \dot{\varepsilon}_0\big)} = \frac{g(T)}{g(T_0)} = g(T). \qquad (3.36)$$

A specific functional form of $g(T)$ can be expressed as (Ohnaka, 1995a)

$$g(T) = \tanh\left[\frac{Q}{R}\left(\frac{1}{T} - \frac{1}{T_1}\right)\right], \qquad (3.37)$$

where T is measured in absolute temperature (K), R is the gas constant, and Q and T_1 are constants. The strain rate effect $h'(\dot{\varepsilon})$ on the shear failure strength in the brittle regime is represented by a logarithmic law (Masuda et al., 1987, 1988; Kato et al., 2003b), which can be written as (Ohnaka, 1995a)

$$h'(\dot{\varepsilon}) = 1 + \alpha \ln\left(\frac{\dot{\varepsilon}}{\dot{\varepsilon}_0}\right), \qquad (3.38)$$

where α is a dimensionless coefficient. By contrast, the strain rate effect $h'(\dot{\varepsilon})$ on the shear resistance to flow in the plastic regime is represented by a power law (e.g., Carter, 1976), which can be written as

$$h'(\dot{\varepsilon}) = \left(\frac{\dot{\varepsilon}}{\dot{\varepsilon}_0}\right)^{\frac{1}{n}}, \qquad (3.39)$$

where n is constant. The functional forms of Eqs. (3.38) and (3.39) have been expressed in such a way that $h'(\dot{\varepsilon}_0) = 1$. The quantitative effect of strain rate on shear strength in the brittle–plastic transition regime is unknown. It is obvious, however, that the strain

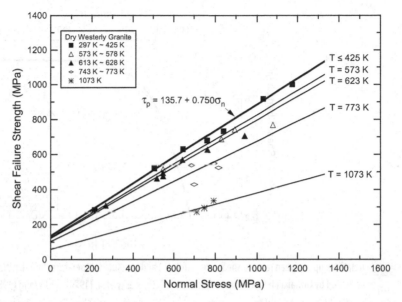

Fig. 3.31 A plot of the shear failure strength τ_p against the normal stress σ_n for dry Westerly granite. Data compiled by Ohnaka (1995a) from Griggs et al. (1960), Stesky et al. (1974), and Wong (1982a, 1982b).

rate effect is more pronounced in the brittle–plastic transition regime than in the brittle regime, but less pronounced in the brittle–plastic transition regime than in the plastic regime.

Let us first examine the specific effect of the normal stress σ_n on the shear failure strength τ_p of dry granite (tested under conditions with no pore water). Figure 3.31 shows a plot of shear failure strength against normal stress for dry Westerly granite data compiled from Griggs et al. (1960), Stesky et al. (1974), and Wong (1982a, 1982b). In the temperature range from room temperature to 150 °C, the mechanical properties of dry Westerly granite are highly insensitive to temperature, and the normal stress dependence of τ_p at 150 °C is almost identical to that at room temperature (Wong, 1982b). Hence, the specific functional form of $f(\sigma_n)$ for dry Westerly granite in the purely brittle regime can be examined by using the data tested in the range from room temperature to about 150 °C. Figure 3.31 indicates that the data points for the shear failure strength of dry Westerly granite tested at temperatures ranging from room temperature to 152 °C align along a single straight line expressed as Eq. (3.35) with $c_0 = 135.7$ MPa and $c_1 = 0.750$. Note that σ_n^{eff} in Eq. (3.35) is replaced with σ_n for dry rock.

Thus, we can conclude from Figure 3.31 that the shear failure strength τ_{p0} for dry Westerly granite at room temperature ($T = T_0$) increases linearly with an increase in the normal stress σ_n, in accordance with (Ohnaka, 1995a)

$$\tau_{p0} = 135.7 + 0.750\sigma_n. \tag{3.40}$$

The linear relation between τ_p and σ_n (Coulomb equation) has been widely documented for geological materials (e.g., see Paterson, 1978).

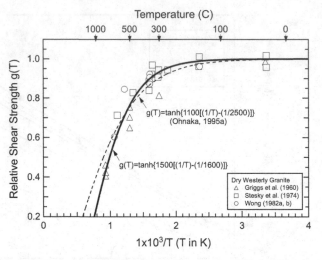

Fig. 3.32   Relationship between relative shear strength $g(T)$ and absolute temperature T for dry Westerly granite. Data compiled by Ohnaka (1995a) from Griggs *et al.* (1960), Stesky *et al.* (1974), and Wong (1982a, 1982b).

Let us abbreviate $\tau_p(\sigma_n^{\text{eff}}, T, \dot{\varepsilon}_0)/\tau_p(\sigma_n^{\text{eff}}, T_0, \dot{\varepsilon}_0)$ as τ_p/τ_{p0}. Figure 3.32 shows a plot of the relative shear failure strength τ_p/τ_{p0} against $1/T$ (where T is measured in absolute temperature K) for the compiled shear failure strength data set of dry Westerly granite. An earlier paper (Ohnaka, 1995a) showed that the data set plotted in Figure 3.32 is empirically well fitted to Eq. (3.37) with $Q/R = 1100$ K and $T_1 = 2500$ K; however, this has been modified because it has since been found that the same data set is better fitted to Eq. (3.37) with $Q/R = 1500$ K and $T_1 = 1600$ K (see Figure 3.32); that is,

$$g(T) = \tanh\left[1500\left(\frac{1}{T} - \frac{1}{1600}\right)\right]. \tag{3.41}$$

Figure 3.32 indicates that the effect of temperature is insignificant below 300 °C for dry Westerly granite, suggesting that dry granite in this temperature range is in the brittle regime. Above 300 °C, the temperature effect is greater (Figure 3.32). This is due to the development of localized plastic deformation in the shear failure zone above 300 °C. In fact, the deformation of Westerly granite was achieved by shear localization in an incipient mylonitization zone, and the fabric in the shear zone was most affected by deformation (Griggs *et al.*, 1960). Biotite crystals in the shear zone showed kink banding of several types, suggesting strong localized plastic deformation (Griggs *et al.*, 1960). Plastic deformation of quartz was optically observed above 500 °C (Griggs *et al.*, 1960; Stesky, 1978), and also observed with a transmission electron microscope at 700 °C, by a marked increase in dislocation density (Stesky, 1978). It has been reported, however, that feldspar minerals did not deform plastically under the experimental conditions employed (Stesky *et al.*, 1974; Stesky, 1978). With these observations in mind, the operative mechanism for the temperature effect on $g(T)$ may be attributed to quartz plasticity, as well as biotite plasticity. Above 500 °C, frictional resistance at the interface of Westerly granite conformed to the shear failure strength of the intact granite, and "welding" over the entire fault surfaces was

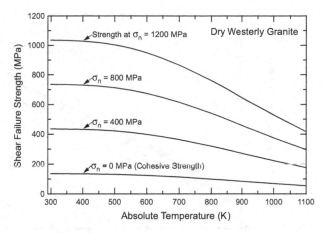

Fig. 3.33 Relationships between the shear failure strength τ_p of dry Westerly granite and absolute temperature T at given constant values of normal stress σ_n.

observed for Westerly granite specimens in frictional sliding tests (Stesky *et al.*, 1974; Stesky, 1978). These facts provide circumstantial evidence for partial plasticity. This is because crystal plastic deformation near the fault surfaces is required to make the real area of contact equal to the nominal area, and to bring about "welding" to conform frictional resistance to the shear failure strength of intact granite. Additional circumstantial evidence for partial plasticity comes from the fact that the fracture angle increased from 30 ° to 40 ° with increasing T above about 500 °C (Ohnaka, 1995a), and that the localized shear zone became thicker at a higher temperature (Wong, 1982b).

From Eqs. (3.33), (3.40), and (3.41), we have

$$\tau_p(\sigma_n, T, \dot{\varepsilon}_0) = (135.7 + 0.750\sigma_n) \tanh\left[1500\left(\frac{1}{T} - \frac{1}{1600}\right)\right]. \qquad (3.42)$$

This equation represents the quantitative effects of normal stress and temperature on the shear failure strength of dry Westerly granite, from the brittle regime through the brittle–plastic transition regime.

The temperature-dependence of the shear failure strength τ_p for dry Westerly granite in the brittle regime through the brittle–plastic transition regime at a given normal stress can be calculated from Eq. (3.42). Specifically, Figure 3.33 shows relationships between τ_p and T at given constant values for σ_n, plotted by using Eq. (3.42). From Figure 3.33, one can see that τ_p is insensitive to temperatures below 423 K ($\cong 150$ °C), and only slightly sensitive to temperatures up to 573 K ($\cong 300$ °C), whereas τ_p becomes more sensitive to temperatures in the brittle–plastic transition regime ($T > 573$ K).

Now, let us examine the effects of σ_n^{eff} and T on τ_p for wet granite. Figure 3.34 shows a plot of τ_p against σ_n^{eff} for wet Tsukuba granite data compiled from Odedra *et al.* (2001), and Kato *et al.* (2004), with additional, unpublished data digitally preserved in our laboratory (Ohnaka, 2003). In Figure 3.34, black circles denote the data points of wet Tsukuba granite tested at temperatures ranging from room temperature to 300 °C ($\cong 573$ K), white squares denote the data points of wet Tsukuba granite tested at temperatures of 400 °C ($\cong 673$ K)

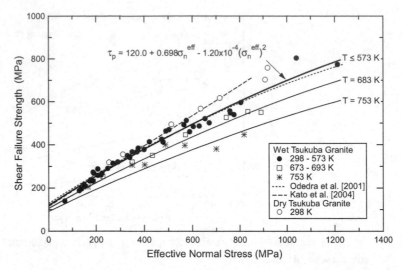

Fig. 3.34   A plot of the shear failure strength τ_p against the effective normal stress σ_n^{eff} for wet or dry Tsukuba granite. Data from Odedra *et al.* (2001) and Kato *et al.* (2004), together with additional unpublished data.

and 420 °C ($\cong$ 693 K), and asterisks denote the data points of wet Tsukuba granite tested at a temperature of 480 °C ($\cong$ 753 K). Data of the shear failure strength for dry Tsukuba granite tested at room temperature have also been displayed as white circles in Figure 3.34 for comparison.

From a set of limited data on τ_p of wet Tsukuba granite tested within the ranges of $T \leq 480$ °C and $\sigma_n^{eff} < 600$ MPa, Kato *et al.* (2004) empirically derived the following linear relationship between τ_p and σ_n^{eff}:

$$\tau_p\left(\sigma_n^{eff}, T\right) = 115 + c_1(T) \times \sigma_n^{eff} \qquad \left(\sigma_n^{eff} < 600\,\text{MPa}\right), \tag{3.43}$$

where

$$c_1(T) = \begin{cases} 0.7 & (T \leq 300\,°\text{C}) \\ 0.7 \times [1 - 1.5 \times 10^{-3}(T - 300)] & (300\,°\text{C} < T \leq 480\,°\text{C}) \end{cases}. \tag{3.44}$$

The linear relationship, $\tau_p = 115 + 0.7\sigma_n^{eff}$, for wet Tsukuba granite in the brittle regime is displayed as a broken line in Figure 3.34 for reference. From Figure 3.34, it can be confirmed that the plotted data points of wet Tsukuba granite in the brittle regime within the range of the effective normal stresses below 600 MPa are well fitted to the linear broken-line relationship. However, the data points of the same wet Tsukuba granite in the brittle regime over the range of the effective normal stresses beyond 600 MPa systematically deviate downward from the linear relationship, although all data points for dry Tsukuba granite in the brittle regime, even when tested at the (effective) normal stresses beyond 600 MPa, lie along the same line (see Figure 3.34), suggesting that a linear relationship between τ_p and σ_n holds for dry Tsukuba granite in the brittle regime. All data points for wet Tsukuba granite in the brittle regime (tested at $T \leq 300$ °C) over a broad range of the effective normal stress up to 1200 MPa, shown in Figure 3.34, are best fitted to the following second-order polynomial:

$$\tau_p = 120.0 + 0.698\sigma_n^{\text{eff}} - 1.20 \times 10^{-4}\left(\sigma_n^{\text{eff}}\right)^2. \tag{3.45}$$

This is consistent with, though slightly different from, the following second-order polynomial:

$$\tau_p = 128.5 + 0.694\sigma_n^{\text{eff}} - 1.39 \times 10^{-4}\left(\sigma_n^{\text{eff}}\right)^2, \tag{3.46}$$

which was derived by Odedra et al. (2001) from some of the data points of wet Tsukuba granite at room temperature, plotted in Figure 3.34. Equations (3.45) and (3.46) have been displayed as a thick, solid curve and as a dotted curve, respectively, in Figure 3.34.

As noted above, the data points of the shear failure strength for wet Tsukuba granite in the brittle regime deviate downward from the linear relationship in the range of σ_n^{eff} higher than 600 MPa. This may be due to the chemical effect of pore water, because the chemical effect of pore water is enhanced at higher stresses. In fact, the data points of the shear failure strength of dry Tsukuba granite in the brittle regime (tested at room temperature) lie slightly higher than those of wet Tsukuba granite in the brittle regime (tested at conditions of $T \le 300\,^\circ\text{C}$), at effective normal stresses higher than 600 MPa. Figure 3.34 suggests that the difference between the strength of dry Tsukuba granite and the strength of wet Tsukuba granite slightly increases with an increase in the effective normal stress.

Under the assumption that the shear failure strength τ_{p0} of wet Tsukuba granite at room temperature is represented by Eq. (3.45), the relative shear failure strength τ_p/τ_{p0} has been calculated, and the calculated τ_p/τ_{p0} is plotted against $1/T$ in Figure 3.35. Odedra et al. (2001) empirically derived the following relation:

$$g(T) = \tanh\left[3200\left(\frac{1}{T} - \frac{1}{950}\right)\right], \tag{3.47}$$

from their own data set shown in Figure 3.35. From the entire data set, shown in Figure 3.35, into which the data from Kato et al. (2004) have been incorporated, $g(T)$ with slightly different parameter values is derived; that is,

$$g(T) = \tanh\left[2800\left(\frac{1}{T} - \frac{1}{1030}\right)\right]. \tag{3.48}$$

Figure 3.35 indicates that temperature has no significant effect on $g(T)$ below 300 °C for wet Tsukuba granite, suggesting that wet granite in this temperature range is in the purely brittle regime. Above 300 °C, wet Tsukuba granite is in the semi-brittle regime (or brittle–plastic transition regime). In fact, complete plastic deformation of biotite and a little plastic deformation of quartz were both observed in the shear failure zone in the presence of pore water, and the coexistence of such plastic deformation with brittle cracking was confirmed in the zone (Kato et al., 2003a). It has been inferred from the data on the shear failure of wet Tsukuba granite that the operative mechanism for plastic deformation in the brittle–plastic transition regime above 300 °C is the dislocation glide derived from the Peierls stress model (Kato et al., 2004).

Constitutive relations for shear failure

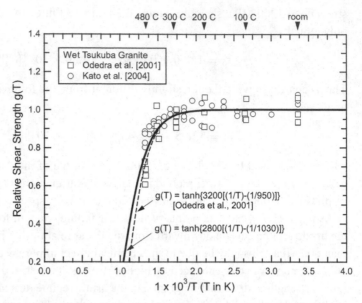

Fig. 3.35 Relationship between relative shear strength $g(T)$ and absolute temperature T for wet Tsukuba granite. Data from Odedra *et al.* (2001) and Kato *et al.* (2004).

From Eqs. (3.33), (3.45), and (3.48), we have

$$\tau_p\big(\sigma_n^{\text{eff}}, T, \dot{\varepsilon}_0\big) = \big[120.0 + 0.698\sigma_n^{\text{eff}} - 1.20 \times 10^{-4}\big(\sigma_n^{\text{eff}}\big)^2\big] \tanh\left[2800\left(\frac{1}{T} - \frac{1}{1030}\right)\right].$$
(3.49)

This equation represents the quantitative effects of effective normal stress and temperature on the shear failure strength of wet Tsukuba granite within the brittle regime through the brittle–plastic transition regime.

The temperature-dependence of τ_p for wet Tsukuba granite within the brittle regime through the brittle–plastic transition regime at a given effective normal stress can be calculated from Eq. (3.49). Figure 3.36 shows relationships between τ_p and T at given constant values for σ_n^{eff}, plotted by using Eq. (3.49). Using Eqs. (3.43) and (3.44) derived by Kato *et al.*, τ_p at $\sigma_n^{\text{eff}} = 0$ and τ_p at $\sigma_n^{\text{eff}} = 400$ MPa have also been plotted against T, displayed as broken lines for comparison, in Figure 3.36. From Figure 3.36, one can see that τ_p of wet Tsukuba granite is quite insensitive to temperatures in the brittle regime ($T \leq 573$ K $\cong$ 300 °C), but becomes sensitive to temperatures in the brittle–plastic transition regime (573 K < T).

We have learned that the shear failure strength of dry Tsukuba granite is lower than that of dry Westerly granite. This difference may be due to a difference in rock texture between Westerly granite and Tsukuba granite, such as the proportion of rock-forming minerals, and the sizes and shapes of constituent grains. Tsukuba granite has grain sizes ranging from 0.2 to 1.3 mm (Kato *et al.*, 2003a), which are coarser than those of Westerly granite. We have

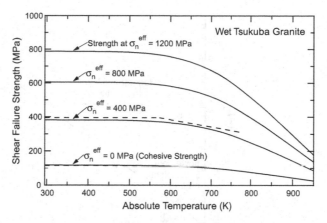

Fig. 3.36 Relationships between the shear failure strength τ_p of wet Tsukuba granite and absolute temperature T at given constant values for the effective normal stress σ_n^{eff}.

also learned that the effect of temperature on the shear failure strength for wet Tsukuba granite in the semi-brittle (or brittle–plastic transition) regime is significantly different from that for dry Westerly granite. This difference may be partly due to the difference in rock texture between Westerly granite and Tsukuba granite, and partly due to the chemical effect of pore water, which is enhanced at higher stresses.

Let us refer to the effect of strain rate on shear failure strength. As noted earlier, the strain rate effect in the brittle regime obeys a logarithmic law. Laboratory experiments (Kato *et al.*, 2003b) indicate that the coefficient α in Eq. (3.38) has a value of 0.01 ($= 0.025 \times \log e$) for the shear failure strength of dry granite in the brittle regime, and 0.02 ($= 0.048 \times \log e$) for the shear failure strength of wet granite in the brittle regime. Thus, the strain rate effect on the shear failure strength of granite is greater in wet environments than in dry environments. This may be attributed to the bond-breaking reaction promoted in wet environments (i.e., stress-aided corrosion). Even so, the effect of strain rate on the shear failure strength is very small if compared to the effect of effective normal stress on the shear failure strength.

On the other hand, the strain rate effect in the plastic regime obeys a power law, and the parameter n in Eq. (3.39) has a value ranging from 1.8 to 4.0 for typical rocks in the plastic regime, depending on rock type, and whether tested in wet or dry environments (e.g., Carter and Ave'Lallemant, 1970; Chopra and Paterson, 1981; Hansen and Carter, 1982; Kronenberg and Tullis, 1984). Thus, the strain rate effect on the shear failure strength in the brittle regime is very small in comparison to that on the shear resistance to flow in the plastic regime, as shown in Figure 3.37. Although the quantitative effect of strain rate on the strength in the brittle–plastic transition regime is unknown, it is clear that it should be between the effect in the brittle regime and the effect in the plastic regime. The quantitative effect of strain rate in the brittle–plastic transition regime remains to be studied.

As noted earlier in this subsection, the chemical effect of pore water on the shear failure strength of rock is exerted through the chemical reaction of some rock-forming minerals with water molecules. Water molecules operate as a corrosive agent, and the chemical reaction is promoted at high stresses to produce weakened products (stress-aided

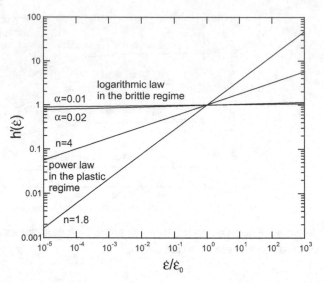

Fig. 3.37 Comparisons of the strain rate effect $h'(\dot{\varepsilon})$ on the shear failure strength in the brittle regime to the strain rate effect $h'(\dot{\varepsilon})$ on the shear resistance to flow in the plastic regime (Ohnaka and Matsu'ura, 2002).

corrosion). For example, strong Si-O bonds of silica react with water molecules (H_2O) at stress concentrations such as crack-tips to produce weaker hydrogen bonded hydroxyl groups linking the silicon atoms; that is (e.g., Scholz, 1972; Atkinson, 1979; Michalske and Freiman, 1982),

$$-\overset{|}{\underset{|}{Si}}-O-\overset{|}{\underset{|}{Si}}-+\;H-O-H \;\rightarrow\; -\overset{|}{\underset{|}{Si}}-OH \cdot HO-\overset{|}{\underset{|}{Si}}-$$

The activity of the corrosive agent plays an important role in controlling the rate of stress corrosion. For more information about the chemical effect of water on the strength of rock, refer to Atkinson (1987).

3.4.3 Dependence of breakdown stress drop on environmental factors

In a triaxial test for the shear failure of intact rock under confining pressure, both the resolved shear stress τ along the fault plane and the resolved normal stress σ_n across the fault plane change with slip displacement D during the breakdown process, as shown in Figure 3.38(a). During this breakdown process, the resolved normal stress σ_n changes with slip displacement D in proportion to the resolved shear stress τ. This can be proved theoretically. From Eqs. (3.1) and (3.2), we have

$$\sigma_n = \sigma_3 + \tau \times \tan\theta. \tag{3.50}$$

In the triaxial test, the minimum principal stress σ_3 equals confining fluid pressure P_c, which is held constant during individual test runs, and the fracture angle θ is also held constant during the breakdown process. Therefore, one can see from Eq. (3.50) that the resolved normal stress σ_n changes in proportion to a change in the resolved shear stress τ during the breakdown process. Nevertheless, the ratio of the resolved shear stress τ to the

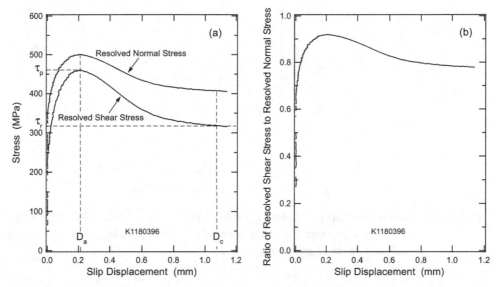

Fig. 3.38 (a) The resolved shear stress and the resolved normal stress against slip displacement for the shear failure of an intact Tsukuba granite sample tested under a constant confining pressure. (b) The ratio of the resolved shear stress to the resolved normal stress against slip displacement for the data in Figure 3.38(a).

resolved normal stress σ_n decreases with ongoing slip displacement during the breakdown process (see Figure 3.38(b)). This slip-weakening relation in Figure 3.38(b) corresponds to the slip-dependent constitutive relation observed under constant normal stress.

In general, the angle between the maximum principal stress axis (σ_1-axis) and the fault plane is non-zero in the case of the shear failure of intact rock under combined compressive stress environments in a triaxial test. In this case, as noted above, the resolved normal stress σ_n decreases in proportion to the resolved shear stress τ during the breakdown process. Thus, the breakdown stress drop $\Delta\tau_b$ (defined as $\Delta\tau_b = \tau_p - \tau_r$) for such a case becomes necessarily greater than $\Delta\tau_b$ for cases where normal stress is held constant during the breakdown process.

The magnitude of $\Delta\tau_b$ under the condition that the normal stress is held constant during the breakdown process can be estimated from data on the shear failure of intact rock obtained in a triaxial test. In order to estimate the magnitude of $\Delta\tau_b$ under the normal stress held constant during the breakdown process, it is necessary to express not only the peak shear strength (i.e., shear failure strength) τ_p as a function of the normal stress σ_n to which the normal stress value at $D = D_a$ is assigned, but also the residual friction stress τ_r as a function of the normal stress σ_n to which the normal stress value at $D = D_c$ is assigned (see Figure 3.38(a)).

Let us assume that the peak shear strength τ_p of intact granite in the brittle regime is expressed as a second-order polynomial function of σ_n^{eff} to which the effective normal stress value at $D = D_a$ is assigned, as follows:

$$\tau_p\left(\sigma_n^{\text{eff}}\right) = c_0 + c_1\sigma_n^{\text{eff}} + c_2\left(\sigma_n^{\text{eff}}\right)^2. \tag{3.51}$$

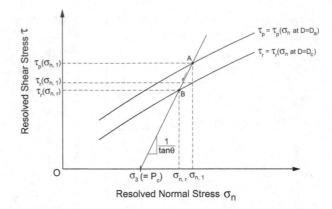

Fig. 3.39 A comparison between the breakdown stress drop $\tau_p(\sigma_{n,l}) - \tau_r(\sigma_{n,r})$ during the shear failure process under a constant confining pressure P_c in a triaxial test and the breakdown stress drop $\tau_p(\sigma_{n,l}) - \tau_r(\sigma_{n,l})$ during the shear failure process under a constant normal stress σ_n.

Likewise, let us assume that the residual friction stress τ_r in the brittle regime is expressed as a second-order polynomial function of σ_n^{eff} to which the effective normal stress value at $D = D_c$ is assigned, as follows:

$$\tau_r\left(\sigma_n^{\text{eff}}\right) = c_{0,r} + c_{1,r}\sigma_n^{\text{eff}} + c_{2,r}\left(\sigma_n^{\text{eff}}\right)^2. \tag{3.52}$$

From Eqs. (3.51) and (3.52), we can express the breakdown stress drop $\Delta\tau_b$ at $\sigma_n^{\text{eff}} = \sigma_{n,1}^{\text{eff}}$ as

$$\Delta\tau_b\left(\sigma_{n,1}^{\text{eff}}\right) = \tau_p\left(\sigma_{n,1}^{\text{eff}}\right) - \tau_r\left(\sigma_{n,1}^{\text{eff}}\right). \tag{3.53}$$

The breakdown stress drop $\Delta\tau_b(\sigma_{n,1}^{\text{eff}})$ defined by Eq. (3.53) equals the magnitude of $\tau_p(\sigma_{n,1}) - \tau_r(\sigma_{n,1})$ in Figure 3.39 (note that $\sigma_{n,1}^{\text{eff}}$ is replaced with $\sigma_{n,1}$ in a triaxial test on dry rock sample). Thus, the magnitude of the breakdown stress drop at a constant normal stress during the breakdown process can be estimated from graphs of Eqs. (3.51) and (3.52), as in Figure 3.39.

Equation (3.50) is displayed as straight line AB in Figure 3.39. The breakdown proceeds from the peak shear strength point A (i.e., $\tau_p(\sigma_{n,1})$) on a line, denoting equation $\tau_p = \tau_p(\sigma_n \text{ at } D = D_a)$ in Figure 3.39, along the straight line to the residual friction stress point B (i.e., $\tau_r(\sigma_{n,r})$) on another line, denoting equation $\tau_r = \tau_r(\sigma_n \text{ at } D = D_c)$ in Figure 3.39. Thus, the breakdown stress drop produced by the shear failure of intact rock under compressive stress environments in a triaxial test equals the magnitude of $\tau_p(\sigma_{n,1}) - \tau_r(\sigma_{n,r})$ in Figure 3.39. It is obvious from Figure 3.39 that $\tau_p(\sigma_{n,1}) - \tau_r(\sigma_{n,r})$ is greater than $\tau_p(\sigma_{n,1}) - \tau_r(\sigma_{n,1})$.

Let us specifically evaluate the magnitude of $\tau_p(\sigma_{n,1}) - \tau_r(\sigma_{n,1})$ for the shear failure of granite. Figure 3.40 shows a plot of the residual friction stress τ_r against the effective normal stress σ_n^{eff} to which the normal stress value at $D = D_c$ is assigned, for the data set

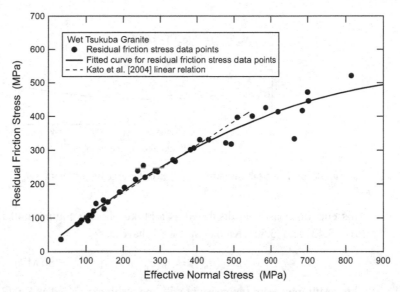

The residual friction stress τ_r against the effective normal stress σ_n^{eff} to which the normal stress value at $D = D_c$ is assigned, for the data set on the shear failure of wet Tsukuba granite in the brittle regime. Data from Odedra *et al.* (2001) and Kato *et al.* (2004), with additional unpublished data.

of the shear failure of wet Tsukuba granite in the brittle regime, plotted in Figure 3.34. The best-fit polynomial curve to the data points in Figure 3.40 is expressed as

$$\tau_r\left(\sigma_n^{\text{eff}}\right) = 19.4 + 0.884\sigma_n^{\text{eff}} - 3.96 \times 10^{-4}\left(\sigma_n^{\text{eff}}\right)^2. \tag{3.54}$$

Equation (3.54) has been displayed as a solid line in Figure 3.40. From Eqs. (3.45) and (3.54), we have

$$\Delta\tau_b\left(\sigma_n^{\text{eff}}\right) = \tau_p\left(\sigma_n^{\text{eff}}\right) - \tau_r\left(\sigma_n^{\text{eff}}\right) = 100.6 - 0.186\sigma_n^{\text{eff}} + 2.76 \times 10^{-4}\left(\sigma_n^{\text{eff}}\right)^2. \tag{3.55}$$

This equation indicates that the breakdown stress drop depends on the effective normal stress in the brittle regime. Specifically, $\Delta\tau_b$ only slightly decreases with an increase in σ_n^{eff} within the range of $\sigma_n^{\text{eff}} < 337$ MPa, and has a value of 69.3 MPa at $\sigma_n^{\text{eff}} = 337$ MPa. Beyond $\sigma_n^{\text{eff}} = 337$ MPa, $\Delta\tau_b$ increases with an increase in σ_n^{eff} (see Figure 3.41).

For a set of limited data on the shear failure of intact Tsukuba granite in wet environments, tested within the ranges of $T \leq 300°C$ and $\sigma_n^{\text{eff}} < 600$ MPa, τ_p has been expressed approximately as a linear function of effective normal stress σ_n^{eff} to which the effective normal stress value at $D = D_a$ is assigned (Eq. (3.43)). Likewise, τ_r can be expressed approximately as a linear function of the effective normal stress σ_n^{eff} to which the effective normal stress value at $D = D_c$ is assigned, and its functional form has been described by Kato *et al.* (2004) as

$$\tau_r\left(\sigma_n^{\text{eff}}\right) = 35 + 0.7\sigma_n^{\text{eff}} \quad (\sigma_n^{\text{eff}} < 600\,\text{MPa}). \tag{3.56}$$

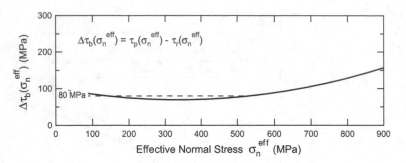

$$\Delta\tau_b(\sigma_n^{\text{eff}}) = \tau_p(\sigma_n^{\text{eff}}) - \tau_r(\sigma_n^{\text{eff}})$$

Fig. 3.41　The dependence of the breakdown stress drop $\Delta\tau_b$ on the effective normal stress σ_n^{eff}.

This equation has been displayed as a broken line in Figure 3.40 for comparison. From Eqs. (3.43) and (3.56), Kato *et al.* (2004) have

$$\Delta\tau_b(\sigma_n^{\text{eff}}) = \tau_p(\sigma_n^{\text{eff}}) - \tau_r(\sigma_n^{\text{eff}}) = 80\,\text{MPa}. \qquad (3.57)$$

This result may seem inconsistent with the result expressed as Eq. (3.55); however, it does not necessarily contradict the result expressed in Eq. (3.55) (see Figure 3.41), if we take into account both experimental error and the fact that Eq. (3.57) has been approximately derived from a limited number of data obtained within a limited range of σ_n^{eff} below 600 MPa.

According to Wong (1986), linear relationships between τ_p and σ_n, and between τ_r and σ_n, approximately hold in a limited range of normal stress for data on the shear failure of Fichtelbirge granite and San Marcos gabbro. Wong (1986) obtained the following relationships:

$$\tau_p(\sigma_n) = 92 + 0.67\sigma_n \qquad (243\,\text{MPa} < \sigma_n < 715\,\text{MPa})$$
$$\tau_r(\sigma_n) = 39 + 0.63\sigma_n \qquad (136\,\text{MPa} < \sigma_n < 608\,\text{MPa}) \qquad (3.58)$$

for the shear failure of Fichtelbirge granite tested at room temperature under confining pressures up to 300 MPa (Rummel *et al.*, 1978), and

$$\tau_p(\sigma_n) = 189 + 0.49\sigma_n \qquad (500\,\text{MPa} < \sigma_n < 784\,\text{MPa})$$
$$\tau_r(\sigma_n) = 77 + 0.57\sigma_n \qquad (439\,\text{MPa} < \sigma_n < 732\,\text{MPa}) \qquad (3.59)$$

for the shear failure of San Marcos gabbro tested at room temperature under confining pressures ranging from 250 to 400 MPa. From Eqs. (3.58) and (3.59), we have

$$\Delta\tau_b(\sigma_n) = \tau_p(\sigma_n) - \tau_r(\sigma_n) = 53 + 0.04\sigma_n \qquad (3.60)$$

and

$$\Delta\tau_b(\sigma_n) = \tau_p(\sigma_n) - \tau_r(\sigma_n) = 112 - 0.08\sigma_n, \qquad (3.61)$$

respectively. These results also indicate that the breakdown stress drop in the brittle regime only slightly depends on the (effective) normal stress.

The breakdown stress drop for real earthquakes in general occurs under combined compressive stress environments in the field. Note, therefore, that the upper limit of the breakdown stress drop for real earthquakes is given, not by the breakdown stress drop under the effective normal stress held constant, but by the breakdown stress drop under combined compressive stress environments. It is therefore worthwhile to know to what extent the breakdown stress drop under combined compressive stress environments is greater than the breakdown stress drop when the effective normal stress is held constant. The regression analysis of the data points for wet Tsukuba granite in the brittle regime, in the range of $\sigma_n^{\mathrm{eff}} < 600$ MPa, gives the following average value for $\Delta \tau_b$ (defined by $\Delta \tau_b = \tau_p(\sigma_{n,1}^{\mathrm{eff}}) - \tau_r(\sigma_{n,r}^{\mathrm{eff}})$) with its standard deviation: $\Delta \tau_b = 119.6 \pm 27.6$ MPa. This indicates that the breakdown stress drop under the effective normal stress held constant in the brittle regime is roughly 67% of the breakdown stress drop observed under combined compressive stress environments in a triaxial test.

Experimental data obtained in the laboratory indicate that the breakdown stress drop produced by the shear failure of intact rock under ambient crustal conditions has a value of about 120 ±28 MPa, which is only slightly affected by the effective normal stress. This does not mean that the breakdown stress drop for a real earthquake is unaffected by the effective normal stress, but simply means that the upper limit of the breakdown stress drop for earthquakes is only slightly affected by the effective normal stress. A real earthquake occurs on a preexisting fault embedded in the seismogenic layer, and such a preexisting fault is in general heterogeneous, consisting of local strong areas highly resistant to rupture growth, and the rest of the fault with low (or little) resistance to rupture growth. The breakdown stress drop produced by the shear failure of intact rock in seismogenic environments provides the upper limit of the breakdown stress drop at such local strong areas on a fault, and this upper limit is only slightly affected by the effective normal stress. When the breakdown stress drop at such a local strong area on a fault is lower than the upper limit provided by the breakdown stress drop produced by the shear failure of intact rock in seismogenic environments, the breakdown stress drop at the local strong area is more likely to be affected by the effective normal stress than the only slightly affected upper limit.

Although the breakdown stress drop does not depend on temperature in the brittle regime, it does become dependent on temperature in the semi-brittle regime. According to Kato *et al.* (2004), the effects of the effective normal stress σ_n^{eff} and temperature T on the breakdown stress drop $\Delta \tau_b(\sigma_n^{\mathrm{eff}}, T)$ for wet Tsukuba granite in the semi-brittle regime can be described as the following linear expression:

$$\Delta \tau_b(\sigma_n^{\mathrm{eff}}, T) = \tau_p(\sigma_n^{\mathrm{eff}}, T) - \tau_r(\sigma_n^{\mathrm{eff}}, T)$$
$$= 80 - 5 \times 10^{-4}(T - 300)\sigma_n^{\mathrm{eff}}, \tag{3.62}$$

for data obtained in a limited range of σ_n^{eff} from about 100 to 600 MPa, and a limited range of T from 300 °C to 480 °C. Equation (3.62) indicates that $\Delta \tau_b$ decreases with increasing temperature in the semi-brittle regime (300 °C < $T \leq$ 480 °C), and that within the temperature range of 300 °C < $T \leq$ 480 °C, $\Delta \tau_b$ at a higher σ_n^{eff} decreases more sharply with increasing temperature than $\Delta \tau_b$ at a lower σ_n^{eff}.

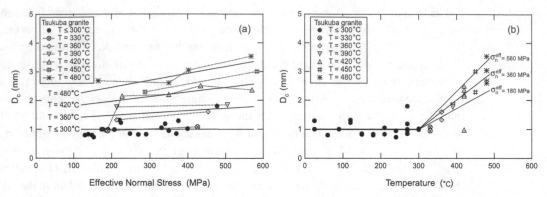

The dependence of the breakdown displacement D_c on the effective normal stress and temperature. Data from Kato *et al.* (2004).

3.4.4 Dependence of breakdown displacement on environmental factors

The actual rupture surfaces of heterogeneous materials such as rock are not flat or smooth but geometrically irregular. As noted previously, the breakdown process of shear rupture is greatly affected by the geometric irregularity of shear-rupture surfaces. In shear rupture, the relative displacement (or slip) between the mating shear-rupturing surfaces progresses along the rupturing surfaces during the breakdown process, and the shear-rupturing surfaces are in mutual contact and interactive during the breakdown process. For this reason, the breakdown process is strongly influenced by the geometric irregularity of shear-rupturing surfaces.

Shear-rupturing surfaces of rock are not self-similar at all scales but self-similar within a finite scale range. The characteristic length λ_c representing the geometric irregularity of such shear-rupturing surfaces can be defined as the corner wavelength that separates the neighboring two self-similar bands with different fractal dimensions, or as the critical wavelength beyond which the geometric irregularity of the rupturing surfaces no longer exhibits self-similarity (see Section 3.1). The characteristic length λ_c thus defined represents the predominant wavelength contained in the geometric irregularity of shear-rupturing surfaces. As noted in Section 3.3, the breakdown displacement D_c scales with the characteristic length λ_c. In other words, D_c changes in proportion to λ_c when other conditions are equal. The characteristic length λ_c representing the geometric irregularity of shear-fracture surfaces of an intact rock sample is prescribed by the structural heterogeneity of rock fabric when the rock sample size is fixed.

Let us focus attention on whether or not the breakdown displacement is affected by ambient temperature and effective normal stress. Figure 3.42(a) shows a plot of the breakdown displacement D_c against the effective normal stress σ_n^{eff} for a set of data (Table 1 in Kato *et al.* (2004)) on the shear failure of intact Tsukuba granite in wet environments under combined compressive stress environments in a triaxial test, and Figure 3.42(b) shows a plot of D_c against ambient temperature T for the same data set. The systematic effect of σ_n^{eff} on D_c may be only slightly recognizable from Figure 3.42(a), even if experimental

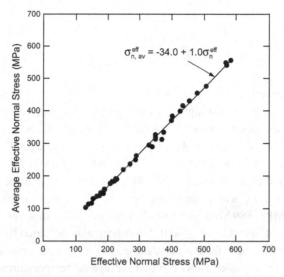

The relation between the average effective normal stress $\sigma_{n,av}^{eff}$ and the effective normal stress σ_n^{eff}. Data points (black circles) are from Kato *et al.* (2004).

errors are taken into account. By contrast, the effect of T on D_c is clearly recognizable from Figures 3.42(a) and (b); that is, D_c is not affected by T below 300 °C, but affected by T above 300 °C. Based on this data set, Kato *et al.* (2004) have empirically derived the following relationship:

$$D_c\left(T, \sigma_{n,av}^{eff}\right) = D_{c0}\rho\left(T, \sigma_{n,av}^{eff}\right), \qquad (3.63)$$

where

$$\rho\left(T, \sigma_{n,av}^{eff}\right) = \begin{cases} 1 & (T \le 300\,°\mathrm{C}) \\ 1 + 5 \times 10^{-3}\left(1 + 3 \times 10^{-3}\sigma_{n,av}^{eff}\right)(T - 300) & (300\,°\mathrm{C} < T \le 480\,°\mathrm{C}). \end{cases}$$
$$(3.64)$$

In the above equations, D_{c0} denotes an average value (1.0 mm) of the breakdown displacement observed in the range of $T \le 300\,°\mathrm{C}$, $\sigma_{n,av}^{eff}$ denotes the average effective normal stress defined as an average value of σ_n^{eff} at $D = D_a$ and at $D = D_c$.

For the shear failure of an intact rock sample in a triaxial test, it is customary to adopt the normal stress σ_n to which the normal stress value at $D = D_a$ is assigned, as representative of the normal stresses during the breakdown process. Accordingly, the effective normal stress value at $D = D_a$ has been assigned to σ_n^{eff} in Figure 3.42. The average effective normal stress $\sigma_{n,av}^{eff}$ changes in proportion to the effective normal stress σ_n^{eff} to which the effective normal stress value at $D = D_a$ is assigned (see Figure 3.43). The straight line in Figure 3.43 denotes the linear relation between $\sigma_{n,av}^{eff}$ and σ_n^{eff}, expressed as $\sigma_{n,av}^{eff} = -34.0 + 1.0\sigma_n^{eff}$. By using this relation, Eqs. (3.63) and (3.64) can be rewritten as

$$D_c\left(T, \sigma_n^{eff}\right) = D_{c0}\rho\left(T, \sigma_n^{eff}\right) \qquad (3.65)$$

and

$$\rho\left(T, \sigma_{\mathrm{n}}^{\mathrm{eff}}\right) = \begin{cases} 1 & (T \le 300\,°\mathrm{C}) \\ 1 + 4.5 \times 10^{-3}\left(1 + 3.3 \times 10^{-3}\sigma_{\mathrm{n}}^{\mathrm{eff}}\right)(T - 300) & (300\,°\mathrm{C} < T \le 480\,°\mathrm{C}), \end{cases}$$

$$(3.66)$$

respectively.

The four solid straight lines in Figure 3.42(a) denote the linear relations between D_{c} and $\sigma_{\mathrm{n}}^{\mathrm{eff}}$ described by Eqs. (3.65) and (3.66) under the conditions of $T \le 300\,°\mathrm{C}$, $T = 360\,°\mathrm{C}$, $T = 420\,°\mathrm{C}$, and $T = 480\,°\mathrm{C}$, respectively. In Figure 3.42(b), the solid line in parallel with the abscissa axis denotes the relation between D_{c} and T described by Eqs. (3.65) and (3.66) under the condition of $T \le 300\,°\mathrm{C}$, and the other three solid lines denote the relations between D_{c} and T, described by Eqs. (3.65) and (3.66) under the conditions of $\sigma_{\mathrm{n}}^{\mathrm{eff}} = 180$ MPa, 380 MPa, and 580 MPa, respectively. Although the effect of T on D_{c} in a limited range of $300\,°\mathrm{C} \le T \le 480\,°\mathrm{C}$ has been approximated by a linear relation between D_{c} and T for the shear failure of intact granite in wet environments (Figure 3.42(b)), the effect of T on D_{c} over a much broader range of temperatures in the brittle–plastic transition regime (above $300\,°\mathrm{C}$) may be expressed as a nonlinear relationship between D_{c} and T, such as $D_{\mathrm{c}}/D_{\mathrm{c}0} = A \exp(-B/T)$ or $D_{\mathrm{c}}/D_{\mathrm{c}0} = A/(B - T)$ (where A and B are constants), for the shear failure of intact granite (Ohnaka, 1992, 1995b). This issue remains to be fully elucidated.

Constitutive laws for earthquake ruptures

4.1 Basic foundations for constitutive formulations

The constitutive law that governs the behavior of earthquake ruptures provides the basis of earthquake physics, and the governing law plays a fundamental role in accounting for the entire process of an earthquake rupture, from its nucleation to its dynamic propagation to its arrest, quantitatively, in a unified and consistent manner. Without a rational law to govern real earthquake rupture processes, the physics of earthquakes cannot be a quantitative science in the true sense. It is therefore of great urgency and importance to establish a rational law strictly formulated for real earthquake ruptures. In this chapter, we will review the constitutive formulations so far proposed for earthquakes, and will rigorously and thoroughly discuss what the governing law for earthquake ruptures ought to be, and how it should be formulated, on the basis of positive facts, from a comprehensive viewpoint. This is a necessary step toward the strict formulation of a constitutive law for real earthquake ruptures, and cannot be avoided if the physics of earthquakes is to be a quantitative science in the true sense.

A shallow earthquake source at crustal depths is a shear rupture instability taking place on a preexisting fault, in geological and tectonic settings, embedded in the seismogenic crust composed of rocks. As described in Chapter 1, the seismogenic crust and preexisting faults embedded therein are inherently heterogeneous. In particular, individual faults contain local strong areas (called "asperities" or "barriers") that are highly resistant to rupture growth, with the rest of the fault having low (or little) resistance to rupture growth. Some local stress drops at these strong areas on faults are high enough to equal the breakdown stress drop of intact rock tested under seismogenic crustal conditions simulated in the laboratory (for example, see Figure 3.23). Hence, it is obvious that the earthquake rupture process at crustal depths is not a simple process of frictional slip failure on a uniformly precut weak fault, but a more complex process, including the fracture of initially intact rock at some local strong areas on a heterogeneous fault. This is indeed corroborated by geological observations of structural heterogeneity and geometric irregularity for real fault zones (see Chapter 1).

The above-mentioned local strong areas on a fault play a much more important role than do the remaining weak areas on the fault in accumulating elastic strain energy as the driving force to bring about an earthquake or to radiate strong motion seismic waves. As stated in Section 3.3, therefore, the constitutive law for real earthquake ruptures must be formulated as a unifying law that governs not only the frictional slip failure at precut interface (or frictional contact) areas on a fault, but also the shear failure of intact rock at some of the local strong areas on the fault. This is the first prerequisite for a rational constitutive formulation for real earthquake ruptures.

Rupture phenomena, including earthquakes, are scale-dependent; in other words, some of the physical quantities inherent in the rupture exhibit scale-dependence. The scale-dependence of scale-dependent physical quantities inherent in the rupture, such as nucleation zone length and slip acceleration, is directly attributed to the scale-dependence of the breakdown displacement (for details, see Chapters 5 and 6). The breakdown displacement D_c is defined as the critical slip displacement required for the shear strength to degrade to a residual friction stress level during the breakdown process; in other words, D_c is the slip displacement at the end of the breakdown zone behind the rupture front (see Figure 2.10). Hence, D_c is directly related to the coherent zone length X_c of the rupture breakdown (see Eq. (3.28)). In addition, the fundamental cause of the scale-dependence of D_c lies in the geometric irregularity of the rupturing surfaces (see Chapters 3 and 6). In light of these facts, it is essential to formulate the constitutive law in such a way that a proper scaling parameter such as D_c, which represents the scaling property intrinsic to the rupture breakdown, is incorporated into the law; otherwise, the scale-dependent physical quantities inherent in the rupture over a broad scale range cannot be treated consistently and quantitatively in terms of a single constitutive law. Thus, the incorporation of a scaling parameter intrinsic to the rupture breakdown into the constitutive law is the second prerequisite for the strict constitutive formulation for real earthquake ruptures.

Once shear rupture instability occurs in the brittle regime, the rupture propagates dynamically at a high speed close to elastic wave velocities. This is referred to as brittle rupture, and a typical large-scale example is the earthquake rupture instability taking place in the brittle layer of the Earth's crust. However, the shear rupture occurring on an inhomogeneous fault cannot begin to propagate abruptly at high speeds close to elastic wave velocities immediately after the onset of the rupture. As presented in detail in Chapter 5, high-resolution laboratory experiments have conclusively demonstrated that shear rupture on a preexisting fault cannot begin to propagate abruptly at high speeds close to elastic wave velocities immediately after the onset of the rupture, even in the brittle regime, when the resistance to rupture growth is heterogeneously distributed on the fault. A shear rupture initially grows stably and quasi-statically in a localized zone on the fault, and subsequently grows spontaneously at accelerating speeds up to the terminal phase of the rupture propagation at a high-speed close to elastic wave velocities. This experimental result has been confirmed by computer numerical simulations based on a constitutive law for shear failure (Shibazaki and Matsu'ura, 1998). The transition process from stable, quasi-static rupture growth to unstable, high-speed rupture in a localized zone is what is referred to as the *nucleation process* (see Figures 5.16 and 5.17). The entire sequential process of an earthquake rupture – from its stable, quasi-static nucleation to its unstable, dynamic propagation, to its arrest – must be accounted for quantitatively in a unified and consistent manner in terms of a single constitutive law. This is the third prerequisite to be met when we rationally formulate the constitutive law for real earthquake ruptures.

As described by Stephen Hawking in his book (Hawking, 1988), in general, a theory is a good theory if it accurately describes a large class of observations on the basis

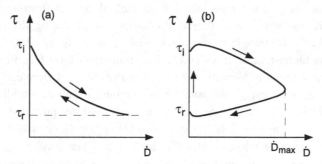

Fig. 4.1 (a) An assumed slip rate-weakening property, and (b) a real relationship between τ and $\dot{D}$ observed during a stick-slip cycle in the laboratory, based on laboratory data from Ohnaka *et al.* (1987a) and Ohnaka and Yamashita (1989).

of a simple model that contains only a few physical elements, and if it makes definite predictions about the results of future observations. In this regard, a constitutive law formulation for earthquake ruptures is a good formulation if the law is formulated as simply as possible in terms of only a few physical elements, in such a way that it meets the aforementioned three prerequisites, and if it enables one to make definite predictions about the results of future observations of an earthquake generation process from its stable, quasi-static nucleation to its unstable, dynamic propagation, by taking into account seismogenic environmental conditions and fault zone properties such as fault surface heterogeneity.

The constitutive formulations for earthquake ruptures so far proposed can be broadly classified into two categories: the rate-dependent formulations (see Section 4.2), and the slip-dependent formulations (see Section 4.3). In the following two sections, we will discuss which formulations are most proper as the governing law for real earthquake ruptures, keeping the aforementioned prerequisites in mind.

4.2 Rate-dependent constitutive formulations

Let us first assume that the constitutive law is expressed as the shear traction τ on shear-rupturing surfaces being a simple function f of the slip rate (or slip velocity) $\dot{D}$ alone; that is,

$$\tau = f(\dot{D}), \tag{4.1}$$

with its functional form being characterized by a slip rate-weakening property, as shown in Figure 4.1(a). A constitutive formulation of this type was employed by a number of authors in their numerical simulations for a simple stick-slip model of earthquakes (e.g., Carlson and Langer, 1989; Carlson *et al.*, 1991; Nakanishi, 1992; Langer *et al.*, 1996). This formulation, however, cannot be a self-consistent constitutive law for frictional stick-slip failure on a fault (Rice and Ruina, 1983; Ruina, 1985). It predicts that since $d\tau/d\dot{D} < 0$, the shear stress τ on the fault unstably decreases from an initial value τ_i to a residual friction stress level τ_r with an increase in $\dot{D}$ during the breakdown process, and that the fault strength recovers reversibly from τ_r to τ_i after the slip velocity has attained its highest value at the residual friction stress level (see Figure 4.1(a)). This contradicts the common observation that frictional stick-slip failure can stably proceed under certain conditions even in the purely brittle regime, and also contradicts the observational fact that the shear traction during the dynamic breakdown process is not a single-valued function of the slip rate (Ohnaka *et al.*, 1987a, 1997; Ohnaka and Yamashita, 1989), as shown in Figure 4.1(b) (see also Figure 3.19(b)). The breakdown process and the subsequent re-strengthening process during a stick-slip cycle are completely irreversible. Thus, this formulation does not meet the indispensable prerequisite for a frictional stick-slip cycle on a fault.

To resolve these contradictions, Dieterich (1978a, 1979, 1981) and Ruina (1983, 1985) introduced an evolving state variable as a measure of the quality of surface contact, and proposed rate- and state-dependent constitutive formulations for frictional stick-slip on a precut fault. This formulation assumes that the slip rate $\dot{D}$ and at least one evolving state variable θ are independent and fundamental variables, and that the transient response of the shear traction τ to $\dot{D}$ is essentially important. In this formulation, therefore, the role of $\dot{D}$ is emphasized, and τ is expressed as an explicit function of $\dot{D}$ and θ. The rate- and state-dependent constitutive law is in general expressed as follows (Ruina, 1983):

$$\tau = \sigma_n F(\dot{D}, \theta), \tag{4.2}$$

$$d\theta/dt = G(\dot{D}, \theta), \tag{4.3}$$

where σ_n denotes the normal stress across the fault, and F and G denote functions of slip rate $\dot{D}$ and state variable θ. Since the state of surface contact may not be described by a single variable, θ is in general expressed as a collection of state variables θ_i $(i = 1, 2, \ldots)$; that is, $\theta = \theta_1, \theta_2, \ldots$ (Ruina, 1983).

This rate- and state-dependent constitutive formulation for frictional stick-slip is based on laboratory tests on frictional sliding on a precut rock interface, illustrated in Figures 4.2(a) and (b) (Dieterich, 1978a, 1979, 1981; Tullis and Weeks, 1986; Kilgore *et al.*, 1993; and others). Figures 4.2(a) and (b) show the response of friction to a stepwise change in slip rate. When the imposed slip rate is suddenly increased from a steady-state slip rate $\dot{D}_1$ (for example, 10^{-4} mm/s) to another steady-state slip rate $\dot{D}_2$ (for example, 10^{-3} mm/s), as shown in Figure 4.2(a), the consequent frictional stress τ changes with slip displacement D (as shown in Figure 4.2(b)) under the condition of a constant normal stress σ_n. In other words, the frictional resistance suddenly increases from its steady-state value $\tau_1^{ss}(= \mu_1^{ss}\sigma_n$, where μ^{ss} is the frictional coefficient at steady-state) at slip rate $\dot{D}_1$ in response to the stepwise change in slip rate from $\dot{D}_1$ to $\dot{D}_2$, and then decays approximately exponentially with slip displacement D to a new steady-state value $\tau_2^{ss}(= \mu_2^{ss}\sigma_n)$ at the slip rate $\dot{D}_2$. The critical slip distance δ_c in Figure 4.2(b) is defined as the characteristic slip required for frictional resistance to reach a steady-state level, when a step change in the slip rate is imposed. Figure 4.2(b) indicates that a new steady-state frictional stress level at fast slip rate $\dot{D}_2$ is lower than the initial steady-state frictional stress at slow slip rate $\dot{D}_1$.

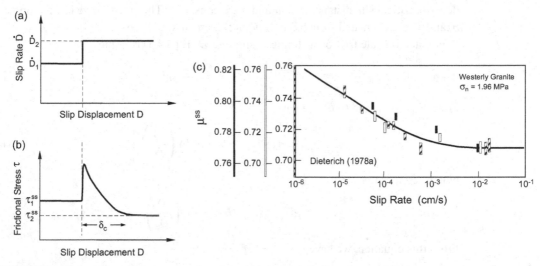

Fig. 4.2 (a) A step change in slip rate from $\dot{D}_1$ to $\dot{D}_2$, (b) The response of friction to the step change in slip rate shown in Figure 4.2(a). (c) An example of slip rate effects on friction, obtained in laboratory experiments. Reproduced from Dieterich (1978a).

Such slip-rate effects on friction were examined in laboratory experiments (e.g., Dieterich, 1978a; Tullis and Weeks, 1986; Kilgore et al., 1993; Marone, 1998b). A typical example is shown in Figure 4.2(c), which is the experimental result obtained in an earlier study by Dieterich (1978a).

Based on the experimental data for extremely slow slip rates less than 1 mm/s (Dieterich, 1978a, 1981; for other references, see Dieterich and Kilgore, 1996), and based on several interpretations of that data, the specific formulation of the rate- and state-dependent constitutive law was proposed originally by Dieterich (1979) and Ruina (1983), and subsequently by Perrin et al. (1995) and other authors (for others see Marone, 1998b; Bizzarri, 2011), with some additional experimental data. Thus, many different forms of functions (4.2) and (4.3) have so far been proposed (see Bizzarri, 2011). Of the many functional forms proposed, the following expression formulae:

$$\tau = \sigma_n \left[\mu_0 + a \ln \left(\frac{\dot{D}}{\dot{D}_0} \right) + b \ln \left(\frac{\dot{D}_0 \theta}{\delta_c} \right) \right], \tag{4.4}$$

$$\frac{d\theta}{dt} = 1 - \frac{\dot{D}\theta}{\delta_c}, \tag{4.5}$$

have been widely adopted as representative expression formulae of the rate- and state-dependent constitutive law, under the assumption that the state of surface contact is described by a single state variable θ representing the average time of asperity contacts. The state variable θ has dimensions of time. In the above equations, $\dot{D}_0$ is a constant slip rate, μ_0 denotes a constant reference value of frictional coefficient at steady-state when $\dot{D} = \dot{D}_0$, δ_c

denotes critical slip distance, and a and b are constants. The constitutive law parameters a, b, and δ_c are estimated from laboratory measurements.

For steady-state frictional sliding at slip rate $\dot{D}$, Eq. (4.5) becomes

$$\frac{d\theta}{dt} = 1 - \frac{\dot{D}\theta}{\delta_c} = 0. \tag{4.6}$$

From Eqs. (4.4) and (4.6), we have

$$\tau^{ss} = \sigma_n \left[\mu_0 - (b - a) \ln \left(\frac{\dot{D}}{\dot{D}_0} \right) \right], \tag{4.7}$$

which reduces to

$$\mu^{ss} = \mu_0 - (b - a) \ln \left(\frac{\dot{D}}{\dot{D}_0} \right). \tag{4.8}$$

From this equation, we have

$$\mu_2^{ss} - \mu_1^{ss} = -(b - a) \ln \left(\frac{\dot{D}_2}{\dot{D}_1} \right). \tag{4.9}$$

When $b > a$, Eq. (4.8) indicates that the frictional coefficient at steady-state μ^{ss} decreases with an increase in slip rate $\dot{D}$. This accounts for the experimental result shown in Figure 4.2(c).

The specific functional forms of the rate- and state-dependent constitutive laws were based on the results of frictional sliding tests under quasi-static conditions. Therefore, they have no experimental foundation applicable to unstable, dynamic slip processes during a stick-slip cycle. Laboratory experiments demonstrate that frictional slip failure is independent of slip rate $\dot{D}$ at high slip rates (or velocities) in the process of dynamic shear rupture instabilities arising during the stick-slip cycle (Okubo and Dieterich, 1986; Ohnaka et al., 1987a), and also independent of state variable θ in the range of sufficiently small θ (Okubo and Dieterich, 1986). This suggests that the rate- and state-dependent constitutive formulation should be modified to achieve the rate and state independence at high slip velocities while maintaining its applicability at low slip rates (Dieterich, 1986; Okubo and Dieterich, 1986; Rice and Tse, 1986). With this in mind, Eq. (4.4) can be rewritten as follows (Dieterich, 1986; Okubo, 1989; Linker and Dieterich, 1992):

$$\tau = \sigma_n \left[\mu_0 - a \ln \left(\frac{\dot{D}_0}{\dot{D}} + 1 \right) + b \ln \left(\frac{\dot{D}_0 \theta}{\delta_c} + 1 \right) \right]. \tag{4.10}$$

The fault-healing (or re-strengthening) process and the fault-weakening (or breakdown) process on a precut fault at a compressive normal load can both be treated by a single law under the framework of the rate- and state-dependent constitutive formulation. This provides a great advantage to simulation researchers, and indeed facilitates computer simulations of earthquake cycles in a virtual world. Thus, a great number of computer numerical simulations have been performed within the framework of the rate- and state-dependent constitutive law since the 1980s (e.g., Tse and Rice, 1986; Rice, 1993; Ben-Zion and Rice, 1995; Stuart and Tullis, 1995; Rice and Ben-Zion, 1996; Tullis, 1996; and others). In particular, computer numerical simulations of dynamic breakdown processes at high slip rates of the order of 1 m/s on a preexisting fault have been done by using Eqs. (4.10) and (4.5) (e.g., Okubo, 1989; Bizzarri *et al.*, 2001; Cocco and Bizzarri, 2002; Bizzarri and Cocco, 2003).

However, this rate- and state-dependent constitutive law poses serious problems when the law is applied to real earthquake rupture processes. As described in Chapter 1, real faults embedded in the Earth's crust are inherently heterogeneous. The process of an earthquake rupture at crustal depths is not a simple process of frictional slip failure on a uniformly precut weak fault, but a more complex process including the fracture of initially intact rock at some local strong areas on a heterogeneous fault. If this is the case, the constitutive law for real earthquake ruptures must be formulated as a unifying law that governs not only frictional slip failure at precut interface (or frictional contact) areas on faults but also the shear fracture of intact rock at some local strong areas on the faults, as previously noted. However, the rate- and state-dependent law, formulated as the governing law for pure frictional stick-slip on precut interface observed in the laboratory, is a function of both slip rate and state variable, and cannot be a constitutive law governing the instability or stability of the shear failure of intact rock. Thus, the rate- and state-dependent law cannot be a unifying constitutive law that governs not only frictional slip failure at precut interface (or frictional contact) areas on a fault, but also the shear failure of intact rock at some local strong areas on the fault. Hence, the rate- and state-dependent law does not meet the first prerequisite described in Section 4.1.

As noted earlier in this section, the rate- and state-dependent constitutive law parameters, a, b, and δ_c, are defined as the parameters directly involved in the response of frictional resistance to the imposed step change in slip rate from a set value to another set value in the laboratory. These parameters are thus definable only in the laboratory, and are not physical quantities associated directly with the breakdown process or the breakdown zone behind the front of a propagating shear rupture along a fault. Therefore, the rate- and state-dependent constitutive law cannot be useful for the purpose of estimating the source parameters of real earthquake ruptures from seismological data actually observed in the field. Indeed, it has not been used successfully for this purpose, although seismologists have estimated earthquake-source parameters from observed near-field or far-field seismic waves by waveform inversion. Thus, the rate- and state-dependent law is limited to use within computer numerical simulations in a virtual world. This is a serious disadvantage for real earthquake ruptures.

Rupture phenomena, including earthquakes, are inherently scale-dependent. Indeed, some of the physical quantities inherent in shear rupture exhibit scale-dependence (see Chapters 5 and 6). To quantitatively account, in a unified and consistent manner, for scale-dependent physical quantities inherent in the rupture over a broad scale range, it is critically important to formulate the governing law in such a way that the scaling property inherent in the rupture breakdown is incorporated into the law. However, any physically meaning-ful parameter representing the scaling property inherent in the rupture breakdown is not incorporated into the rate- and state-dependent constitutive law. The only possible scaling parameter incorporated into the rate- and state-dependent law is the critical slip distance δ_c, defined as the slip distance required for frictional resistance to reach a steady-state value when a step change in the slip rate is suddenly imposed from a set value $\dot{D}_1$ (e.g., 10^{-4} mm/s) to another set value $\dot{D}_2$ (e.g., 10^{-3} mm/s). Therefore, it is obvious that the parameter δ_c thus defined is not directly related to the geometric length of the coherent zone of the rupture breakdown. Accordingly, δ_c cannot be a physically meaningful parameter representing the scaling property inherent in the rupture breakdown. Therefore, the rate- and state-dependent law also does not meet the second prerequisite described in Section 4.1.

According to Bizzarri and Cocco (2003), the quantity D_c^{eq} equivalent to the breakdown displacement D_c defined in the framework of the slip-dependent constitutive law can be expressed in the framework of the rate- and state-dependent constitutive law as follows:

$$D_c^{eq} \cong \delta_c \ln \left(\frac{\theta_{init} \dot{D}_c^{eq}}{\delta_c} \right), \tag{4.11}$$

where θ_{init} is the initial value of state variable, and $\dot{D}_c^{eq}$ is the slip rate (or slip velocity) at the end of the breakdown process when the slip displacement is D_c^{eq}. Cocco and Bizzarri (2002) and Bizzarri and Cocco (2003) reported that $D_c^{eq}/\delta_c \cong 15$ under further assumptions (see Bizzarri and Cocco, 2003). Even if δ_c is related to the breakdown displacement D_c through $D_c^{eq}/\delta_c \cong 15$ under certain assumptions, it is clear that δ_c is not an optimum parameter for quantitative scaling of the scale-dependent physical quantities inherent in the rupture breakdown, because D_c^{eq}/δ_c rigorously does not have a universal constant value of 15. This is obvious from the definition of δ_c and from Eq. (4.11). In addition, since δ_c is a parameter definable only in the laboratory, it is impossible to estimate δ_c for real earthquakes directly from observed seismological data. Indeed, Guatteri et al. (2001) had no choice but to estimate δ_c for the 1995 Kobe earthquake under certain assumptions, indirectly from the breakdown displacement D_c estimated by Ide and Takeo (1997). This also suggests that the rate- and state-dependent formulation is not suitable for the constitutive law for real earthquake ruptures.

Equations (4.4) and (4.5) were derived by considering the effects of $\dot{D}$ and θ alone, based on laboratory data on rock friction under the condition that the direct effect of slip displacement D on friction is negligible ($\partial \mu / \partial D \cong 0$). In this formulation, therefore, the direct effect of slip displacement is not incorporated into the law, despite the fact that the effect of slip rate is secondary to the principal effect of slip displacement during the breakdown process. This fact has already been established in laboratory experiments, and it is common knowledge among experimentalists. In fact, the slip rate effect can be measured only after an adequate amount of slip displacement where the effect of slip displacement on friction (or $\partial \mu / \partial D$) is reduced (Dieterich, 1979, 1981), and laboratory experiments show that the parameter a in Eq. (4.4) has a very small value ranging from 0.003 to 0.015 for rock (e.g., Dieterich, 1981; Gu et al., 1984; Tullis and Weeks, 1986; Blanpied et al., 1987). This indicates that the quantitative effect of slip rate is very small. Thus, it is clear that the condition stated above can be attained only after an adequate amount of slip displacement, and that this condition cannot be met during the breakdown process of shear rupture.

As described previously, real shear-rupture surfaces of inhomogeneous rock are not flat planes, but exhibit geometric irregularity with band-limited self-similarity. The re-strengthening on such irregular fault surfaces can be attained by a displacement-induced increase in friction, due to an increase in the sum of asperity junction areas, such as asperity interlocking and asperity ploughing, on the fault surfaces with progressive displacement under compressive normal stress. This direct effect is dominant as the physical mechanism of fault re-strengthening, and therefore this must be incorporated into the constitutive formulation for the fault-healing (or re-strengthening) process. In the framework of the rate- and state-dependent formulation, however, it is assumed that the fault re-strengthening is caused by the slip rate and state (or contact-time) effects alone. In this formulation, therefore, the dominant effect of interlocking asperities with progressive displacement under compressive normal stress on the fault-healing (or re-strengthening) is overlooked, despite the fact that the slip rate and state (or contact-time) effects on the fault-healing may be masked completely by the re-strengthening effect of interlocking asperities with progressive displacement under compressive normal stress. The fault re-strengthening may also be reinforced by a gradual increase in the effective normal stress with tectonic loading during the interseismic period. Thus, the fault re-strengthening can easily be attained without having to assume the effect of slip rate, so that we do not find any compelling reason to emphasize the slip rate effect in the constitutive formulation for a real earthquake rupture on a heterogeneous fault and for the fault re-strengthening during the interseismic period.

4.3 Slip-dependent constitutive formulations

As understood from the basic fact that three fundamental modes (modes I, II, and III) of fracture are defined in terms of crack-tip displacement in fracture mechanics, the relative displacement between fracture surfaces plays a fundamental and primary role in the fracture process. Consistent with this, and based on the laboratory-proven fact that slip-dependency is more fundamental to shear rupture than any other property, including slip rate-dependency, the slip-dependent constitutive law is formulated in such a manner that the shear traction τ along shear-rupturing surfaces is a function of slip displacement D; that is,

$$\tau = f(D), \tag{4.12}$$

where f denotes a functional form representing the constitutive relation between τ and D. When the shear traction τ is specifically given as a function of slip displacement D, as shown in Figure 2.13 or Figure 6.9, the energy G_c required for shear fracture can be calculated from Eq. (2.44), and the calculated result is equal to the vertical-lined area in Figure 2.13 or Figure 6.9. From this, we can understand intuitively that the slip-dependent constitutive law automatically satisfies the Griffith energy balance fracture criterion.

In the slip-dependent constitutive formulation, slip displacement is an independent variable, and the transient response of the shear traction to the slip displacement is fundamentally important, whereas the effect of slip rate is secondary compared to the principal effect of the slip displacement. This is a significant point to be emphasized, and as noted in Chapter 3, the slip-dependent constitutive law is self-consistent as the governing law for shear rupture. This is in contrast to Eq. (4.1) in the rate-dependent constitutive formulation.

The slip-dependent constitutive law has been adopted not only for theoretical modeling of earthquake ruptures and numerical simulations of earthquake generation processes using computers (e.g., Ida, 1972; Andrews, 1976a, 1976b, 1985; Campillo and Ionescu, 1977; Day, 1982; Matsu'ura et al., 1992; Yamashita and Ohnaka, 1992; Madariaga et al., 1998; Shibazaki and Matsu'ura, 1998; Madariaga and Olsen, 2000, 2002; Fukuyama and Madariaga, 2000; Hashimoto and Matsu'ura, 2000, 2002; Campillo et al., 2001; Fukuyama and Olsen, 2002; Fukuyama et al., 2002; Aochi and Madariaga, 2003), but also for estimating constitutive law parameters for real earthquakes from seismological data (e.g., Papageorgiou and Aki, 1983b, 1983b; Ide and Takeo, 1997; Mikumo and Yagi, 2003; Mikumo et al., 2003; Ruiz and Madariaga, 2011). For these purposes, a simplified, linear slip-weakening model, as illustrated in Figure 2.14, has often been adopted. The linear slip-weakening model illustrated in Figure 2.14 can be expressed as (e.g., Ida, 1972; Andrews, 1976a, 1976b)

$$\tau(D) = \begin{cases} \tau_p - (\tau_p - \tau_r)\dfrac{D}{D_c} = \tau_p - \Delta\tau_b \dfrac{D}{D_c} & (D < D_c) \\ \tau_r & (D_c \leq D), \end{cases} \tag{4.13}$$

where $\Delta\tau_b = \tau_p - \tau_r$, τ_p denotes the peak shear strength, τ_r denotes the residual friction stress level, and D_c denotes the breakdown displacement. Such a simplified, linear slip-weakening model is certainly useful in most cases, since the fundamental property of the slip-dependent constitutive relation revealed in high-resolution laboratory experiments (see Chapter 3) is incorporated into the model.

The slip-dependent constitutive formulation has very definite advantages. First, the slip-dependent formulation is the only one that makes it possible to govern the stability or instability of the shear fracture, which possibly occurs at some local strong areas (called "asperities" in the field of earthquake seismology) on faults. Hence, the slip-dependent constitutive law is a unifying law that governs both frictional slip failure at precut interface areas on a fault and the shear fracture of intact rock at some local strong areas on the fault (Section 3.3). This can also be justified from the standpoint of microcontact physics (subsection 2.1.4). Thus, the slip-dependent constitutive formulation meets the first prerequisite described in Section 4.1.

Second, scale-dependent physical quantities inherent in shear rupture directly scale with the breakdown displacement D_c (see Chapter 6) defined in the framework of the slip-dependent constitutive formulation. By definition, D_c is the slip displacement at the end of the breakdown zone behind the front of a propagating shear rupture (see Figure 2.10), and therefore D_c is directly related to the geometric length of the coherent zone of rupture breakdown (see Eq. (3.28)). Since D_c is one of the slip-dependent constitutive law parameters, it is obvious that the slip-dependent formulation is most suitable to account consistently in quantitative terms for scale-dependent physical quantities inherent in earthquake ruptures over a broad scale range (see Section 3.3, and Chapters 5 and 6). Thus, the slip-dependent constitutive formulation also meets the second prerequisite described in Section 4.1.

Third, both stable, quasi-static rupture growth and unstable, dynamic high-speed rupture propagation can be accounted for quantitatively in a unified and consistent manner within the framework of the slip-dependent constitutive law (see Chapter 5). In particular, scale-dependent strong motion source parameters, such as slip acceleration, during dynamically propagating shear ruptures over a broad scale range, can be accounted for quantitatively in terms of a single slip-dependent constitutive law. This also supports the slip-dependent formulation as the constitutive formulation for earthquake ruptures. Thus, the slip-dependent constitutive formulation meets the third prerequisite for the rational constitutive formulation for real earthquake ruptures, described in Section 4.1.

Thus, it can be concluded that the slip-dependent constitutive law meets all three prerequisites described in Section 4.1. In addition, the slip-dependent constitutive law is a generalized, more universal law than the Griffith energy balance fracture criterion, in the sense that the physical scaling property is incorporated into the slip-dependent constitutive law.

However, the aforementioned simplified linear slip-weakening model (illustrated in Figure 2.14 and expressed in Eq. (4.13)) definitely leads to a singularity of slip acceleration near the front of a shear rupture dynamically propagating on a fault (Ida, 1973), which is physically unrealistic. To avoid such an unrealistic singularity of slip acceleration, the function form of the slip-dependent constitutive law needs to be more faithful to the laboratory-derived slip-dependent constitutive relation. The slip-dependent constitutive relation derived from high-resolution laboratory experiments is illustrated in Figure 3.22, which shows that there are two phases involved in the entire breakdown process: the slip-strengthening phase, and the slip-weakening phase. The shear strength increases with ongoing slip during the slip-strengthening phase, and decreases with ongoing slip during the slip-weakening phase after the strength has attained its peak value. With this in mind, the slip-dependent constitutive law must be formulated in such a way that the slip-strengthening phase is incorporated into the constitutive equation (Ida, 1973; Ohnaka and Yamashita, 1989). This is particularly important when strong motion source parameters such as peak slip acceleration and peak slip velocity in the dynamic rupture regime of high slip velocities are discussed in quantitative terms.

In order to avoid an unrealistic singularity of slip acceleration, the specific constitutive equation

$$\tau = (\tau_i - \tau_r)g(D)\exp(-\eta D) + \tau_r \tag{4.14}$$

has been proposed, based on laboratory-derived constitutive relations (Ohnaka and Yamashita, 1989). In the above equation, τ_i denotes the initial stress on the verge of slip, τ_r denotes the residual friction stress, η is a constant, and $g(D)$ is a function of D, representing the slip-strengthening effect, where $g(D)$ must be chosen so as to meet the following conditions:

$$g(0) = 1 \qquad (4.15)$$

and

$$g(D)\exp(-\eta D) \to 0 \qquad (4.16)$$

at adequate values of D subsequently after the breakdown (or slip-weakening). The form of the exponential in Eq. (4.14) has been employed to represent the slip-weakening phase of the constitutive relations observed. An exponential form similar to Eq. (4.14) was assumed by Stuart (1979a, 1979b) and Stuart and Mavko (1979) to model the breakdown process of earthquake instabilities.

The amount of elastic deformation has been removed when the constitutive relation is derived from laboratory data on shear fracture (cf. Figures 3.3 and 3.5), and hence the slip-strengthening phase is intrinsically inelastic. Given this, the following logarithmic equation:

$$g(D) = 1 + \alpha \log(1 + \beta D), \qquad (4.17)$$

may be suitable for $g(D)$ to describe the slip-strengthening phase (Ohnaka and Yamashita, 1989), where α and β are constants. Equation (4.17) satisfies the conditions (4.15) and (4.16), and hence it is a possible solution. The following constitutive equation:

$$
\begin{aligned}
\tau(D) &= \tau_{\mathrm{p}} \left(\frac{\beta D}{D_{\mathrm{c}}} \right) \exp \left(1 - \frac{\beta D}{D_{\mathrm{c}}} \right) \\
&= \Delta\tau_{\mathrm{b}} \left(\frac{\beta D}{D_{\mathrm{c}}} \right) \exp \left(1 - \frac{\beta D}{D_{\mathrm{c}}} \right) + \tau_{\mathrm{r}} \qquad (4.18)
\end{aligned}
$$

has also been proposed under assumption of $\beta = 1$ by Matsu'ura *et al.* (1992), and practically employed by Shibazaki and Matsu'ura (1998) to conduct numerical simulations of shear rupture generation processes using computers, under the assumption that $\beta = 5$ and $\tau_{\mathrm{r}} = 0$.

In Section 5.3, slip velocity and slip acceleration are theoretically derived using Eqs. (4.14) and (4.17), and strong motion source parameters such as peak slip velocity and peak slip acceleration in the dynamic rupture regime of high velocities are discussed in quantitative terms. Therefore, see Section 5.3 for the theoretical derivation of the slip acceleration based on the laboratory-derived constitutive equation to avoid an unrealistic singularity of the slip acceleration.

As noted in Section 3.3, the specific functional form f is dictated by constitutive law parameters such as τ_{p}, $\Delta\tau_{\mathrm{b}}$, and D_{c}, and these constitutive law parameters can, in general, be affected by the seismogenic environment and the properties of the fault zone. In fact, as presented in Section 3.4, each constitutive law parameter τ_{p}, $\Delta\tau_{\mathrm{b}}$, or D_{c} is affected by some of the following factors: the effective normal stress $\sigma_{\mathrm{n}}^{\mathrm{eff}}$, which is defined as the difference between normal stress σ_{n} and interstitial pore water pressure P, ambient temperature T, slip rate $\dot{D}$ (or strain rate $\dot{\varepsilon}$), the characteristic length λ_{c} representing the geometric irregularity of shear rupture surfaces, and the chemical effect of interstitial pore water CE. In light of this, the slip-dependent constitutive law expressed as Eq. (4.12) may in general be rewritten as (Ohnaka, 1996, 2004a; Ohnaka et al., 1997):

$$\tau = f\left(D; \sigma_n^{\text{eff}}, T, \dot{D}, \lambda_c, \text{CE}\right). \tag{4.19}$$

When the constitutive law for earthquake ruptures is formulated as simply as possible in terms of a few physical variables, it is important to clearly distinguish the most fundamental and primary variable D from the other variable parameters. Equation (4.19) is expressed as such, and therefore may be a suitable representation of the function to express the constitutive law for real earthquake ruptures.

The functional form f representing the fault constitutive relation between τ and D can be particularly affected by ambient environmental conditions such as the effective normal stress σ_n^{eff}, pore water pressure P, and temperature T in the seismogenic layer at crustal depths, and in general σ_n^{eff}, P, and T vary with depth. The constitutive law parameters such as τ_p and $\Delta\tau_b$ are functions of σ_n^{eff}, P, and T, which, in turn, are functions of position r in the seismogenic crust. Accordingly, the constitutive law parameters are functions of position r. In light of this, the constitutive Eq. (4.14) at position r can be rewritten as follows (Ohnaka, 1995c, 1996; Ohnaka and Kato, 2007):

$$\tau(D, r) = \tau_p(r) - (\tau_p(r) - \tau_{r\infty}(r))\left\{1 - h(D, r)\exp\left[-A(r)B(r)\left(\frac{D}{D_a(r)} - 1\right)\right]\right\}, \tag{4.20}$$

where

$$h(D, r) = 1 + A(r)\log\left(1 + B(r)\left(\frac{D}{D_a(r)} - 1\right)\right). \tag{4.21}$$

The above equations have been expressed mathematically in such a way that the exponential equals 1 at $D = D_a$. In these equations, $\tau_{r\infty}$ is the residual friction stress as $D \to \infty$, and $A(r)$ and $B(r)$ are dimensionless parameters, which depend on environmental conditions such as σ_n^{eff}, P, and T at position r in the seismogenic layer.

Here, we introduce a small fixed numerical value χ (for instance, 0.1), to define the realistic breakdown displacement D_c^{eff} (hereafter referred to as the effective breakdown displacement). The effective breakdown displacement D_c^{eff} is defined as the slip displacement at which τ decreases to stress level $\chi(\tau_p - \tau_{r\infty}) + \tau_{r\infty}$ (see Figure 4.3). In this case, the corresponding effective breakdown stress drop $\Delta\tau_b^{\text{eff}}$ is given by $\Delta\tau_b^{\text{eff}} = (1 - \chi)(\tau_p - \tau_{r\infty})$, and Eq. (4.20) can be rewritten as

$$\tau(D, r) = \tau_p(r) - \frac{\Delta\tau_b^{\text{eff}}(r)}{(1 - \chi)}\left\{1 - h(D, r)\exp\left[-A(r)B(r)\left(\frac{D}{D_a(r)} - 1\right)\right]\right\}. \tag{4.22}$$

Substituting D_c^{eff} for D and $\tau_p - \Delta\tau_b^{\text{eff}}$ for τ in Eqs. (4.22) and (4.21), we have

$$\left\{1 + A(r)\log\left[1 + B(r)\left(\frac{D_c^{\text{eff}}(r)}{D_a(r)} - 1\right)\right]\right\}\exp\left[-A(r)B(r)\left(\frac{D_c^{\text{eff}}(r)}{D_a(r)} - 1\right)\right] = \chi. \tag{4.23}$$

Substituting zero for D and τ_i for τ in Eqs. (4.22) and (4.21), we have

$$1 - [1 + A(r)\log(1 - B(r))]\exp[A(r)B(r)] = R(r), \tag{4.24}$$

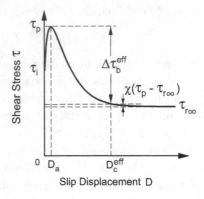

Fig. 4.3   A slip-dependent constitutive relation.

where

$$R(r) = \frac{\tau_p(r) - \tau_i(r)}{\Delta\tau_b^{eff}(r)/(1-\chi)}. \tag{4.25}$$

If the constitutive law parameters τ_p, τ_i, $\Delta\tau_b^{eff}$, D_a, and D_c^{eff} are given as functions of position r, the parameters $A(r)$ and $B(r)$ can be determined by solving the simultaneous equations (4.23) and (4.24). This will be done in Section 4.4.

Let us define the dimensionless shear stress σ by $\sigma = \tau/\tau_p(r)$, and the dimensionless slip displacement d by $d = D/D_a(r)$. In this case, Eq. (4.20) or (4.22) can be expressed in terms of σ and d as

$$\sigma(d, r) = 1 - S(r)\{1 - [1 + A(r)\log(1 + B(r)(d-1))]\exp[-A(r)B(r)(d-1)]\}, \tag{4.26}$$

where

$$S(r) = \frac{\tau_p(r) - \tau_{r\infty}(r)}{\tau_p(r)} = \frac{\Delta\tau_b^{eff}(r)/(1-\chi)}{\tau_p(r)}. \tag{4.27}$$

Figure 4.4 shows a dimensionless constitutive relation between σ and d. The parameter d_c in the figure denotes the dimensionless breakdown displacement defined by $d_c = D_c^{eff}/D_a$. If $A(r)$, $B(r)$, and $S(r)$ are evaluated at any position r on a fault embedded in the seismogenic crust, Eq. (4.26) completely specifies the dimensionless constitutive relation between σ and d on the fault (see Section 4.4), and hence the physical constitutive relation between τ and D on the same fault can also be uniquely determined from Eq. (4.26), given that $\sigma = \tau/\tau_p(r)$ and $d = D/D_a(r)$. The dimensionless expression of the constitutive equation may be useful for computer simulations of earthquake generation processes.

Figures 4.5(a) and (b) show two specific examples of dimensionless constitutive relations calculated from Eq. (4.26). Figure 4.5(a) shows a typical example of the constitutive relation for the shear failure of intact rock, for which the following parameter values have been used: $S = 0.5347$, $R = 1.1655$, $A = 2.3603$, $B = 0.3645$, and $d_c = D_c^{eff}/D_a = 5$ for a preset value of $\chi = 0.1$. Figure 4.5(b) shows an extreme example of the constitutive relation for frictional stick-slip failure on a uniformly pre-cut weak fault with very smooth surfaces,

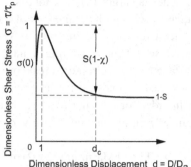

Fig. 4.4   A slip-dependent constitutive relation expressed in terms of dimensionless shear stress $\sigma(d)$ and dimensionless slip displacement d. $\sigma(0)$ is the normalized initial shear stress at the onset of slip, d_c is the normalized breakdown displacement, and $S(1-\chi) = \Delta\tau_b^{\text{eff}}/\tau_p$ (see Eq. (4.27) in text).

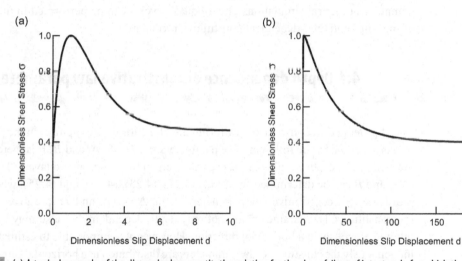

Fig. 4.5   (a) A typical example of the dimensionless constitutive relation for the shear failure of intact rock, for which the following parameter values have been used: $S = 0.5347$, $R = 1.1655$, $A = 2.3603$, $B = 0.3645$, and $D_c^{\text{eff}}/D_a = d_c = 5$ for a preset value of $\chi = 0.1$. (b) An extreme example of the dimensionless constitutive relation for frictional slip failure on a uniformly precut weak fault with very smooth surfaces, for which the following parameter values have been used: $S = 0.6000$, $R = 0.2850$, $A = 0.03112$, $B = 0.999962$, $D_c^{\text{eff}}/D_a = d_c = 100$ for a preset value of $\chi = 0.05$.

for which the following parameter values have been used: $S = 0.6000$, $R = 0.2850$, $A = 0.03112$, $B = 0.999962$, and $d_c = D_c^{\text{eff}}/D_a = 100$ for a preset value of $\chi = 0.05$. In this extreme example, d_c has a large value of 100 because of a near-zero small value for D_a.

The aforementioned slip-dependent constitutive law is formulated for the shear-rupture breakdown process. Therefore, in order to simulate earthquake cycles in a virtual world using computers, the constitutive law for the fault re-strengthening during the interseismic period needs to be formulated separately from the constitutive law for earthquake ruptures.

In this respect, one may argue that it is not easy to conduct the computer numerical simulation of earthquake cycles in the framework of the slip-dependent constitutive formulation. However, the fault re-strengthening process on a heterogeneous fault with "asperities" and/or "barriers" (see Chapter 1) during an interseismic period is a physical process that is completely different from the rupture breakdown (or fault-weakening) process. Therefore, there are no physical grounds for the insistence that both processes must be governed by a single law. In order for the physics of earthquakes to become a complete, quantitative science, the underlying physics must be given priority over the convenience of computer simulation.

Nonetheless, Aochi and Matsu'ura (2002) proposed a slip- and time-dependent constitutive law derived from theoretical modeling by integrating effects of microscopic interactions (abrasion and adhesion) between statistically self-similar fault surfaces. This constitutive law, though its function expression is not simple, consistently explains not only the slip-weakening at high slip rates, but also the logarithmic time strengthening in stationary contact and the slip rate-weakening in steady-state slip (Aochi and Matsu'ura, 2002). Thus, computer numerical simulations of earthquake cycles can be performed in the framework of this slip- and time-dependent constitutive formulation.

4.4 Depth dependence of constitutive law parameters

The slip-dependent constitutive law formulated for earthquake ruptures in the previous section is specified by the following five parameters: τ_p, D_a, S, A, and B. It is therefore critical to identify the depth profiles of these parameters in the seismogenic layer. The parameters S, A, and B can be determined from Eqs. (4.27), (4.23), (4.24), and (4.25), when the depth profiles of the constitutive law parameters τ_i, τ_p, $\Delta\tau_b$, D_a, and D_c are given. In general, the constitutive law parameters are not only depth-dependent, but also may be distributed inhomogeneously in a horizontal direction. However, it is impossible to estimate how inhomogeneously the constitutive law parameters are distributed in a horizontal direction, unless inhomogeneous distributions of temperature, lithostatic pressure, and pore water pressure in a horizontal direction are specifically given, together with structural inhomogeneity of seismogenic layer (and preexisting faults therein). In this section, therefore, we only present depth profiles of the parameters S, A, and B, which have been estimated by Ohnaka and Kato (2007) based on experimental data obtained under seismogenic crustal conditions simulated in the laboratory (Kato *et al.*, 2003a). The symbols $\Delta\tau_b^{\mathrm{eff}}$ and D_c^{eff} are used in place of $\Delta\tau_b$ and D_c, respectively, in this section.

Depth distributions of temperature, lithostatic pressure, and pore water pressure in the seismogenic crust were specifically assumed in the laboratory simulation as shown in Figure 4.6, in which both temperature (30 °C/km) and lithostatic pressure (30 MPa/km) increase linearly with depth, in the regimes of hydrostatic (10 MPa/km) and suprahydrostatic (24 MPa/km) pore water pressure gradients (Kato *et al.*, 2003a). These distributions at crustal depths were simulated using the sophisticated high-pressure testing apparatus described in Section 3.1.

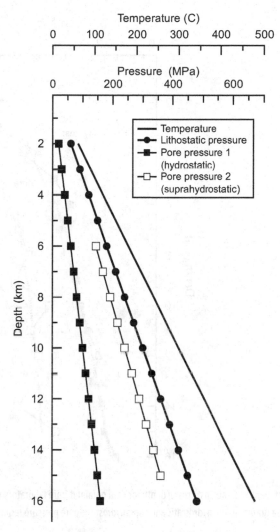

Fig. 4.6 Depth distributions, assumed in the laboratory simulation with a high-pressure testing apparatus, of temperature, lithostatic pressure, and pore water pressure in the seismogenic crust. Reproduced from Ohnaka and Kato (2007).

As emphasized in previous chapters, a real fault embedded in the seismogenic crust is inherently inhomogeneous, with local strong areas highly resistant to rupture growth, and the rest of the fault having low (or little) resistance to rupture growth. For example, the sites of fault segment stepover zones and/or interlocking asperities are potential candidates for strong areas highly resistant to rupture growth, and these sites act as "barriers" or "asperities" (see Chapter 1). Such strong areas on a fault play a critical role in accumulating an adequate amount of elastic strain energy in the medium at and around those areas, until they break. Once they break, the elastic strain energy accumulated is released to act as the driving force of a resulting earthquake. Hence, local strong areas highly resistant to rupture growth on a fault play a much more important role in prescribing the size (or magnitude)

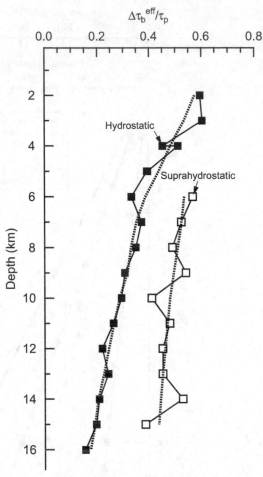

Fig. 4.7   Plots of $\Delta\tau_b^{\text{eff}}/\tau_p$ against crustal depth for data obtained from laboratory experiments on the shear failure of intact Tsukuba granite in the hydrostatic and suprahydrostatic pore pressure regimes (Ohnaka and Kato, 2007).

of the resulting earthquake than does the rest of the fault with low (or little) resistance to rupture growth. Given that the strength of some local strong areas on a fault is equal to the shear failure strength of intact rock (see Chapter 1 and Section 3.3), and that the shear failure strength of intact rock is the upper limit of frictional strength, it is critically important to comprehend the depth profiles of S, A, and B for the shear failure of intact rock under seismogenic crustal conditions. Hence, they will be presented below, as a typical example.

Equation (4.27) indicates that the dimensionless parameter S is substantially represented by $\Delta\tau_b^{\text{eff}}/\tau_p$, because χ is a fixed numerical value. Thus, we present the depth profile of $\Delta\tau_b^{\text{eff}}/\tau_p$, in place of the depth profile of S. Figure 4.7 shows how $\Delta\tau_b^{\text{eff}}/\tau_p$ varies with depth, under the simulated environmental conditions at crustal depths shown in Figure 4.6. The broken lines in Figure 4.7 indicate five-point moving averages for each pore pressure

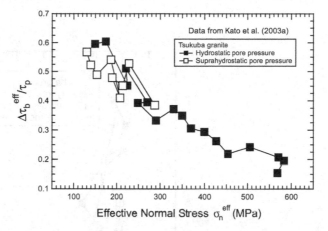

Plots of $\Delta\tau_b^{eff}/\tau_p$ against the effective normal stress σ_n^{eff} for data obtained from laboratory experiments on the shear failure of intact Tsukuba granite in the hydrostatic and suprahydrostatic pore pressure regimes (Ohnaka and Kato, 2007).

regime. From Figure 4.7, one can see that $\Delta\tau_b^{eff}/\tau_p$ decreases with depth, and is greater in the suprahydrostatic pore pressure regime than in the hydrostatic pore pressure regime. This difference is attributed to the difference in the effective normal stress σ_n^{eff} due to different pore water pressures (Figure 4.8). Figure 4.8 shows a plot of $\Delta\tau_b^{eff}/\tau_p$ against σ_n^{eff} for two different sets of data obtained under the two different pore pressure regimes. From Figure 4.8, one can clearly see that the two different sets of data are aligned with each other in spite of the differences in pore water pressure and temperature.

A set of numerical solutions for $A(r)$ and $B(r)$ at a depth r can be obtained by solving simultaneous equations (4.23) and (4.24). In order to do so, however, we have to know the depth distributions of the parameters $R = (\tau_p - \tau_i)/[\Delta\tau_b^{eff}/(1-\chi)]$ and D_c^{eff}/D_a. Since χ is a fixed numerical value, R is substantially represented by $(\tau_p - \tau_i)/\Delta\tau_b^{eff}$. Thus, the depth profile of $(\tau_p - \tau_i)/\Delta\tau_b^{eff}$, in place of R, is shown in Figure 4.9. The data points plotted in Figure 4.9 were obtained under the simulated crustal conditions shown in Figure 4.6. The broken lines in Figure 4.9 indicate five-point moving averages for each pore water pressure regime. From Figure 4.9, one can see that $(\tau_p - \tau_i)/\Delta\tau_b^{eff}$ increases with depth in the hydrostatic pore pressure regime, while it is insensitive to depth in the suprahydrostatic pore pressure regime. One can also see that $(\tau_p - \tau_i)/\Delta\tau_b^{eff}$ is greater in the hydrostatic pore pressure regime than in the suprahydrostatic pore pressure regime. The difference in $(\tau_p - \tau_i)/\Delta\tau_b^{eff}$ between the two pore pressure regimes is attributed to the difference in σ_n^{eff} between the two regimes, due to different pore water pressures (Figure 4.10). Figure 4.10 shows a plot of $(\tau_p - \tau_i)/\Delta\tau_b^{eff}$ against σ_n^{eff} for two different sets of data obtained in the two different pore pressure regimes. From this figure, one can confirm that these different sets of data are aligned with each other in spite of the differences in pore water pressure and temperature.

In order to evaluate how D_c^{eff}/D_a varies with depth, we first determine to what extent the ratio D_c^{eff}/D_a is affected by σ_n^{eff}, P, and T. Since D_{wc} is defined by $D_{wc} = D_c^{eff} - D_a$,

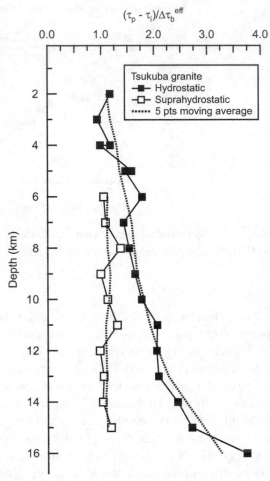

Fig. 4.9 Plots of $(\tau_p - \tau_i)/\Delta\tau_b^{eff}$ against crustal depth for data obtained from laboratory experiments on the shear failure of intact Tsukuba granite in the hydrostatic and supra-hydrostatic pore pressure regimes (Ohnaka and Kato, 2007).

the ratio D_c^{eff}/D_a is uniquely related to the ratio D_{wc}/D_c^{eff} ($=1 - D_a/D_c^{eff}$), and hence the investigation of how the ratio D_{wc}/D_c^{eff} is affected by σ_n^{eff}, P, and T suffices for the present purpose. Figure 4.11 shows a plot of $c_1 = D_{wc}/D_c^{eff}$ against the peak shear strength τ_p for data on frictional slip failure on precut granite interface and on the shear failure of intact granite tested in the brittle regime (where temperature ranges from room temperature to roughly 300 °C) (Ohnaka and Kato, 2007). This figure indicates that D_{wc}/D_c^{eff} does not depend on τ_p. Since τ_p is an increasing function of σ_n^{eff} in the brittle regime, one can see from Figure 4.11 that the ratio D_c^{eff}/D_a is not only unaffected by τ_p but also unaffected by σ_n^{eff} in the brittle regime. Figure 4.11 also shows that the ratio D_{wc}/D_c^{eff} has an average value of roughly 0.8 (which corresponds to $D_c^{eff}/D_a = 5$). In particular, it is worth noting that the average value of roughly 0.8 is commonly shared by both data on the shear failure of intact rock and on frictional slip failure on a precut rock interface. This is consistent

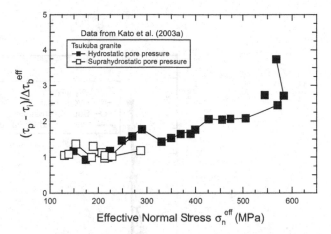

Fig. 4.10   Plots of $(\tau_p - \tau_i)/\Delta\tau_b^{\mathrm{eff}}$ against the effective normal stress σ_n^{eff} for data obtained from laboratory experiments on the shear failure of intact Tsukuba granite in the hydrostatic and suprahydrostatic pore pressure regimes (Ohnaka and Kato, 2007).

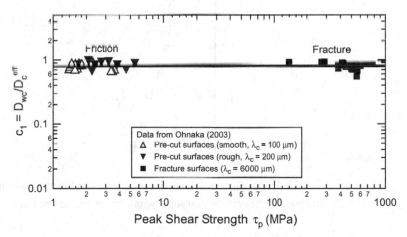

Fig. 4.11   A plot of $c_1 = D_{wc}/D_c^{\mathrm{eff}}$ against the peak shear strength τ_p for data obtained from laboratory experiments on frictional stick-slip failure and the shear failure of intact granite in the brittle regime (Ohnaka and Kato, 2007).

with the view that the shear failure of intact rock and frictional slip failure on a precut rock interface can both be unified in a consistent manner in the framework of slip-dependent constitutive formulation (Ohnaka, 2003).

Figure 4.12 shows how $D_c^{\mathrm{eff}}/D_a(= 1/c_2)$ varies with depth. This figure indicates that D_c^{eff}/D_a does not significantly vary with depth at depths shallower than roughly 10 km, whereas it increases at greater depths. From Figure 4.12, one can see that D_c^{eff}/D_a at depths shallower than 10 km is significantly unaffected by pore water pressure within the range of experimental error. At greater depths, however, D_c^{eff}/D_a is affected by pore water pressure; specifically, D_c^{eff}/D_a at depths greater than 10 km is affected more

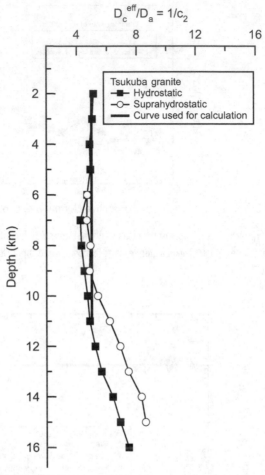

$$D_c^{\text{eff}}/D_a = 1/c_2$$

Fig. 4.12 Plots of $D_c^{\text{eff}}/D_a(= 1/c_2)$ against crustal depth for data obtained from laboratory experiments on the shear failure of intact Tsukuba granite in the hydrostatic and suprahydrostatic pore pressure regimes (Ohnaka and Kato, 2007).

sensitively in the supra-hydrostatic pore pressure regime than in the hydrostatic pore pressure regime. Note that the ratios of one displacement parameter D_a, D_{wc}, or D_c^{eff} to another are scale-independent (Section 3.3). This can easily be confirmed from Figure 4.13, which shows a plot of $c_1 = D_{\text{wc}}/D_c^{\text{eff}}$ against the effective breakdown displacement D_c^{eff} for data on frictional slip failure on precut granite interface and on the shear failure of intact granite tested in the brittle regime (Ohnaka and Kato, 2007). From this figure, one can see that $c_1 = D_{\text{wc}}/D_c^{\text{eff}}$ does not depend on D_c^{eff}. This means that the ratio $D_{\text{wc}}/D_c^{\text{eff}}$ is scale-independent, although each displacement parameter D_c^{eff} or D_{wc} is scale-dependent, as described in Section 3.3. Likewise, the ratio $D_c^{\text{eff}}/D_a(= D_c^{\text{eff}}/(D_c^{\text{eff}} - D_{\text{wc}}))$ is also scale-independent.

We can thus summarize these facts about the constitutive law displacement parameters D_a, D_{wc}, and D_c^{eff} as follows:

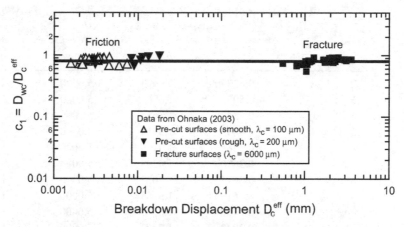

Fig. 4.13   A plot of $c_1 = D_{wc}/D_c^{eff}$ against the effective breakdown displacement D_c^{eff} for data obtained from laboratory experiments on frictional stick-slip failure and the shear failure of intact granite in the brittle regime (Ohnaka and Kato, 2007).

(1) The ratios of one displacement parameter D_a, D_{wc}, or D_c^{eff} to another are neither scale-dependent nor dependent on ambient conditions such as the effective normal stress and temperature in the brittle regime, and

(2) D_c^{eff}/D_a has an average value of roughly 5 within the range of experimental error in the brittle regime (corresponding to a depth range down to roughly 10 km), and the average value of roughly 5 is commonly shared by both sets of data on frictional slip failure on a precut rock interface and on the shear failure of intact rock.

These two facts play an important role in evaluating the depth profiles of D_c^{eff}/D_a. The value of roughly 5 for D_c^{eff}/D_a in the brittle regime is in agreement with the value determined using a different approach (see Section 3.3).

For the hydrostatic pore pressure regime, we thus assume that D_c^{eff}/D_a has a constant value of 5 within a depth range down to 11 km, and increases at greater depths, as shown in Table 4.1 (see also Figure 4.12). For the suprahydrostatic pore pressure regime, we take into account the fact that D_c^{eff}/D_a is more sensitive to pore water pressure even at shallower depths, and hence we assume that D_c^{eff}/D_a has the same value of 5 within a depth range down to 9 km, and increases at greater depths, as shown in Table 4.2 (see also Figure 4.12).

Using values listed in Tables 4.1 and 4.2 for D_c^{eff}/D_a and R as a function of depth, simultaneous equations (4.23) and (4.24) were solved to find a set of numerical solutions for A and B at individual depths. The results are listed in Tables 4.1 and 4.2, and plotted as a function of depth in Figure 4.14. In this calculation, $\chi = 0.1$ has been assumed. The possible errors in estimating A and B primarily come from the errors involved in the parameters $(\tau_p - \tau_i)/\Delta\tau_b^{eff}$ and D_c^{eff}/D_a, and the errors in these parameters are due to experimental error. To minimize the errors in a set of numerical solutions for A and B, we used the values for $(\tau_p - \tau_i)/\Delta\tau_b^{eff}$ and D_c^{eff}/D_a smoothed by removing the experimental

Table 4.1 Changes in dimensionless constitutive law parameters with depth in the hydrostatic pore pressure regime (Ohnaka and Kato, 2007)

Depth (km)	S	R	D_c^{eff}/D_a	A	B
2	0.6379	0.9943	5.0000	4.9634	0.1821
3	0.5931	1.0567	5.0000	3.3733	0.2617
4	0.5347	1.1655	5.0000	2.3603	0.3645
5	0.4849	1.2114	5.0000	2.1333	0.4001
6	0.4285	1.3196	5.0000	1.7816	0.4720
7	0.3896	1.4209	5.0000	1.5758	0.5280
8	0.3674	1.4652	5.0000	1.5075	0.5498
9	0.3519	1.5163	5.0000	1.4400	0.5732
10	0.3180	1.6279	5.0000	1.3238	0.6188
11	0.2942	1.7273	5.0000	1.2455	0.6541
12	0.2722	1.8718	5.2073	1.0097	0.7520
13	0.2504	2.0424	5.6314	0.7843	0.8582
14	0.2264	2.3440	6.3809	0.5940	0.9488
15	0.2120	2.6479	6.8876	0.5194	0.9784
16	0.1925	2.9456	7.4705	0.4605	0.9921

For the definition of symbols S, R, D_c^{eff}/D_a, A, and B used, see text.

Table 4.2 Changes in dimensionless constitutive law parameters with depth in the suprahydrostatic pore pressure regime (Ohnaka and Kato, 2007)

Depth (km)	S	R	D_c^{eff}/D_a	A	B
6	0.5936	1.0074	5.0000	4.4725	0.2009
7	0.5814	0.9967	5.0000	4.8619	0.1857
8	0.5661	1.0093	5.0000	4.4089	0.2036
9	0.5491	1.0549	5.0000	3.4015	0.2597
10	0.5327	1.0355	5.3993	1.0262	0.6294
11	0.5222	0.9777	6.1837	0.6370	0.7875
12	0.5116	0.9840	6.8782	0.6459	0.7842
13	0.5011	0.9950	7.4397	0.4961	0.8659
14	0.4922	0.9732	8.3092	0.3983	0.9149
15	0.4840	1.0106	8.5930	0.3996	0.9196

For the definition of symbols S, R, D_c^{eff}/D_a, A, and B used, see text.

error superimposed on the trends of the parameters $(\tau_p - \tau_i)/\Delta\tau_b^{eff}$ and D_c^{eff}/D_a with depth (Tables 4.1 and 4.2). In particular, five-point moving average values were used for $(\tau_p - \tau_i)/\Delta\tau_b^{eff}$ (see Figure 4.9) to remove fluctuating measurement values caused by the experimental error.

The parameters A and B in the suprahydrostatic pore pressure regime sharply change at depths ranging from 8 to 10 km (see Figure 4.14). To figure out why these sharp changes

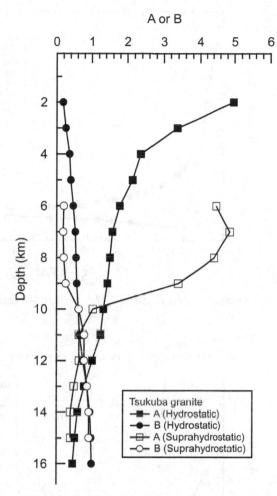

Fig. 4.14 Plots of *A* and *B* against crustal depth for data obtained from laboratory experiments on the shear failure of intact Tsukuba granite in the hydrostatic and suprahydrostatic pore pressure regimes (Ohnaka and Kato, 2007).

occur at those depths, the effects of D_c^{eff}/D_a and $(\tau_p - \tau_i)/\Delta\tau_b^{\mathrm{eff}}$ were examined to see which one has more influence on *A* and *B* (Ohnaka and Kato, 2007). Figure 4.15 shows how *A* and *B* depend on $(\tau_p - \tau_i)/\Delta\tau_b^{\mathrm{eff}}$, when D_c^{eff}/D_a is held constant ($D_c^{\mathrm{eff}}/D_a = 5$ has been assumed), and Figure 4.16 shows how *A* and *B* depend on D_c^{eff}/D_a, when *R* is held constant (two possible extreme examples are shown: $R = 1.0000$ and 2.9456). One can see from Figure 4.15 that *A* and *B* react more and more sensitively as $(\tau_p - \tau_i)/\Delta\tau_b^{\mathrm{eff}}$ approaches unity. In particular, *A* in the suprahydrostatic pore pressure regime is greatly influenced by a slight change in $(\tau_p - \tau_i)/\Delta\tau_b^{\mathrm{eff}}$, because $(\tau_p - \tau_i)/\Delta\tau_b^{\mathrm{eff}}$ has a value close to unity in the suprahydrostatic pore pressure regime (see Figure 4.9), and because a slight change in $(\tau_p - \tau_i)/\Delta\tau_b^{\mathrm{eff}}$ around unity causes a sizable amount of the change in *A* due to a steeper slope (Figure 4.15). In addition, one can see from Figure 4.16 that when *R* (or $(\tau_p - \tau_i)/\Delta\tau_b^{\mathrm{eff}}$) has a value close to unity, the parameter *A* increases sharply toward 5 as D_c^{eff}/D_a decreases to 5.

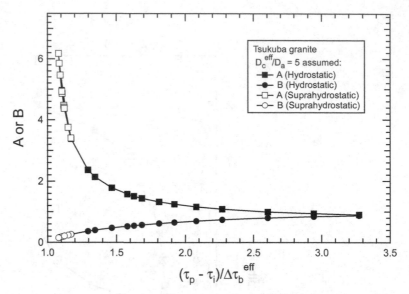

Fig. 4.15 Relation of A or B to $(\tau_p - \tau_i)/\Delta\tau_b^{\text{eff}}$, where $D_c^{\text{eff}}/D_a = 5$ has been assumed (Ohnaka and Kato, 2007).

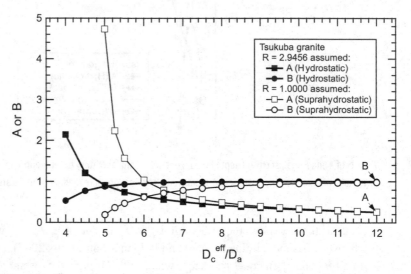

Fig. 4.16 Relation of A or B to D_c^{eff}/D_a, where $R = 1.0000$ or $R = 2.9456$ has been assumed in the hydrostatic and suprahydrostatic pore pressure regimes (Ohnaka and Kato, 2007).

In other words, A is greatly influenced by a slight change in D_c^{eff}/D_a around $D_c^{\text{eff}}/D_a = 5$, which occurs at depths ranging from 8 to 10 km in the case of the suprahydrostatic pore pressure regime (see Table 4.2). Figures 4.15 and 4.16 also suggest that the parameter A for the suprahydrostatic pressure regime is very sensitive to depth, and this may possibly

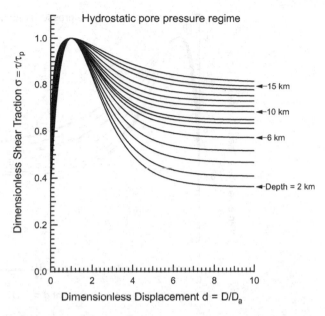

Fig. 4.17 Changes in the normalized constitutive relation between σ and d with crustal depth in the hydrostatic pore pressure regime (Ohnaka and Kato, 2007).

cause a large fluctuation of A with depth, as a slight change in $(\tau_p - \tau_i)/\Delta\tau_b^{\text{eff}}$ occurs with depth.

We have estimated depth distributions of the dimensionless constitutive law parameters S, A, and B at strong areas highly resistant to rupture growth on faults. These estimates enable us to specify depth distributions of the dimensionless constitutive relation between σ and d from Eq. (4.26). Figures 4.17 and 4.18 show how constitutive relations thus derived between σ and d change with depth, in the hydrostatic and suprahydrostatic pore pressure regimes, respectively. These constitutive relations were calculated using the values for S, A, and B listed in Tables 4.1 and 4.2. From Figures 4.17 and 4.18, it is clear that slip-weakening behavior, after the peak strength has been attained, is observed at all depths to 15–16 km; however, this slip-weakening behavior (or the constitutive relation between σ and d) systematically changes with depth, depending upon variations of lithostatic pressure, pore water pressure, and temperature.

The dimensionless shear traction σ is converted to a dimensional shear traction τ by the relation $\tau = \tau_p\sigma$, and the dimensionless slip displacement d is converted to a dimensional slip displacement D by the relation $D = D_a d$. In order to convert the dimensionless form of the constitutive relation to its dimensional form at individual depths, therefore, we need to know the depth profiles of τ_p and D_a under seismogenic, crustal conditions. Figure 4.19 plots τ_p, D_a, and D_c^{eff} against depth for laboratory data on the shear fracture of intact granite obtained under the simulated crustal conditions of temperature and lithostatic pressure in

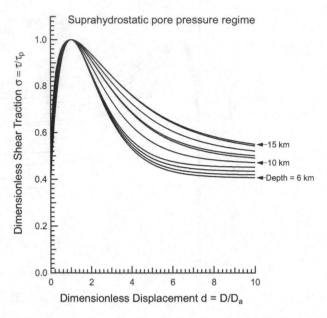

Fig. 4.18 Changes in the normalized constitutive relation between σ and d with crustal depth in the suprahydrostatic pore pressure regime (Ohnaka and Kato, 2007).

the hydrostatic and suprahydrostatic pore pressure regimes shown in Figure 4.6. Since D_a and D_c^{eff} are scale-dependent, it is meaningless to directly compare their absolute values obtained from the shear fracture of small rock samples (circular cylinders 40 mm long and 16 mm in diameter) in the laboratory, with those of earthquake ruptures that occur on faults of much larger size in the seismogenic crust. Therefore, both D_a and D_c^{eff} were normalized to the average value $\langle D_c^{eff}\rangle$ ($= 1.0$mm) for D_c^{eff} obtained in the laboratory, under the simulated conditions corresponding to shallow crustal depths to 11 km (brittle regime) in the hydrostatic pore pressure regime, and the dimensionless $D_a/\langle D_c^{eff}\rangle$ and $D_c^{eff}/\langle D_c^{eff}\rangle$ were plotted in Figure 4.19.

Since τ_p shown in Figure 4.19 is the shear failure strength (peak shear strength) of intact granite obtained under the simulated seismogenic crustal conditions, it represents the highest possible value for the strength of strong areas (highly resistant to rupture growth) on faults embedded in the seismogenic crust. The highest possible value for the strength at strong areas on faults can be greatly affected by pore water pressure. Therefore, the dimensional shear traction τ, to which the dimensionless shear traction σ is converted by $\tau = \tau_p\sigma$, can also be greatly affected by the mechanical effect of pore water pressure. Indeed, τ_p in the hydrostatic pore pressure regime is much higher than that in the suprahydrostatic pore pressure regime (Figure 4.19).

The constitutive law parameter D_a plays a key role in converting the dimensionless parameter d to the corresponding slip displacement D, which is scale-dependent. Hence, it is important to comprehend how and to what extent D_a is affected by

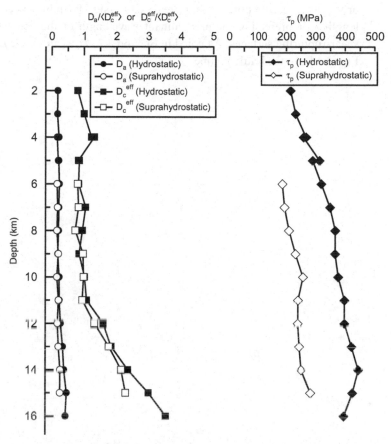

Fig. 4.19   Plots of $D_a/\langle D_c^{\mathrm{eff}}\rangle$ and $D_c^{\mathrm{eff}}/\langle D_c^{\mathrm{eff}}\rangle$ against crustal depth for data obtained from laboratory experiments on the shear failure of intact Tsukuba granite in the hydrostatic and suprahydrostatic pore pressure regimes. The shear strengths τ_p in the hydrostatic and suprahydrostatic pore pressure regimes are also plotted for reference. Data from Kato *et al.* (2003a).

seismogenic, crustal conditions. Figure 4.19 indicates that D_a can be regarded as virtually constant over a depth range to 16 km; in other words, it is neither appreciably affected by effective normal stress or temperature, nor affected by whether failure occurs in the hydrostatic pore pressure regime or in the suprahydrostatic pore pressure regime. This fact is important and noteworthy because D_c^{eff} exhibits contrasting behavior in a depth range roughly from 10 to 16 km, where D_c^{eff} increases with depth (Figure 4.19). It has been demonstrated that the increase in D_c^{eff} with depth in the range from 10 to 16 km is primarily due to the effect of temperatures above 300 °C (see Figure 3.42).

Although D_a is neither influenced by such crustal conditions as the effective normal stress and temperature, nor influenced by whether failure occurs in the hydrostatic or suprahydrostatic pore pressure regime, D_a significantly depends on the characteristic length λ_c representing the predominant wavelength of the geometric irregularity of the shear-rupturing surfaces. As noted in Section 3.3, D_a scales with D_c^{eff} in accordance with Eq. (3.14), in the brittle regime.

Earthquake generation processes

5.1 Shear failure nucleation processes observed in the laboratory

5.1.1 Introduction

When an external stress is applied at a slow and steady rate to an inhomogeneous material in the brittle regime, fracture eventually begins to occur locally somewhere in the material, and the elastic strain energy stored in the medium surrounding the rupture front is released. If the released strain energy is much greater than the resistance to rupture growth, the released energy is expended not only as the driving force for the rupture to grow further but also as kinetic energy. As a result, the rupture extends spontaneously beyond external control, and dynamically at high speeds close to elastic wave velocities (unstable, dynamic rupture in the brittle regime). We often come across such brittle fractures in everyday life: for example, the fracture of fragile bodies such as glass. However, a typical large-scale example of brittle fractures would be the earthquake rupture instability that takes place in the brittle layer of the Earth's crust.

Most earthquakes occur at shallow depths in the brittle regime, and shallow focus earthquakes are caused by slip-failure mechanical instabilities that give rise to dynamically propagating shear ruptures on preexisting faults in the seismogenic layer. In the past, such an earthquake rupture in the brittle regime has been thought to begin to propagate abruptly at high speeds close to elastic wave velocities immediately after the onset of the rupture.

Strictly speaking, however, shear rupture on a preexisting fault cannot begin to propagate abruptly at high speeds close to elastic wave velocities immediately after the onset of the rupture, even in the brittle regime, when the resistance to rupture growth is heterogeneously distributed on the fault. The elastic strain energy released immediately after the onset of the rupture cannot be much greater than the resistance to rupture growth. The shear rupture initially begins to nucleate at a location of minimum resistance to rupture growth on a heterogeneous fault, and grows stably and quasi-statically at a slow speed toward the surrounding area of somewhat higher resistance to rupture growth on the fault. When the rupture growth distance reaches a critical distance (L_{sc}), the released elastic strain energy becomes greater than the resistance to rupture growth, and the rupture begins to extend spontaneously at accelerating speeds (see subsections 5.1.3 and 5.1.4). After the rupture growth distance eventually attains another critical distance (L_c), the rupture propagates at a high speed close to elastic wave velocities. Thus, the shear-rupture nucleation process prior to its dynamic high-speed propagation in the brittle regime consists of two phases: an initial phase of stable, quasi-static rupture growth, and the subsequent phase of spontaneous rupture extension at accelerating speeds (see subsection 5.1.3). This was elucidated and demonstrated in the 1990s by well-prepared, high-resolution laboratory experiments (Ohnaka and Shen, 1999).

If, however, temporal and spatial observation resolutions are not sufficiently high, it is not possible to observe such a sequence of the initial phase of stable, quasi-static rupture growth, and the subsequent phase of spontaneous rupture extension at accelerating speeds. In other words, high temporal and spatial resolution observations are essential for revealing the nucleation process. In laboratory experiments on stick-slip rupture on a preexisting fault, the experimental method can be properly devised for the purpose intended, and high temporal and spatial resolution measurements can be made along the entire fault. Indeed, the nucleation process of slip failure on a preexisting fault was first revealed in the 1970s (Dieterich, 1978; Dieterich et al., 1978), and in considerable detail in the 1980s (Ohnaka et al., 1986, 1987b; Ohnaka and Kuwahara, 1990). Since then, a series of high-resolution laboratory experiments have been performed to elucidate in more detail the physical nature of the shear rupture nucleation process on a preexisting fault in the brittle regime.

By contrast, it has not been possible to closely observe the ongoing nucleation process of a real earthquake rupture, because the deployment of a series of instruments (sensors or meters) for measuring local shear stresses (or strains) and slip displacements along the fault on which a pending earthquake is expected to occur, is required to closely observe the ongoing nucleation process, and because the amounts of slip and stress change during the nucleation process are very small in comparison to those of the slip and of the stress drop during the dynamic mainshock breakdown process. Hence, in order to understand the earthquake nucleation process, it is critically important to specifically understand it in terms of the underlying physics, based on the results of high-resolution laboratory experiments. The underlying physics explains how a shear rupture nucleates, spontaneously begins to grow at accelerating speeds, and eventually propagates at a high speed close to elastic wave velocities on an inhomogeneous fault in the brittle regime.

In this section, therefore, we present the results of laboratory experiments on frictional stick-slip rupture nucleation on preexisting faults in the brittle regime, elucidated and demonstrated by well-prepared, high-resolution laboratory experiments (Ohnaka and Shen, 1999).

5.1.2 Experimental method

For the experiments on stick-slip rupture nucleation, Tsukuba granite from Ibaraki Prefecture, central Japan, was chosen. The mechanical and physical properties of Tsukuba granite and its modal analysis have been described in Chapter 3 (see subsection 3.1.1). The testing apparatus utilized and the experimental configuration employed have also been described in Chapter 3 (see subsection 3.2.1).

As described previously, the shear rupture surfaces are in mutual contact and interact during the breakdown process under a compressive normal load, so that the shear rupture process is strongly affected by the geometric irregularity (or roughness) of the rupturing surfaces. This suggests that the nucleation process is also strongly affected by the geometric irregularity. In order to investigate to what extent the nucleation process is affected by the geometric irregularity of fault surfaces, precut faults with different surface roughnesses were prepared by grinding precut flat fault surfaces with carborundum grit of different grain sizes, ranging from coarse grit (#60, representative particle sizes of 210–250 µm) to fine grit (#2000, representative particle sizes of 7–9 µm). The profiles of these fault surfaces were measured with a diamond stylus profilometer with a tip radius of 2 µm, and the resolution of the elevation data collected was 0.0084 µm. The fault surface roughnesses prepared with grits #60, #600, and #2000 are hereafter referred to as rough, smooth, and extremely smooth faults, respectively, and the three-dimensional surface profiles of the faults with these three representative geometric irregularities are shown in Figure 5.1 for comparison.

Figures 5.2(a) and (b) show plots of the logarithm of topographical length $L(r)$ against the logarithm of ruler length r for the three representative examples of the fault surfaces with different roughnesses shown in Figure 5.1. $L(r)$ in Figure 5.2 was measured with a ruler length r along the slip direction on the fault surfaces. From Figure 5.2, we find that the relation between $\log L(r)$ and $\log r$ can be represented by segments of linear line with different slopes, suggesting that the surface roughnesses of the faults exhibit band-limited self-similarity. In this case, the relation between $L(r)$ and r within each bandwidth can be expressed by the following power law (Ohnaka and Shen, 1999):

$$L(r) = A_i r^{-a_i} \quad (\lambda_{ci-1} < r < \lambda_{ci}), \tag{5.1}$$

where A_i and a_i are constants in a bandwidth bounded by lower corner length λ_{ci-1} and upper corner length λ_{ci} ($i = 1, 2, \ldots, k$). The corner length λ_{ci} is defined as the length that separates the neighboring two bands with different segment slopes a_i and a_{i+1}. Note that the corner length thus defined is a characteristic length representing the fault surface irregularity. Fault surface roughnesses may, in general, be quantified and characterized in terms of these two parameters: a_i and λ_{ci}.

Specifically, the parameters characterizing the fault surface roughnesses shown in Figure 5.1 were evaluated from the data shown in Figure 5.2; that is, $\lambda_{c1} = 10^{1.83} = 67.6$ µm, $\lambda_{c2} = 10^{2.3} = 199.5$ µm, $a_1 = 0.0229$, and $a_2 = 0.0085$ for the rough surface (grit #60); $\lambda_{c1} = 10^{1.25} = 17.8$ µm, $\lambda_{c2} = 10^{1.66} = 45.7$ µm, $a_1 = 0.0076$, and $a_2 = 0.0028$ for the smooth surface (grit #600); and $\lambda_c = 10^1 = 10$ µm, and $a = 0.0017$ for the extremely smooth surface (grit #2000) (Ohnaka and Shen, 1999). These values for λ_{ci} suggest that λ_{ci} is primarily characterized by the grain size of carborundum grit used for grinding the fault surfaces.

Fault surface roughnesses can be characterized in terms of two parameters: a_i and λ_{ci}. Of these two parameters, however, only the corner length λ_{ci} straightforwardly represents a characteristic length scale in the slip direction on the fault. Hence, attention will hereafter be paid to λ_{ci} alone, and λ_{ci} will be used in a later analysis (subsection 5.1.4). We will see later that λ_{ci}, in fact, plays a critical role in scaling the shear rupture nucleation zone size and its duration.

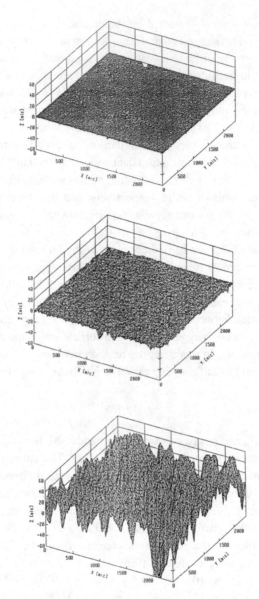

Fig. 5.1   Three representative examples of fault surface geometric irregularities: extremely smooth, smooth, and rough surfaces. Reproduced from Ohnaka and Shen (1999).

5.1.3 Nucleation phases observed on faults with different surface roughnesses

Rough fault

When the tangential force along the fault surfaces is remotely applied by the vertical ram of a biaxial testing apparatus, shown in Figure 3.8, at a constant rate, under a constant normal force across the fault surfaces, the shear stress along the fault builds up at a steady

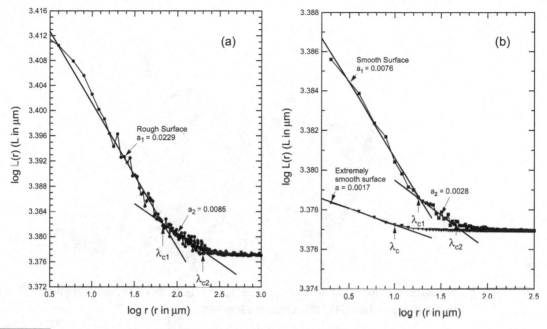

Plots of the logarithm of topographical length $L(r)$ against the logarithm of ruler length r (Ohnaka and Shen, 1999): (a) for the rough fault surface, and (b) for the smooth and extremely smooth fault surfaces, shown in Figure 5.1.

rate with time. Such an example of the shear stress built up with time along the rough fault has been shown in Figure 3.12. As noted in Chapter 3 (see Section 3.2), the remotely applied shear load increased with time at a constant rate until shear failure occurred at $t = 195.08$ s, and no strength degradation or no slip acceleration preceding overall shear failure was discernible from Figure 3.12. For the event T4111805 shown in Figure 3.12, local shear stresses measured at a series of 12 different positions along the fault during the shear load application are shown in Figure 5.3.

Figure 5.3 specifically shows how local shear stresses at the series of 12 positions along the fault were nonuniformly induced and built up with time, when the shear load was remotely applied at the constant rate of deformation (6.7×10^{-4} mm/s). Differential shear stress (τ_{d0}) in Figure 5.3 is defined by (Ohnaka and Shen, 1999)

$$\tau_{d0} = \tau - \tau_{t0}, \tag{5.2}$$

where τ represents the local shear stress measured at a position along the fault, and τ_{t0} represents the initial local shear stress induced at the same position before the shear load application by the normal load that was remotely applied and held constant. Note that the initial shear stress τ_{t0} is not necessarily zero (see Figure 5.4). This is because the shear stress on a fault with topographic irregularities on the surfaces can be induced locally by the application of normal load alone. Thus, the differential shear stress τ_{d0} defined by Eq. (5.2) is the net local shear stress at each position along the fault accumulated during the shear load application. Figure 5.4 indicates the non-uniform shear-stress distribution along the fault for the stick-slip failure event (T4111805) shown in Figure 5.3.

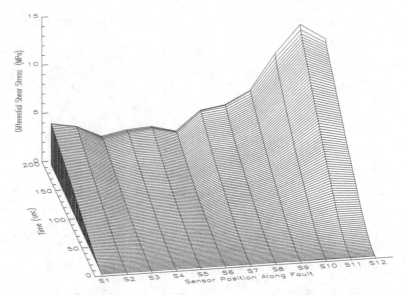

Fig. 5.3 Time and space variations of differential shear stresses (τ_{d0}) recorded locally at 12 positions along the rough fault for the stick-slip rupture event (T4111805) (Ohnaka and Shen, 1999).

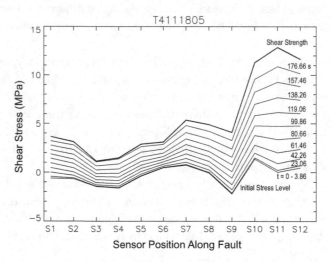

Fig. 5.4 Non-uniform distributions of the fault shear strength (top thick line) and stresses at 12 positions along the rough fault (event T4111805), induced and accumulated with time during the shear load application (Ohnaka and Shen, 1999). The bottom thick line represents the non-uniform distribution of the initial shear stress induced by the application of the normal load (6.2 MPa) alone.

A rough fault has topographic irregularities on the surfaces, and the stress distribution is structure-sensitive. Hence, the non-uniform distribution of shear stress along the fault is inevitably induced and developed with time, despite the fact that remotely applied, normal and shear loads are uniform. It can be clearly seen from Figure 5.4 how the shear stress was nonuniformly induced along the fault and accumulated with time during the shear load application. The top thick line in Figure 5.4 represents the spatial distribution of the peak shear strength τ_p along the fault. The peak shear strength is defined as the maximum shear strength during the local breakdown process behind the front of a propagating rupture at a position on the fault. The bottom thick line in the figure represents the non-uniform distribution of the initial shear stress, induced by the application of the normal load (of 6.2 MPa) alone. Positive and negative shear stresses in Figure 5.4 indicate clockwise and anti-clockwise shear, respectively. A series of thin lines in Figure 5.4 indicates levels of local shear stresses accumulated with time on the fault, and these local stress levels have been plotted at 19.2 s intervals from the time of $t = 3.86$ s after the shear load application.

From Figure 5.4, it is clearly seen, not only that the fault strength and shear stress are both non-uniformly distributed along the fault, but also that the higher the local strength at a position on the fault, the higher the rate of shear stress accumulation at the position during the shear load application. This suggests that the strength distribution along a fault could be inferred from the stress distribution along the fault, or vice versa, because of a strong correlation between the two. However, this does not mean that shear rupture begins to nucleate at a location of highest shear stress on a heterogeneous fault. The stress cannot build up beyond the strength, but builds up more only at the location where the fault strength is locally higher. Frictional stick-slip shear rupture nucleation begins to occur at a location where the resistance to rupture growth is lowest on a fault (see subsection 5.1.4), and the strength is generally low at the location of low resistance to rupture growth.

In general, the amounts of stress degradation and slip during the nucleation process are very small, and this makes it difficult to discern the nucleation phase. In the experiment on the stick-slip failure event (T4111805) shown in Figures 3.12, 5.3, and 5.4, overall slip failure occurred at $t = 195.08$ s; however, the nucleation phase for this event is not significantly discernible prior to the time $t = 195.08$ s in those figures. This is because the amounts of stress degradation and slip for the nucleation phase are very small in comparison to those for the overall slip failure event.

In order to make it possible to discern the nucleation phase, it is necessary to enlarge the scales of the stress and time axes. Figure 5.5 shows the degradation of shear stress associated with frictional stick-slip rupture nucleation along the fault on enlarged scales of the stress and time axes. This figure makes it possible to identify the position from which the rupture began to nucleate, and how the nucleation grew with time. Differential shear stress (τ_{d180}) in Figure 5.5 is defined by

$$\tau_{d180} = \tau - \tau_{t180}, \tag{5.3}$$

where τ represents the local shear stress at a time t recorded at a position along the fault, and τ_{t180} represents the local shear stress at $t = 180$ s recorded at the same position.

From Figure 5.5, we find that local shear stresses at nine positions (S1 through S9) along the fault decreased gradually with time after they had attained peak values (peak shear strength τ_p) at these individual positions. A gradual decrease in local shear stress at a position along the fault indicates that the rupture has begun to occur stably and slowly at that position. Shear rupture is defined, in terms of fracture mechanics, as the process

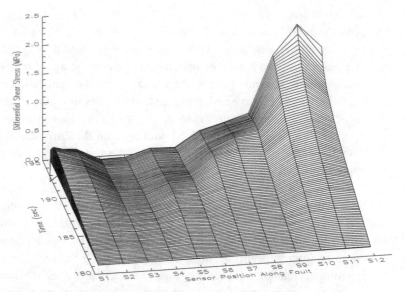

Fig. 5.5   Time and space variations of differential shear stresses (τ_{d180}) recorded locally at 12 positions along the rough fault during the nucleation of the stick-slip rupture event T4111805 (Ohnaka and Shen, 1999).

during which the shear strength degrades to a residual friction stress level with ongoing slip on the rupturing surfaces. Therefore, whether or not shear rupture has indeed occurred at the position of shear stress degradation can be verified by observing the slip-weakening relation at that position.

An observed slip-weakening relation for the nucleation phase of the frictional stick-slip rupture event shown in Figure 5.5 is exemplified in Figure 5.6. Figure 5.6 shows a plot of the local shear stress measured at the position S4 against the local slip displacement measured at the same position during the nucleation process for this particular event. Figure 5.6 demonstrates that frictional stick-slip rupture has actually proceeded locally at the position of shear stress degradation on the fault. We can thus conclude that the rupture began to nucleate roughly at $t = 185$ s somewhere around positions S3 and S4 on the fault, and that the rupture grew slowly and bi-directionally along the fault (see Figure 5.5). This slow extension of a localized slip-weakening zone along the fault with time, prior to unstable, overall frictional stick-slip rupture at $t = 195.08$ s, is what has been referred to as the rupture nucleation phase, and this nucleation phase is an integral part of the subsequent dynamic rupture.

In Figure 5.6, the slip rate (or velocity) is also plotted against the slip displacement. The slip velocity in Figure 5.6 is the time derivative of the slip displacement smoothed by a moving average of three points (Ohnaka and Shen, 1999). From Figure 5.6, one can see that the slip velocity remains very low during the stable, quasi-static nucleation process on the rough fault.

Where on the fault the frictional stick-slip rupture began to nucleate, and how it grew bi-directionally along the fault may be more clearly seen from Figure 3.13. From Figure 3.14, one can see that local slips at positions D1, D2, and D3 proceeded slowly prior to the imminent, overall dynamic slip rupture. These local slow pre-slips correspond

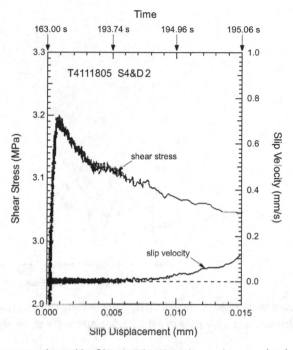

Fig. 5.6 A plot of the shear stress measured at position S4 against the slip displacement measured at the same position during the nucleation for the stick slip rupture event (T4111805) (Ohnaka and Shen, 1999). The slip velocity is also plotted against the slip displacement. The time axis is added to facilitate comparison with Figures 5.5 and 5.7, and Figures 3.13 and 3.14.

to local shear stress degradations at the same positions, as noted in subsection 3.2.3. We can confirm from Figures 3.13 and 3.14 that the slip rupture nucleation proceeded under the condition that both ends of granite sample blocks A and B (shown in Figure 3.8) were servo-controlled to move at a constant rate (6.7×10^{-4} mm/s) in the fault length direction. In fact, the overall displacement versus time record presented in the bottom frame of Figure 3.14 indicates that both ends of the fault were well servo-controlled up to the time $t \cong 195$ s, when dynamic slip rupture occurred over the entire fault length. This overall dynamic slip rupture occurrence was induced immediately after the rupture front reached either end of the fault. This is because the overall rupture was triggered by the breakout of either end of the fault. The event T4111805 generated on the rough fault, shown in Figures 3.12 to 3.14, and Figures 5.3 to 5.6, is a typical example of the stable, quasi-static shear rupture nucleation phase for stick-slip rupture.

In Figure 5.7, the time when the rupture began to occur locally at individual positions S1 to S12 is plotted against the distance along the fault for the event T4111805. The vertical-lined area in this figure indicates the zone in which the breakdown (or slip-weakening) proceeded with time. Figure 5.7 clearly shows how progressively the rupture nucleation grew. A local rupture growth rate (or rupture velocity) V can be estimated from the slope of a segment connecting the neighboring two points in Figure 5.7. Figure 5.8 shows plots of the logarithm of the rupture growth rate V against the logarithm of the rupture growth

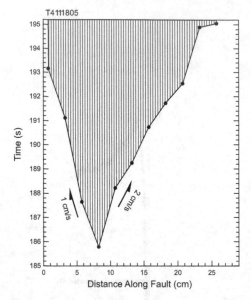

Fig. 5.7 Spatio-temporal evolution of a stick-slip rupture nucleation on the rough fault (Ohnaka and Shen, 1999). The vertical-lined area indicates the zone in which the breakdown (or slip-weakening) proceeds with time.

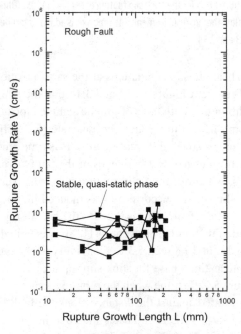

Fig. 5.8 Plots of the logarithm of the rupture growth rate V against the logarithm of the rupture growth length L during the nucleation for stick-slip rupture events on the rough fault (Ohnaka and Shen, 1999).

length L for several stick-slip rupture events that occurred on the rough fault. The experiments conducted under the condition that the remotely applied normal stress was 6.2 MPa and the deformation rate was 6.7×10^{-4} mm/s led to the observational result that the rupture nucleation on the rough fault grew stably at slow rates of 1 to 10 cm/s (Figure 5.8) over the entire fault length of 29 cm.

The phase of spontaneous rupture growth at accelerating speeds and the subsequent phase of dynamic rupture propagation at a high velocity close to elastic wave velocities were never observed for any event on the rough fault, despite the fact that these phases were clearly observed for events on smoother faults. This is because the fault length of 29 cm was too short to enable observation of the accelerating phase and the subsequent phase of high-speed rupture propagation on the rough fault. If the length of this particular rough fault had been sufficiently long, the accelerating phase and the subsequent phase of high-speed rupture propagation would have been observed on the rough fault.

Smooth fault

Figures 5.9 and 5.10 show examples of the series of shear stress versus time records and slip displacement versus time records, respectively, for stick-slip rupture nucleation generated on the smooth fault. Arrows in Figure 5.9 indicate the time when slip failure began to occur locally at individual positions S1 to S12 along the fault. Thus, from Figure 5.9, one can identify the time when slip failure began to occur locally at positions S1 to S12 on the fault, and confirm that the stick-slip rupture nucleation occurred at one end of the fault and grew uni-directionally toward the other end of the fault.

From Figure 5.10, we find that slow slip began to proceed locally at position D1, prior to the imminent overall slip rupture at $t = 106.58$ s. This local, slow slip corresponds to the shear stress degradation at the same position in the nucleation zone. In contrast, however, slip occurred abruptly without any slow preslip in the zone of dynamic rupture propagation (positions D3, D4, and D5 in Figure 5.10). Note that the positions D3, D4, and D5 in Figure 5.10 correspond to the positions S6, S8, and S11, respectively, in Figure 5.9 (see Figure 3.9).

Figure 5.10 (and Figure 3.14) demonstrates that the observation of an abrupt slip rupture without any slow preslip at one position near the fault does not mean that there is no nucleation process preceding an unstable, dynamic high-speed rupture. Figures 5.9 and 5.10 exemplify that in order to detect the shear rupture nucleation that proceeds with time in a localized zone on a fault, it is critically important to monitor local shear stresses and slip displacements continuously at a series of positions along the fault.

The rupture growth rate (or rupture velocity) V for stick-slip rupture events generated on the smooth fault was also estimated in the same manner that was employed to estimate V for stick-slip rupture events generated on the rough fault. Figure 5.11 shows plots of the logarithm of the rupture growth rate V against the logarithm of the rupture growth length L for three stick-slip rupture events that occurred on the smooth fault. From Figure 5.11, we find that each shear rupture nucleation initially proceeded slowly at a steady rate of about 10 cm/s, and extended rapidly at accelerating speeds after the rupture growth length L exceeded the critical length L_{sc}.

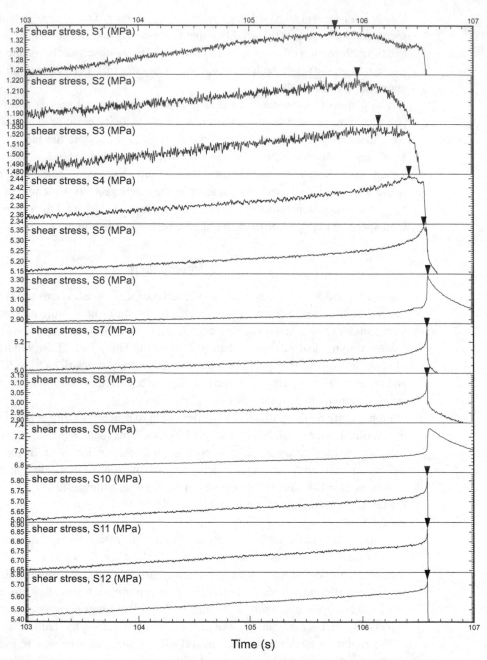

Fig. 5.9 Local shear stress versus time records at 12 positions (S1 to S12) along the fault for a stick-slip rupture event (TS081608) that occurred on the smooth fault (Ohnaka and Shen, 1999). Arrows denote the time when slip failure began to occur locally at individual positions S1 to S12 along the fault.

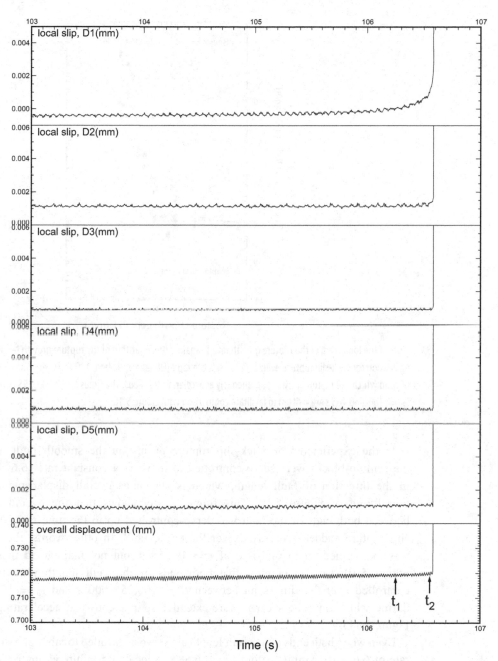

Fig. 5.10 Local slip displacement versus time records at five positions (D1 to D5) along the fault, and a time record of the relative displacement between both ends of the rock sample blocks A and B (bottom frame) for the event T5081608 shown in Figure 5.9 (Ohnaka and Shen, 1999). During the time span between $t_1 (= 106.255000\ \text{s})$ and $t_2 (= 106.560465\ \text{s})$, the rupture locally extended spontaneously at accelerating speeds (see Figure 5.11).

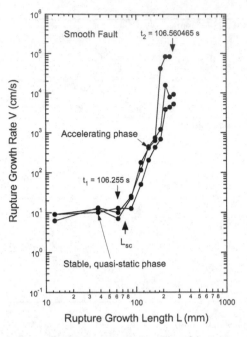

Fig. 5.11 Plots of the logarithm of the rupture growth rate V against the logarithm of the rupture growth length L during the nucleation for stick-slip rupture events on the smooth fault (Ohnaka and Shen, 1999). L_{sc} denotes the critical length beyond which the rupture grows spontaneously at accelerating speeds. The times $t_1 = 106.255\,000$ s and $t_2 = 106.560\,465$ s are written to facilitate comparison with Figure 5.10.

In the experiments on stick-slip rupture events on the smooth fault, both ends of the sample blocks were servo-controlled to move at a constant rate (6.6×10^{-4} mm/s) in the direction of fault length, which is shown as overall displacement in the bottom frame of Figure 5.10. This frame clearly shows that the overall displacement between both ends of the fault was servo-controlled up to the time $t = 106.58$ s, when slip rupture suddenly occurred over the entire fault. In other words, the servo-control was maintained up to the time of $t = 106.58$ s, but not maintained thereafter. From Figure 5.10, we can confirm that both ends of the fault length were indeed servo-controlled over the time span between $t_1 = 106.255\,000$ s and $t_2 = 106.560\,465$ s, during which time the shear rupture extended spontaneously at accelerating speeds (see Figure 5.11).

Even when both ends of the fault length are servo-controlled to move at a slow and steady rate of 6.6×10^{-4} mm/s before $t = 106.58$ s, a local slip failure at any position within a localized zone on the fault cannot be controlled, if the elastic strain energy accumulated in the medium surrounding the localized zone is released, and if the released energy is greater than the resistance to rupture growth. In this case, the frictional slip rupture grows spontaneously and unstably at accelerating speeds in the localized zone (Figure 5.11). Figure 5.11 implicitly indicates that an adequate amount of elastic strain energy was released as the rupture grew beyond the critical length L_{sc}.

Figure 5.11 conclusively demonstrates that the shear rupture nucleation process consists of two phases: an initial, slow and steady growth (phase I), and the subsequent, spontaneous growth at accelerating speeds (phase II). The rupture in phase I grows quasi-statically at a slow and steady rate, whereas the rupture in phase II grows dynamically at accelerating speeds. The presence of this slow and quasi-static rupture growth prior to dynamic rupture growth at accelerating speeds has recently been corroborated by other laboratory experiments (e.g., Nielsen *et al.*, 2010) using a different methodology and using transparent materials not found in rocks constituting the Earth's crust.

The result presented in Figure 5.11 is significant because this is the first successful observation of the transition process from the phase of stable, slow rupture growth to the subsequent accelerating phase of rupture growth for stick-slip rupture events. The high-resolution laboratory experiments under the conditions of a normal stress of 6.2 MPa, and a deformation rate of 6.6×10^{-4} mm/s led to common observations of these two phases on the smooth fault, with good reproducibility (Ohnaka and Shen, 1999).

The terminal phase of dynamic shear rupture propagation at a high speed close to the shear wave velocity was not observed for stick-slip events on the smooth fault, although the terminal phase was commonly observed for stick-slip events on the extremely smooth fault, which will be described below. In order to make it possible to observe the entire process from an initial, slow and steady phase of the nucleation to the subsequent, accelerating phase, and to the terminal phase of high-speed rupture propagation for a single event on the smooth fault, the fault length has to be much longer than 29 cm. The fault length of 29 cm was too short to enable observation of the terminal phase of high-speed rupture propagation on the smooth fault.

Extremely smooth fault

In the experiments on stick-slip rupture on the extremely smooth fault, metallic foil strain gauges with an active gauge length of 2 mm were used to measure local slip displacements along the fault, and amplified signals of local stress and slip displacement were sampled at a frequency of 1 MHz (see subsection 3.2.1). On enlarged scales of both stress and time axes, Figure 5.12 shows a typical example of the series of shear stress records at 12 positions (S1 through S12) during the transition process from shear rupture nucleation to dynamic high-speed rupture for stick-slip events taking place on the extremely smooth fault. In this experiment, the shear load was applied at a constant deformation rate of 4.0×10^{-4} mm/s under the remotely applied normal load of 9.0 MPa. The origin of time t was taken arbitrarily in Figure 5.12. Differential shear stress (τ_{d0}) in Figure 5.12 is defined by

$$\tau_{d0} = \tau - \tau_{t0} + 0.4, \tag{5.4}$$

where τ represents the local shear stress measured at a time t at a position along the fault, and τ_{t0} represents the shear stress recorded at $t = 0$ at the same position. In order to make the nucleation process more clearly visible in Figure 5.12, all data points were shifted upward by 0.4 MPa on the stress axis in Figure 5.12, by adding 0.4 to the right-hand side of Eq. (5.4).

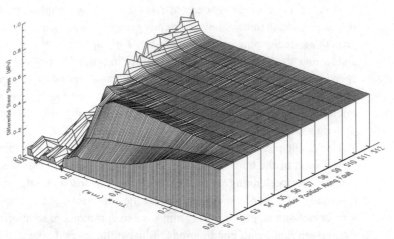

Fig. 5.12   Time and space variations of differential shear stresses (τ_{d0}) recorded locally at 12 positions along the extremely smooth fault during the nucleation for a stick-slip rupture event generated on the fault (Ohnaka and Shen, 1999).

From Figure 5.12, we find that local shear stresses at two positions (S1 and S2) along the fault began to degrade at about $t = 0.1$ ms, and continued to decrease thereafter, while the stresses at other positions remained constant until the passage of a stress pulse accompanying dynamic rupture propagation. This leads to the conclusion that the rupture nucleation began at one end of the fault, and propagated uni-directionally to the other end. Figure 5.13 shows a slip-weakening relation observed at the position S2 in the nucleation zone for the stick-slip rupture event shown in Figure 5.12. From Figure 5.13, it can be confirmed that slip failure actually occurred at a position of stress degradation.

The slip velocity during the nucleation process on the extremely smooth fault is also plotted against the slip displacement in Figure 5.13. The slip velocity in Figure 5.13 is the time derivative of the slip displacement smoothed by a moving average of nine points (Ohnaka and Shen, 1999). One can see from Figure 5.13 that the slip velocity on the extremely smooth fault began to accelerate very rapidly at a slip displacement of about 0.0006 mm. This is contrasted with the slip velocity on the rough fault (see Figure 5.6). These results suggest that the geometric irregularity of fault surfaces has a significant implication for the slip velocity.

From Figures 5.5 and 5.12, one can see that the rupture nucleation zone size and its duration on the extremely smooth fault are small compared with those on the rough fault. As will be shown later, the nucleation zone size and its duration on the extremely smooth fault are also smaller than those on the smooth fault (see Chapter 6).

Figure 5.14 shows a plot of the time when the rupture began to occur locally at positions S1 to S12 against the distance along the fault for the event shown in Figure 5.12. The vertical-lined area in Figure 5.14 indicates the zone in which the breakdown (or slip-weakening) proceeded with time. The rupture growth rate V estimated from the slope of a segment connecting the neighboring two points in Figure 5.14 had the lowest value of 50 m/s in the nucleation zone (see Figure 5.15). The lowest rupture growth rate of 50 m/s is much

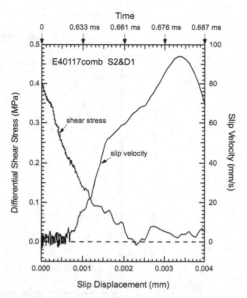

Fig. 5.13 A plot of the local shear stress measured at position S2 against the local slip displacement measured at the same position for the slip failure event generated on the extremely smooth fault shown in Figure 5.12 (Ohnaka and Shen, 1999). The slip velocity at the same position is also plotted against the slip displacement. The time axis is added to facilitate comparison with Figure 5.12.

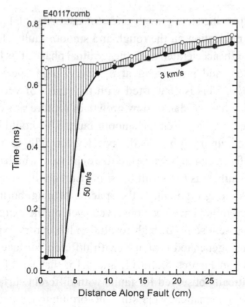

Fig. 5.14 Spatio-temporal evolution of a stick-slip rupture nucleation to dynamic propagation on the extremely smooth fault for the event shown in Figure 5.12 (Ohnaka and Shen, 1999). The vertical-lined area indicates the zone in which the breakdown (or slip-weakening) proceeds with time.

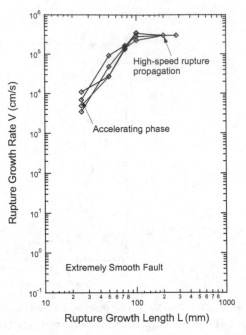

Fig. 5.15 Plots of the logarithm of the rupture growth rate V against the logarithm of the rupture growth length L for stick-slip rupture events on the extremely smooth fault (Ohnaka and Shen, 1999).

higher than the slow growth rates (of the order of 1–10 cm/s) observed during the initial rupture nucleation on the rough and smooth faults. The rupture on the extremely smooth fault rapidly accelerated up to the terminal phase of its high-speed propagation (Figures 5.14 and 5.15), and propagated at a steady, high speed close to the shear wave velocity (2.9 km/s). This is contrasted with the phase of very slow rupture growth on the rough fault. No phase of stable, slow growth of rupture was observed for stick-slip rupture events generated on the extremely smooth fault. This could be because the phase of stable, slow growth of rupture (phase I) did not develop fully to a measurable extent on the extremely smooth fault, and/or because the size of phase I, which might have appeared on the extremely smooth fault, was too small to be measured with strain gauge sensors mounted at 25 mm intervals along the fault. If the spacing of neighboring sensors had been much less than 25 mm, it might have been observed even on the extremely smooth fault.

The present series of high-resolution laboratory experiments on stick-slip shear rupture nucleation generated on faults with different surface roughnesses led to different results as shown in Figures 5.8, 5.11, and 5.15; that is, different phases of the nucleation were systematically observed for faults with different surface roughnesses. These observations could be systematically classified according to the geometric irregularity of fault surfaces. If a scaling parameter representing the geometric irregularity of fault surfaces can be appropriately defined, a unified comprehension and consistent interpretation of the experimental results shown in Figures 5.8, 5.11, and 5.15 can be provided. Let us consider this in the next subsection.

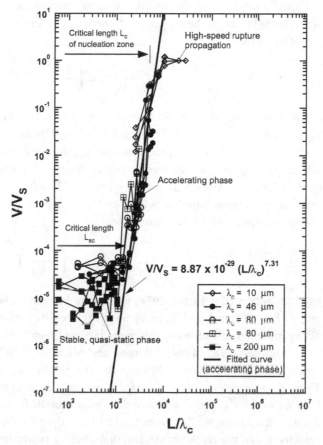

Fig. 5.16 Log–log plots of the rupture growth rate V normalized to the shear wave velocity V_S against the rupture growth length L normalized to the characteristic length λ_c during the nucleation process for stick-slip rupture events generated on faults with different surface geometric irregularities (Ohnaka and Shen, 1999).

5.1.4 Scaling of the nucleation zone size

Let us recall that the corner length λ_{ci} defined in subsection 5.1.2 is a characteristic length representing the geometric irregularity of fault surfaces in the slip direction. If we assume the effective characteristic length λ_c as the upper limit corner length beyond which the fault surface irregularity no longer exhibits self-similarity, then we have $\lambda_c = 200$ μm for the rough fault, $\lambda_c = 46$ μm for the smooth fault, and $\lambda_c = 10$ μm for the extremely smooth fault (see subsection 5.1.2). By using these values for λ_c, the rupture growth velocity V normalized to the shear wave velocity V_S is plotted against the rupture growth length L normalized to λ_c in Figure 5.16 for the data shown in Figures 5.8, 5.11, and 5.15, and for additional data. From Figure 5.16, we find that L scales well with λ_c, and that a unified comprehension and consistent interpretation can be provided for the entire data set on the rupture nucleation on faults with different surface irregularities.

From Figure 5.16, we also find that all data points in the accelerating phase fall around a single straight line on the log–log plot, showing that the accelerating phase obeys a power law of the form

$$\frac{V}{V_S} = \alpha \left(\frac{L}{\lambda_c} \right)^n, \tag{5.5}$$

where α and n are numerical constants. The data set in the accelerating phase shown in Figure 5.16 gives $\alpha = 8.87 \times 10^{-29}$ and $n = 7.31$ (Ohnaka and Shen, 1999).

One may argue that an exponential type of behavior is more physically reasonable for a self-amplifying process, and therefore that a functional form of exponential type would be better fitted to the data set shown in Figure 5.16. This has been examined, and a functional form of exponential type has been found to be a poorer fit with the present data set. In addition, a power law of the form

$$V = V_0 K_I^n \exp \left(-\frac{H}{RT} \right) \tag{5.6}$$

has been proposed (Charles, 1958) to explain the relation between the subcritical growth rate V of mode I crack and the crack tip stress intensity factor K_I, observed for a large body of experimental data on many different classes of materials, including rocks and minerals (e.g., Atkinson, 1982, 1984; Atkinson and Meredith, 1987). In Eq. (5.6), V_0 and n are constants, H is the activation enthalpy, R is the gas constant, and T is absolute temperature. Since K_I is expressed in terms of the remotely applied stress σ and the crack length L as $K_I = m\sigma\sqrt{L}$ (m, a dimensionless modification factor) in the framework of linear fracture mechanics, we find from Eq. (5.6) that the relation between the subcritical crack velocity V and the crack growth length L obeys a power law. In spite of different modes of rupture (modes I and II), Eqs. (5.5) and (5.6) show that there is a clear phase in which the rupture extends at accelerating speeds with an increase in the rupture growth length, prior to the terminal phase of high-speed rupture close to elastic wave velocities, and that this accelerating phase is commonly described by the power law. This suggests that the accelerating phase is governed by a common, underlying physical law (Ohnaka and Shen, 1999).

The rupture nucleation initially begins to proceed stably and quasi-statically (phase I), because no elastic strain energy stored in the medium surrounding the fault is released. The rupture growth in phase I is controlled by the applied loading rate. Beyond the critical length L_{sc}, however, the elastic strain energy stored in the medium surrounding the rupture front is adequately released for the rupture to extend spontaneously and unstably at accelerating speeds, and the rupture extension no longer depends on the applied loading rate. In other words, the behavior of rupture growth changes at $L = L_{sc}$ from the quasi-static phase controlled by the applied loading rate to the self-driven, dynamic phase controlled by inertia. In this respect, the critical length L_{sc} is a physical parameter of paramount importance in the course of shear rupture nucleation. The size of the critical length L_{sc} is affected by not only the applied loading rate, but also the characteristic length λ_c representing the geometric irregularity of the rupturing surfaces.

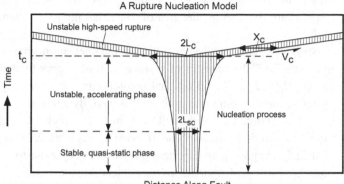

Fig. 5.17 A shear rupture nucleation model (Ohnaka, 2000). The behavior of the nucleation growth changes at $L = L_{sc}$ (half-length) from the stable, quasi-static phase controlled by the applied loading rate to the self-driven, accelerating phase controlled by inertia. The accelerating phase develops into the terminal phase of dynamic rupture propagation at a high speed V_c at the critical time t_c at which $L = L_c$ is attained. X_c denotes the breakdown zone size.

Figure 5.16 indicates that an increase in V/V_S with increasing L/λ_c begins to deviate from the power law (5.5) at $L/\lambda_c = 5.6 \times 10^3$, suggesting that the rupture velocity V converges to a terminal velocity after the rupture growth length attains another critical length L_c. This critical length L_c is important when laboratory data on stick-slip rupture nucleation are compared with seismological data on earthquake rupture nucleation, because the size of the seismic nucleation zone estimated from seismological data does not correspond to the critical length L_{sc} but to the critical length L_c (see Section 5.2). Note that L_c depends on the characteristic length λ_c, though it is not affected by the applied loading rate.

The experimental results have shown that the run-up distance for attaining a high-speed rupture on extremely smooth fault surfaces is very short, whereas a long run-up is required for attaining the high-speed rupture on rough, irregular fault surfaces. This has an important implication for predicting the onset of dynamic, high-speed rupture.

Based on the above and other results of high-resolution laboratory experiments on the stick-slip rupture nucleation, the nucleation process can be specifically modeled as shown in Figure 5.17. It has been clarified where on an inhomogeneous fault a frictional stick-slip shear rupture nucleation begins to occur. The high-resolution laboratory experiments (Ohnaka and Kuwahara, 1990; Ohnaka, 1990, 1996) revealed that a frictional stick-slip rupture begins to nucleate at a location where $\Delta\tau_b$ and D_c are both smallest, and hence the resistance to rupture growth G_c is lowest on a precut fault on which $\Delta\tau_b$ and D_c are nonuniformly distributed. Based on this experimental result, Shibazaki and Matsu'ura (1998) performed a computer numerical simulation by using a theoretical model with a slip-dependent constitutive law for stick-slip failure, and this computer simulation regenerated the transition process from stable, quasi-static nucleation to high-speed rupture propagation observed in the laboratory experiments. Thus, it was demonstrated in the 1990s that a shear rupture nucleation originates at a localized area of minimum resistance to rupture growth on a heterogeneous fault, and grows with time toward the surrounding area of somewhat higher resistance to rupture growth on the fault.

Now, let us consider theoretically how the nucleation zone size scales with λ_c. As shown in Figure 5.17, the behavior of rupture nucleation growth changes at the nucleation zone length $2L = 2L_{sc}$ (L and L_{sc}, half-length) from the stable, quasi-static phase controlled by the applied loading rate to the self-driven, dynamic and accelerating phase controlled by inertia. Thereafter, at the critical time t_c at which $2L = 2L_c$, the accelerating phase develops into the terminal phase of dynamic, high-speed rupture. Beyond the critical time t_c, the rupture propagates at a steady, high speed, V_c, close to elastic wave velocities. In this model, the critical length $2L_c$ of the nucleation zone is related to the breakdown zone size X_c defined in the terminal phase of dynamic, high-speed rupture as follows (Ohnaka and Shen, 1999: Ohnaka, 2000):

$$L < X_c \quad \text{for } t < t_c$$
$$L_c = X_c \quad \text{at } t = t_c. \tag{5.7}$$

In the zone of dynamic high-speed rupture, X_c is directly related to the breakdown displacement D_c by Eq. (3.28). From Eqs. (5.7) and (3.28), we have

$$L_c = \frac{1}{k} \frac{\mu}{\Delta \tau_b} D_c. \tag{5.8}$$

Since k has a value of the order of unity (see Chapter 3), Eq. (5.8) indicates that the critical size $2L_c$ of the nucleation zone is substantially determined by D_c and $\Delta \tau_b$. Since D_c is related to λ_c by Eqs. (3.15) and (3.16), Eq. (5.8) can be rewritten in terms of λ_c as follows:

$$L_c = \frac{1}{k} \frac{\mu}{\Delta \tau_b} \left(\frac{1}{\beta} \right)^{1/M} \left(\frac{\Delta \tau_b}{\tau_p} \right)^{1/M} \lambda_c. \tag{5.9}$$

If appropriate values are substituted for k, μ, β, and M; that is, $k \cong 2.8$, $\mu = 20\,000$ MPa, $\beta = 1.64$, and $M = 1.20$ for the Tsukuba granite used (see Section 3.3), Eq. (5.9) is reduced to

$$L_c = 4.73 \times 10^3 \frac{1}{\Delta \tau_b} \left(\frac{\Delta \tau_b}{\tau_p} \right)^{0.833} \lambda_c. \tag{5.10}$$

In the laboratory experiments on stick-slip failure, $\Delta \tau_b$ had a value ranging from 0.014 to 0.173 MPa, and $\Delta \tau_b / \tau_p$ had a value ranging from 0.01 to 0.07 (Ohnaka, 2003). If we substitute 0.065 MPa for $\Delta \tau_b$ and 0.04 for $\Delta \tau_b / \tau_p$ in Eq. (5.10), Eq. (5.10) is further reduced to

$$L_c \cong 5 \times 10^3 \lambda_c, \tag{5.11}$$

which roughly agrees with the value for L_c / λ_c estimated from Figure 5.16.

5.2 Earthquake rupture nucleation

5.2.1 Seismogenic background

In the past, there was a pervasive hypothesis that the Earth's crust is in a state of perpetual self-organized criticality in which any small earthquake may cascade into a large event, and that earthquakes are unpredictable catastrophes. The Gutenberg–Richter frequency–magnitude power law was cited as evidence that the Earth's crust is in a self-organized critical state. However, it has been substantiated that this hypothesis is far from the truth, at least in the case of large earthquakes (Ohnaka, 2004a). In fact, real data on seismicity including large earthquakes, do not fall on a Gutenberg–Richter frequency–magnitude power law curve (e.g., see Engdahl and Villasenor, 2002).

Furthermore, paleoseismological data indicate that individual faults or fault segments tend to generate characteristic earthquakes having a relatively narrow range of magnitude (e.g., Schwartz and Coppersmith, 1984; Sieh, 1984, 1996), disobeying the Gutenberg–Richter frequency–magnitude power law. In addition, historical records and seismic instrumentation recordings consistently indicate that large earthquakes, in particular those along plate boundaries, have occurred repeatedly, not in clusters or at random but quasi-periodically on a single fault. For those events, average recurrence time intervals are well defined (e.g., Schwartz and Coppersmith, 1984; Sieh, 1984; Ishibashi and Satake, 1998; Utsu, 1998). To give a specific example, large earthquakes occurring repeatedly along the tectonic plate interface in the Nankai region in the southwest of Japan are well documented, and the average recurrence time interval with its standard deviation for those earthquakes is calculated to be 117.1 ± 21.2 years under certain assumptions (Utsu, 1998; for details see Section 7.1). It can thus be concluded that a sequence of large earthquakes along plate boundaries may occur repeatedly on a quasi-periodic basis.

The aforementioned facts show that the Earth's crust contains a number of characteristic length scales, such as the depth of seismogenic layer, fault and/or its segment sizes. In addition, the sizes of local strong areas highly resistant to rupture growth (called "asperities" or "barriers") on faults can also be characteristic length scales (e.g., Aki, 1984).

We have to keep in mind that an earthquake cannot occur anywhere in the Earth's crust, but can occur only in regions where an adequate amount of elastic strain energy has been accumulated as the driving force for a pending earthquake. Once an earthquake occurs in such a region, the elastic strain energy is necessarily released and expended, and hence the stored energy in the region is lowered to a subcritical level. When the earthquake is small, the energy released is restricted within a small region, so that the released energy in the small region may be restored to the critical level of resistance to rupture growth needed for earthquake occurrence, even shortly after the event, by such means as fault–fault interaction and/or dynamic stress transfer. When the earthquake is large, however, a large amount of the elastic strain energy stored in a wide region is expended and lowered to a subcritical level. The large amount of the energy released in a wide region cannot be restored to the critical level shortly after the event, by means of fault–fault interaction and/or dynamic stress transfer. Tectonic loading due to perpetual slow plate motion is necessarily required for this. Therefore, the next large earthquake cannot occur in the region for a long time until an adequate amount of elastic strain energy has accumulated to reach the critical level with tectonic loading.

If the Earth's crust were in a perpetual critical state, it should continue to have the potential to cause the next large earthquake even immediately after the occurrence of a large event, and hence one would expect that large earthquakes would occur more often in the same region, irrespective of the tectonic loading rate. This, however, does not follow from seismological and geotectonic observations. Thus, it is clear that the hypothesis of perpetual self-organized criticality is neither reasonable nor acceptable, at least in regard to large earthquakes.

An alternative, physically rational hypothesis is to assume that the crustal region immediately after the occurrence of a large earthquake is in a subcritical state, and that crustal deformation in the region proceeds toward the critical state of earthquake recurrence with tectonic loading. An essential feature of this model is that the process leading from a subcritical state to the critical state is repeated intermittently on a single fault under perpetual tectonic loading in a crustal region.

As described in previous chapters, the brittle, seismogenic crust and individual faults embedded therein are inherently heterogeneous. Indeed, there is increasing evidence that real faults contain local strong areas highly resistant to rupture growth, with the rest of the faults having low (or little) resistance to rupture growth. Such local strong areas are called "asperities" or "barriers" in the field of earthquake seismology (see Chapter 1). When sites of such strong areas act as "barriers," stress will build up and elastic strain energy will accumulate in the medium at and around these sites until they break. Once a site of such a strong area is broken down, whatever the cause may be, the elastic strain energy accumulated in the medium surrounding the site is released, and the released energy is expended as the driving force for an earthquake. In this case, such a strong area on a fault will be regarded as an "asperity." Thus, it is obvious that local strong areas highly resistant to rupture growth on a fault play a much more important role in the generation of an earthquake than the rest of the fault with low (or little) resistance to rupture growth.

Let us envision a brittle, inhomogeneous crustal region with an embedded preexisting fault that has the potential to cause a large earthquake, and contains local strong areas highly resistant to rupture growth, with the rest of the fault having low (or little) resistance to rupture growth. Such a fault may be called a master fault. We specifically consider the deformation process of such a crustal region, including a master fault such as a plate boundary fault, from a subcritical state toward the critical state of earthquake occurrence with tectonic loading. Shortly after the occurrence of a large earthquake, the preexisting fault begins to heal and to be re-strengthened under a compressive normal load. The re-strengthening on irregular fault surfaces can be attained by a displacement-induced increase in frictional resistance, due to an increase in the sum of the real areas of asperity junction, such as asperity interlocking, on the fault surfaces with progressive displacement under compressive normal stress. Accordingly, the elastic strain energy can again accumulate in the region surrounding the fault, as tectonic stress builds up slowly and steadily. The fault thus regains the potential to cause the next large earthquake.

At an early stage where the tectonic stress is far below the critical level for earthquake occurrence, and where the amount of the elastic strain energy stored in the region is inadequate, seismic activity in the region is quiescent. This quiescent seismicity is called *background seismicity*. As the tectonic stress reaches higher levels, and gradually approaches the critical level, the crustal region begins to behave in-elastically in the brittle regime; consequently, seismicity in the region surrounding the master fault having the potential to cause the next large event becomes progressively active with time. This is because the crust is inherently heterogeneous and includes numerous faults of small-to-moderate size. This later stage can thus be characterized by the activation of seismicity.

Seismic activity is enhanced by fluid–rock interaction, and by the triggering effect of stress transfer at the later stage where the tectonic stress is in close proximity to the critical level. For example, a small stress perturbation due to the occurrence of neighboring earthquakes and/or the Earth's tides may trigger small to moderate earthquakes at this stage. Indeed, it has been demonstrated that a small stress change due to the Earth's tides can trigger such earthquakes in the region where the tectonic stress is in close proximity to the critical level (Tanaka *et al.*, 2002). Activated seismicity at this stage may be recognized as premonitory phenomena for an ensuing large earthquake on the master fault.

In due course, the tectonic stress reaches the critical level, and an adequate amount of the elastic strain energy builds up in the region surrounding the fault. Consequently, a rupture nucleation begins locally at an area of lowest resistance to rupture growth on the master fault, and the nucleation necessarily proceeds toward the surrounding area of higher resistance to rupture growth on the fault. At this stage, the rupture nucleation proceeds stably and quasi-statically. After the elastic strain energy accumulated in the surrounding medium is released over the resistance to rupture growth, the rupture nucleation grows spontaneously at accelerating speeds (see Section 5.1).

The nucleation process eventually leads to a resulting dynamic high-speed rupture (or earthquake) on the fault, accompanied by a rapid stress drop and the dissipation of a great amount of the elastic strain energy, resulting in the radiation of seismic waves. The arrest of the earthquake rupture results in aftereffects that include re-distribution of local stresses on and around the fault, leading to aftershock activity.

5.2.2 Physical modeling and theoretical derivation of the nucleation zone size

In this subsection, we concentrate on how the nucleation process leading up to a large earthquake can be physically modeled in seismogenic environments. We assume that the laboratory-derived rupture nucleation model shown in Figure 5.17 is applicable to the nucleation process of an earthquake rupture. In this model, it is assumed that the rupture nucleates to grow bi-directionally. We can also assume a similar model in which the rupture nucleates to grow uni-directionally. However, whether the rupture grows bi-directionally or uni-directionally is extraneous to the matter at hand. Here we employ the bi-directional rupture nucleation model.

As described in Section 5.1, the shear-rupture nucleation process consists of two phases: an initial, stable and quasi-static phase (phase I), and the subsequent, unstable and accelerating phase (phase II). Phase I is a slow, steady rupture growth controlled by the rate of tectonic loading. On the other hand, phase II is a spontaneous rupture extension driven by the release of the elastic strain energy stored in the surrounding medium. The rupture nucleation initially proceeds stably and quasi-statically at a slow and steady speed V_{st} up to the critical length $2L_{sc}$ (phase I), beyond which the rupture growth velocity V increases with an increase in the rupture growth length L (phase II), in accordance with a power law specified by Eq. (5.5). After the rupture growth length has reached the critical length $2L_c$ at $t = t_c$ (see Figure 5.17), the rupture propagates at a steady, high speed V_c close to the shear wave velocity.

Thus, the earthquake rupture generated on a heterogeneous fault cannot begin to propagate abruptly at a high speed close to elastic wave velocities, immediately after the onset of the spontaneous rupture growth, but begins to develop at accelerating speeds from the phase of quasi-static slow growth to the phase of dynamic high-speed propagation in a finite duration. As described in Section 5.1, the zone size and duration of the nucleation are both scale-dependent, and therefore, the zone size and duration of the nucleation for field-scale earthquake ruptures are much larger than those for laboratory-scale shear ruptures.

Let us examine how specifically the rupture grows with time during the nucleation process. Shibazaki and Matsu'ura (1998) computed the far-field velocity waveform from a circular fault model with the slip-time function obtained from the numerical simulation of a slip failure event governed by a slip-dependent constitutive law, and demonstrated that the seismic nucleation phase, defined by Ellsworth and Beroza (1995), on a seismogram corresponds to the critical nucleation phase at $t = t_c$ shown in Figure 5.17, at the earthquake source. This means that the size of the seismic nucleation zone estimated from seismological data (e.g., Ellsworth and Beroza, 1995; Beroza and Ellsworth, 1996) corresponds not to the critical length L_{sc} but to the critical length L_c at $t = t_c$ defined in the previous section (see Figure 5.17). Hence, we define the critical size of the seismic nucleation zone, not as L_{sc} but as L_c, and hereafter we will focus attention on the critical length L_c. Note that the critical length L_c defined in the previous section is physically not identical to the critical length defined by Andrews (1976a, 1976b).

In the present model, the rupture growth length L during the stable, quasi-static phase (phase I) increases slowly and steadily with time t in accordance with

$$L(t) = L_0 + V_{st}t \quad (0 < t < t_{sc}), \tag{5.12}$$

where $L_0 = L(0)$, which is an initial length of $L(t)$, and V_{st} is the steady, rupture growth rate controlled by the tectonic loading rate. By contrast, the rupture growth length L during the unstable, accelerating phase (phase II) increases more rapidly with time. A specific functional form of $L(t)$ during phase II leading to the critical point will be derived below. The critical point is defined as the critical time t_c at which $L = L_c$ is attained.

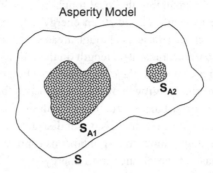

Asperity Model

Fig. 5.18 An "asperity" earthquake-fault model. S denotes entire fault area. S_{A1} and S_{A2} denote "asperity" patch areas highly resistant to rupture growth on the fault.

Noting that $dt = dL/V$, we have from Eq. (5.5)

$$t - t_{sc} = \int_{L_{sc}}^{L} \frac{dL}{V(L)} = \frac{\lambda_c^n}{\alpha V_S} \int_{L_{sc}}^{L} L^{-n}dL = \frac{\lambda_c^n}{\alpha V_S(1-n)}\left(L^{-n+1} - L_{sc}^{-n+1}\right) \quad (n \neq 1)$$

(5.13)

and

$$t_c - t_{sc} = \int_{L_{sc}}^{L_c} \frac{dL}{V(L)} = \frac{\lambda_c^n}{\alpha V_S} \int_{L_{sc}}^{L_c} L^{-n}dL = \frac{\lambda_c^n}{\alpha V_S(1-n)}\left(L_c^{-n+1} - L_{sc}^{-n+1}\right) \quad (n \neq 1).$$

(5.14)

In the above equations, time t has been set such that $t = t_{sc}$ when $L = L_{sc}$. Subtracting Eq. (5.13) from Eq. (5.14), we have (Ohnaka and Matsu'ura, 2002; Ohnaka, 2004b)

$$L(t) = L_c \left(\frac{t_a - t_c}{t_a - t}\right)^{1/(n-1)} \quad (t_{sc} \leq t \leq t_c),$$

(5.15)

where t_a is defined by

$$t_a = \frac{1}{\alpha(n-1)} \frac{\lambda_c}{V_S} \left(\frac{\lambda_c}{L_c}\right)^{n-1} + t_c.$$

(5.16)

Thus, the present model leads to the conclusion that the rupture growth length $L(t)$ during phase II increases with time t in accordance with Eq. (5.15).

The critical length L_c for earthquakes can be obtained from Eq. (5.8), if D_c and $\Delta\tau_b$ can be properly evaluated. Here, we take a different approach in order to examine how large the nucleation zone size of an actual earthquake is in comparison to the size of the resulting earthquake.

For simplicity, we assume an "asperity" earthquake-fault model as shown in Figure 5.18. In this model, the fault has local areas (denoted as S_{A1} and S_{A2} in Figure 5.18) that are highly resistant to rupture growth, with the rest of the fault having low (or little) resistance to rupture growth. Local areas (called "asperities") highly resistant to rupture growth on the fault are so tough that an adequate amount of elastic strain energy can be stored in the medium surrounding the local tough areas, whereas the rest of the fault is so weak or deformable that little elastic strain energy can be sustained. We further assume that the interaction between these local tough areas is negligible. The size of such a local tough area may represent a characteristic length on the fault.

As noted previously, a shear rupture nucleation begins to occur locally at a site of lowest resistance to rupture growth on a heterogeneous fault, and grows towards the surrounding area of higher resistance to rupture growth on the fault. The rupture begins to propagate dynamically when the initial, slow and steady rupture growth controlled by the rate of tectonic loading changes to a spontaneous rupture extension at accelerating speeds driven by the release of the elastic strain energy stored in the surrounding medium. Once any one of the local "asperity" areas on the fault is completely broken down, the elastic strain energy stored in the surrounding medium is released and expended as the driving force for dynamic rupture propagation at a high speed. If a strong and geometrically large "asperity" area is broken down, the amount of the elastic strain energy stored in the surrounding medium is large, so that the resulting earthquake will be large. On the other hand, a large displacement is required for the breakdown of a geometrically large tough area, and the large breakdown displacement D_c necessarily leads to a large nucleation zone size (see Eq. (5.8)). This predicts that the size of an earthquake scales with its nucleation zone size, which will be confirmed theoretically in quantitative terms below.

As described in Chapter 2, the resistance to rupture growth is a distinct physical quantity defined as the energy G_c required for a rupture front to grow, and is expressed as Eq. (2.45); that is,

$$G_c = \frac{1}{2}\Gamma \Delta \tau_b D_c, \qquad (5.17)$$

where Γ is the dimensionless quantity of the order of unity (here we assume that $\Gamma = 1$; see Section 2.2), $\Delta \tau_b$ is the breakdown stress drop, and D_c is the breakdown displacement.

Let us pay attention to the resistance to rupture growth on a heterogeneous fault. The resistance to rupture growth $\overline{G_c}$ averaged over the entire fault area S is defined by

$$\overline{G_c} = \frac{1}{S}\int_S G_c dS. \qquad (5.18)$$

Let S_{A1} be the area of the geometrically largest "asperity" on the fault. Equation (5.18) may be rewritten as follows:

$$\overline{G_c} = \frac{1}{S}\left[\int_{S_{A1}} G_c dS + \int_{S-S_{A1}} G_c dS\right]. \qquad (5.19)$$

The first term of the right-hand side of Eq. (5.19) denotes the integral over the area S_{A1} of the geometrically largest "asperity" on the fault, and the second term denotes the integral over the rest $(S - S_{A1})$ of the fault area. The first term of the right-hand side of Eq. (5.19) is a fraction $a(< 1)$ of $\overline{G_c}$, so that we have

$$\overline{G_c} = \frac{1}{aS}\int_{S_{A1}} G_c dS = \left(\frac{S_{A1}}{aS}\right)\overline{G_c^{A1}}, \qquad (5.20)$$

where

$$\overline{G_c^{A1}} = \frac{1}{2}\Gamma\Delta\tau_b^{A1}D_c^{A1}. \tag{5.21}$$

In Eq. (5.21), $\Delta\tau_b^{A1}$ and D_c^{A1} represent the breakdown stress drop and the breakdown displacement, respectively, required for the geometrically largest "asperity" on the fault to break down.

Let $\overline{D}$ and $\overline{\Delta\tau}$ be the mean values of slip displacement D and stress drop $\Delta\tau$, respectively, averaged over the entire fault area S. The stress drop $\Delta\tau$ is defined as $\Delta\tau = \tau_i - \tau_r$ (for the definition of τ_i and τ_r, see Figure 3.22). The stress drop $\Delta\tau$ thus defined is practically estimated from $\overline{D}$, using the relationship between $\Delta\tau$ and $\overline{D}$ derived from a shear crack model for a planar fault surface, under the assumption that $\Delta\tau$ is uniform over the entire fault area. In this case, $\overline{\Delta\tau} = \Delta\tau$. For example, for a circular crack of radius R, $\Delta\tau$ is theoretically related to $\bar{D}$ by (e.g., Madariaga, 1977; Papageorgiou and Aki (1983a))

$$\Delta\tau = \mu\left(\frac{7\pi}{16}\right)\left(\frac{\overline{D}}{R}\right). \tag{5.22}$$

We have to keep in mind that $\overline{\Delta\tau} < \Delta\tau_b$, because $\Delta\tau_b$ is the breakdown stress drop (defined as $\Delta\tau_b = \tau_p - \tau_r$) in the localized zone (called breakdown zone) behind a rupture front.

The seismic moment M_0, expressed as (Aki, 1966)

$$M_0 = \mu\overline{D}S, \tag{5.23}$$

represents the size of an earthquake, where μ is the rigidity of the elastic medium surrounding the fault. The physical meaning of the seismic moment will be easily understandable if it is derived from the elastic rebound model proposed by Reid (1911), as follows (Ohnaka, 1976). For this purpose, we use a Cartesian coordinate system with the xz plane being the fault plane and the y-axis being normal to the fault plane (see Figure 5.19, where the positive z-axis is down-pointing). We assume a strike-slip plate boundary fault along the x-axis. Tectonic movements of both sides of the plate boundary to opposite directions result in a steady, slow buildup of elastic shear strain in the medium surrounding the fault. Eventually, the shear strain reaches the critical level beyond which the fault cannot sustain its strength, resulting in fault failure. Consequent elastic-rebound movements of the two sides of the fault plane (as shown in Figure 5.19) release the stored strain energy, accompanied by shear dislocation on the fault plane. This causes an earthquake rupture on the fault.

In Figure 5.19, the heavy solid line represents pre-seismic shear deformation of the medium surrounding the fault, and the heavy broken lines represent residual strains of the surrounding medium immediately after the earthquake. The residual strains shown in Figure 5.19 may be maintained by the presence of frictional resistance on the fault surfaces. Let u_+ and u_- be the displacements on the right ($y = +0$) and left ($y = -0$) fault surfaces, respectively, in the x-direction, and $u_{+\eta}$ and $u_{-\eta}$ be the displacements at $y = \eta$ and $y = -\eta$, respectively, in the x-direction, where η denotes the distance between the fault plane and the critical point beyond which relation $\partial u/\partial y \cong 0$ holds. The shear dislocation resulting from the elastic rebounds of both sides of the fault plane is represented by the relative displacement $D \equiv u_+ - u_-$ on the fault plane. The shear force τ_{yx} per unit

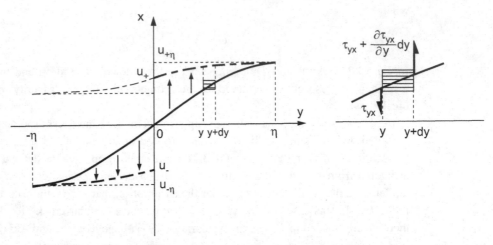

An elastic rebound model. The heavy solid line denotes deformation in the medium surrounding a plate boundary fault due to tectonic movements of both sides of the plate boundary to opposite directions, and the heavy broken lines denote elastic rebound movements of the two sides of the fault plane after the fault failures. The arrows indicate the elastic rebound motion. The moment per unit area on volume element $1 \times dy \times 1$ about the z-axis is $y(\partial \tau_{yx}/\partial y)dy$.

area on the xz plane acts in the x-direction, and the moment of the forces per unit area on volume element $1 \times dy \times 1$ about the z-axis is given by $y(\tau_{yx} + (\partial \tau_{yx}/\partial y)dy - \tau_{yx}) = y(\partial \tau_{yx}/\partial y)dy$. Thus, the total moment of the forces per unit area over the interval $-\eta \le y \le \eta$ immediately before the earthquake is given by (Ohnaka, 1976)

$$\int_{-\eta}^{\eta} y \frac{\partial \tau_{yx}^{(1)}}{\partial y} dy = \left[y\tau_{yx}^{(1)} \right]_{-\eta}^{\eta} - \int_{-\eta}^{\eta} \tau_{yx}^{(1)} dy = - \int_{-\eta}^{\eta} \mu \frac{\partial u^{(1)}}{\partial y} dy = -\mu(u_{+\eta} - u_{-\eta}), \quad (5.24)$$

where superscript (1) denotes "immediately before the earthquake." Similarly, the total moment per unit area over the same interval immediately after the earthquake is given by (Ohnaka, 1976)

$$\int_{-\eta}^{\eta} y \frac{\partial \tau_{yx}^{(2)}}{\partial y} dy = \left[y\tau_{yx}^{(2)} \right]_{-\eta}^{\eta} - \int_{-\eta}^{\eta} \tau_{yx}^{(2)} dy = - \int_{-\eta}^{\eta} \mu \frac{\partial u^{(2)}}{\partial y} dy = -\mu(u_{+\eta} - u_{-\eta} - D), \quad (5.25)$$

where superscript (2) denotes "immediately after the earthquake." Subtracting Eq. (5.25) from Eq. (5.24), we have

$$\int_{-\eta}^{\eta} y \frac{\partial}{\partial y} \left[\tau_{yx}^{(1)} - \tau_{yx}^{(2)} \right] dy = -\mu D. \quad (5.26)$$

The minus sign on the right-hand side of Eq. (5.26) indicates counterclockwise rotation. Thus, the total moment M_0 of an earthquake with fault area S is expressed as $M_0 = \mu DS$. This is what is referred to as the seismic moment, and equal to Eq. (5.23), given that D has been assumed implicitly to be constant over the entire fault area S in this derivation.

Let us consider how M_0 of a mainshock earthquake scales with the breakdown displacement D_c. The earthquake rupture finally arrests when the driving force has become lower than the resistance to rupture growth. If the breakdown zone size X_c behind the rupture front is sufficiently small in comparison to the final fault length L_f, the condition of the rupture arrest may be written as $L_f(\Delta\tau)^2 \leq \kappa\mu G_c$, where κ is a constant. The sign of equality in this equation represents the condition at which the rupture grows quasi-statically, and this may be regarded as the critical condition below which the dynamic rupture must arrest. When the driving force (or $\Delta\tau$) gradually decreases with distance toward the fault end, or when the resistance to rupture growth (or G_c) gradually increases with outward distance from the fault end, the following relation:

$$\left(\overline{\Delta\tau}\right)^2 = \kappa\mu\frac{\overline{G_c}}{L_f}, \tag{5.27}$$

is expected to hold between $\overline{G_c}$, $\overline{\Delta\tau}$, and L_f (Ohnaka, 2000).

If we further assume the following scaling relations:

$$S = c_1 L_f^2 \tag{5.28}$$

and

$$\overline{D} = c_2 L_f, \tag{5.29}$$

where c_1 and c_2 are numerical constants, then we have from Eqs. (5.20), (5.21), (5.23), and (5.27)–(5.29) (Ohnaka, 2000)

$$M_0 = c_1 c_2 \left(\frac{\kappa\Gamma}{2}\right)^3 \left(\frac{S_{A1}}{aS}\right)^3 \left(\frac{\mu}{\overline{\Delta\tau}}\right)^3 \left(\frac{\Delta\tau_b^{A1}}{\overline{\Delta\tau}}\right)^3 \mu\left(D_c^{A1}\right)^3. \tag{5.30}$$

On the right-hand side of Eq. (5.30), the entire fault area S, the "asperity" area S_{A1}, and the breakdown displacement D_c^{A1} are scale-dependent, while the rest of the parameters are scale-independent. If the ratio $S_{A1}/aS(= \overline{G_c}/\overline{G_c^{A1}})$ is scale-independent, the breakdown displacement D_c^{A1} is the only scale-dependent parameter on the right-hand side of Eq. (5.30). If we assume that the stress drop $\overline{\Delta\tau}$ averaged over the entire fault area and the breakdown stress drop $\Delta\tau_b$ in the breakdown zone behind the rupture front are virtually constant (in a statistical sense), Eq. (5.30) predicts theoretically that the mainshock seismic moment M_0 is directly proportional to the third power of the breakdown displacement D_c at the geometrically largest "asperity" area highly resistant to rupture growth on the fault.

The theoretically derived scaling relation $M_0 \propto D_c^3$ is well-grounded, because D_c is by definition the critical slip displacement required for the breakdown of a largest "asperity" highly resistant to rupture growth on a fault ($D_c = D_c^{A1}$), and because the breakdown of the largest "asperity" on the fault causes the release of the largest amount of the elastic strain energy stored in the surrounding medium, resulting in the generation of a mainshock earthquake on the fault.

Likewise, from Eqs. (5.8) and (5.30), we have the following relation between the seismic moment M_0 and the critical size $2L_c$ of the nucleation zone (Ohnaka, 2000):

$$M_0 = c_1 c_2 \left(\frac{k\kappa\Gamma}{4}\right)^3 \left(\frac{S_{A1}}{aS}\right)^3 \left(\frac{\Delta\tau_b^{A1}}{\overline{\Delta\tau}}\right)^6 \mu(2L_c)^3. \tag{5.31}$$

Equations (5.30) and (5.31) were derived theoretically by assuming an "asperity" earthquake-fault model and scaling laws, in the framework of fracture mechanics, based on the laboratory-derived slip-dependent constitutive law (Ohnaka, 2000). These equations lead to the conclusion that the seismic moment is primarily prescribed by the properties of the geometrically largest "asperity" on the fault, and is directly proportional to the third power of the breakdown displacement and the critical size of the nucleation zone, if the assumptions made are physically reasonable.

Equations (5.30) and (5.31) can be reduced to simplified theoretical relations, if appropriate values for the following parameters: c_1, c_2, μ, $S_{A1}/(aS)$, $\overline{\Delta\tau}$, and $\Delta\tau_b^{A1}$, are substituted under the constraint that $\overline{\Delta\tau} < \Delta\tau_b^{A1}$. We here assume $c_1 = 0.5$ and $c_2 = 5 \times 10^{-5}$, in view of the values given by Abe (1975). We further assume $\mu = 30\,000$ MPa, $\overline{\Delta\tau} = 3$ MPa, $\Delta\tau_b^{A1} = 10$ MPa, $S_{A1}/S = 0.4$, and $a = 0.6$ (hence, $S_{A1}/(aS) = 2/3$). Under these assumptions, and given that $k = 3$, $\Gamma = 1$, and $\kappa = 2$, Eq. (5.30) is reduced to

$$M_0 = 0.8 \times 10^{19} D_c^3 \approx 1 \times 10^{19} D_c^3, \tag{5.32}$$

and Eq. (5.31) is reduced to

$$M_0 = 1.0 \times 10^9 (2L_c)^3, \tag{5.33}$$

where M_0 are measured in Nm, and D_c and L_c are both measured in m. To what extent these theoretical relations are justified has been verified by using seismological data on the nucleation phase (Ohnaka, 2000). The results of the comparison of those theoretical relations with seismological data will be shown in the next subsection.

5.2.3 Comparison of theoretical relations with seismological data

Slip-dependent constitutive law parameters for actual earthquakes were estimated in the 1980s to 1990s by leading-edge seismologists (e.g., Papageorgiou and Aki (1983b); Ellsworth and Beroza, 1995; Ide and Takeo, 1997). These data were used in order to verify to what extent the theoretical relations derived in the previous subsection are justified (Ohnaka, 2000).

In particular, Papageorgiou and Aki (1983b) first estimated the breakdown stress drop $\Delta\tau_b$, the breakdown displacement D_c, the breakdown zone size X_c, and the shear rupture energy G_c for moderate to large earthquakes, based on their own specific barrier model for earthquake faulting. Ellsworth and Beroza (1995) analyzed near-source recordings of slow initial P wave for earthquakes with a wide magnitude range from 2.6 to 8.1, and evaluated the breakdown stress drop $\Delta\tau_b$ and the critical length $2L_c$ (L_c, half-length) of the seismic nucleation zone. Ide and Takeo (1997) estimated constitutive relations for the 1995 Kobe earthquake with $M7.3$, from near-field seismic waves by waveform inversion and the solution of elastodynamic equations using a finite difference method.

The basic parameters of the present concern are $\Delta \tau_b$, D_c, and L_c (or X_c), together with the seismic moment M_0. Ellsworth and Beroza (1995) evaluated $\Delta \tau_b$ and L_c from near-source recordings of slow initial P wave for small to large earthquakes by using the same analytical method, and therefore, their data are best suited for the present purpose. However, they did not estimate D_c at the critical stage $L = L_c$ for those earthquakes. Thus, D_c was calculated from Eq. (5.8), as follows (Ohnaka, 2000). In order to evaluate $\Delta \tau_b$ and L_c, Ellsworth and Beroza assumed that the longitudinal wave velocity $V_P = 6$ km/s, the shear wave velocity $V_S = V_P /1.73$, the rupture velocity $V_c = 0.8 V_S$, and the rigidity $\mu = 30\,000$ MPa. Under these assumptions, and assuming that $\Gamma/\chi_c = 3.3$ in Eq. (3.29), we have $k = 2.9$ for an in-plane shear crack (mode II), and $k = 3.5$ for an anti-plane shear crack (mode III). With this in mind, if $k = 3$ is assumed for a circular crack model employed by Ellsworth and Beroza, then D_c for those earthquakes can be calculated from Eq. (5.8).

Although L_c for the Kobe earthquake was not estimated by Ide and Takeo (1997), it can be inferred as follows (Ohnaka, 2000). If we assume that at the hypocenter, which was determined to be 16 km deep, the nucleation reached the critical stage $L = L_c$ (beyond which the rupture propagated at a high speed), then we obtain $L_c = 1700$ m from Eq. (5.8), using the values $\Delta \tau_b = 3$ MPa and $D_c = 0.5$ m estimated by Ide and Takeo (1997). They suggested that 0.5 m is the upper limit of D_c at a deeper part of the fault where the nucleation must have reached the critical stage. In this case, 1700 m would be the upper limit of L_c for the Kobe earthquake. If $\Delta \tau_b = 3$ MPa and, for example, $D_c = 0.3$ m are assumed, then we have $L_c = 1000$ m. These suggest that L_c for the Kobe earthquake was of the order of 1×10^3 m.

Specific values for the parameters $\Delta \tau_b$, D_c, and X_c estimated for moderate to large earthquakes by Papageorgiou and Aki (1983b) are also useful for the present purpose, given that the breakdown zone size X_c is of the order of the critical nucleation zone size L_c (half-length). (Note that $L_c = X_c$ in the present nucleation model.) Hence, their data are also plotted for comparison in the figures shown below.

Figure 5.20 shows a plot of the logarithm of M_0 against the logarithm of $\Delta \tau_b$ for earthquakes. In this figure, white squares denote data points taken from Papageorgiou and Aki (1983b), black circles from Ellsworth and Beroza (1995), and black rectangles from Ide and Takeo (1997). It is found from Figure 5.20 that $\Delta \tau_b$ is independent of M_0, though $\Delta \tau_b$ fluctuates considerably between 1 and 100 MPa. In other words, the breakdown stress drop does not depend on the size of the earthquake. This justifies the premise that the breakdown stress drop is scale-independent. It is seen from Figure 5.20 that the average value for $\Delta \tau_b$ is roughly 10 MPa. This average value has been used in the previous subsection in order to derive the simplified, scaling relation (or Eq. (5.32)) between M_0 and D_c from Eq. (5.30), and the simplified, scaling relation (or Eq. (5.33)) between M_0 and $2L_c$ from Eq. (5.31).

Figure 5.21 shows a plot of the logarithm of the mainshock seismic moment M_0 against the logarithm of the breakdown displacement D_c for earthquake data taken from Papageorgiou and Aki (1983b), Ellsworth and Beroza (1995), and Ide and Takeo (1997). From Figure 5.21, we can see that there is a distinct trend among the plotted data points that M_0 increases with an increase in D_c, although there is considerable fluctuation. This empirical trend is compared with the theoretical scaling relation expressed by Eq. (5.32) in Figure 5.21. The

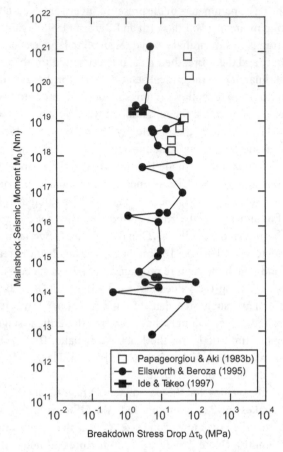

Fig. 5.20 A plot of the seismic moment M_0 against the breakdown stress drop $\triangle\tau_b$ for earthquakes. Reproduced from Ohnaka (2000).

thick straight line in Figure 5.21 indicates the theoretical scaling relation expressed by Eq. (5.32). From Figure 5.21, we can see that the theoretical scaling relation agrees well with the empirical trend in the plotted data points.

Figure 5.22 shows a plot of the logarithm of the mainshock seismic moment M_0 against the logarithm of the critical size $2L_c$ of the nucleation zone for earthquakes taken from Ellsworth and Beroza (1995), and Ide and Takeo (1997). As mentioned above, M_0 for earthquakes analyzed by Papageorgiou and Aki (1983b) has also been over-plotted in Figure 5.22 against $2L_c$ calculated from Eq. (5.8). Figure 5.22 clearly shows that the mainshock seismic moment scales with the critical size of the nucleation zone. This scaling relation was found empirically by Ellsworth and Beroza (1995) and Beroza and Ellsworth (1996). The thick straight line in Figure 5.22 represents the theoretical scaling relation derived in the previous subsection (that is, Eq. (5.33)). From Figure 5.22, we find that there is a good agreement between the theoretical prediction and seismological data on the nucleation. The agreement of both Eqs. (5.32) and (5.33) with seismological data

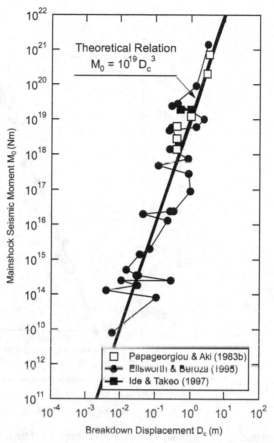

Fig. 5.21 A plot of the seismic moment M_0 against the breakdown displacement D_c for earthquakes. The simplified theoretical scaling relation (Eq. (5.32)) denoted by the thick line in the figure is compared with seismological data. Reproduced from Ohnaka (2000).

on the nucleation suggests that the assumption of $S_{A1}/(aS) = 2/3$ is not unreasonable for typical earthquakes.

From Eq. (5.33), we have, for example, $2L_c = 10$ km for earthquakes with $M_0 = 10^{21}$ Nm, $2L_c = 1$ km for earthquakes with $M_0 = 10^{18}$ Nm, and $2L_c = 100$ m for earthquakes with $M_0 = 10^{15}$ Nm. These are estimates for the nucleation zone size for "normal" earthquakes having an average value of 10 MPa for $\Delta\tau_b$. In reality, however, individual earthquakes do not necessarily have the average value but have different values for $\Delta\tau_b$ (see Figure 5.20). Though $\Delta\tau_b$ is scale-independent, this does not mean that $\Delta\tau_b$ has a constant value (such as 10 MPa) for any earthquake. If a strong area highly resistant to rupture growth (with a strength possibly equal to the shear strength of intact rock mass) is broken down, then $\Delta\tau_b$ will necessarily have a high value of the order of 10 to 100 MPa, depending upon the individual, seismogenic environments (see Chapter 3). By contrast, if frictional slip failure occurs on a preexisting fault having very low resistance to rupture growth, $\Delta\tau_b$ may have

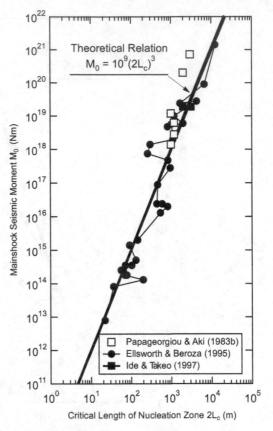

Theoretical Relation
$$M_0 = 10^9 (2L_c)^3$$

Papageorgiou & Aki (1983b)
Ellsworth & Beroza (1995)
Ide & Takeo (1997)

Fig. 5.22 A plot of the seismic moment M_0 against the critical length $2L_c$ of the seismic nucleation zone for earthquakes. The simplified theoretical scaling relation (Eq. (5.33)) denoted by the thick line in the figure is compared with seismological data. Reproduced from Ohnaka (2000).

a low value of the order of 1 MPa or less (see also Chapter 3). It thus follows that $\Delta\tau_b$ can have any value in a wide range from 1 to 100 MPa, depending upon the seismogenic environment and fault zone properties.

Equation (5.8) indicates that L_c is directly proportional to D_c if $\Delta\tau_b$ is constant. However, the linear relationship between L_c and D_c, in general, is severely violated by a fluctuation of $\Delta\tau_b$. Figure 5.23 shows how the linear relationship between L_c and D_c is violated by the fluctuation of $\Delta\tau_b$ for actual earthquakes. The four straight lines in Figure 5.23 represent theoretical linear relations between L_c and D_c, expressed by Eq. (5.8) under the assumption that $\Delta\tau_b$ has a different, constant value of either 0.1 or 1 or 10 or 100 MPa. Thus, in order to properly evaluate L_c for real earthquakes, we have to keep in mind that L_c depends on not only D_c but also $\Delta\tau_b$, as Eq. (5.8) indicates.

Equations (5.30) and (5.31) have been derived under the assumption that the interaction between "asperities" on a fault is negligible. If, however, the interaction is not negligible, the scaling relation between, for example, M_0 and $2L_c$ may no longer hold. Let us consider

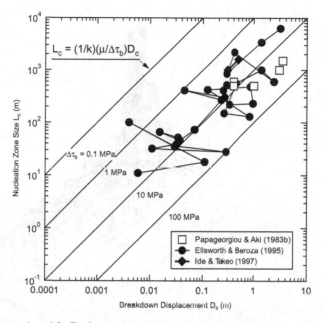

Fig. 5.23 Scaling relation between L_c and D_c. The four straight lines in the figure represent theoretical scaling relations between l_c and D_c with $\triangle\tau_b$ having a value of either 0.1 or 1 or 10 or 100 MPa. The scaling relation between L_c and D_c is violated by the fluctuation of $\triangle\tau_b$ for actual earthquakes. Reproduced from Ohnaka (2000).

a case where the amount of the elastic strain energy released by the breakdown of a small "asperity" (for example, A2 in Figure 5.18) is adequate enough to break down a neighboring large "asperity" (A1 in Figure 5.18). In this case, the size of the ensuing mainshock earthquake will be determined by the amount of the elastic strain energy released by the breakdown of this large "asperity" (A1 in Figure 5.18). On the other hand, the nucleation zone size of the mainshock earthquake is prescribed by the size of the small "asperity" (A2 in Figure 5.18) which has been initially broken down. Hence, the eventual size of the mainshock earthquake may not necessarily scale with its nucleation zone size, when "asperity"–"asperity" interaction is not negligible. Such a case may be an earthquake of multiple shock type, to which the physical scaling relations (5.32) and (5.33) are unlikely to be applicable.

We will describe in Chapter 6 the duration of the nucleation for typical earthquakes.

5.2.4 Foreshock activity associated with the nucleation process

Jones and Molner (1979) and Scholz (1988) pointed out that typical foreshocks occurring within about 10 days before the occurrence of a mainshock tend to concentrate close to the mainshock hypocenter. Such foreshocks may be called *immediate foreshocks*. Using the Japan Meteorological Agency (JMA) earthquake catalogue data from 1977 through 1997, Maeda (1999) examined temporal and spatial distributions of representative foreshocks as functions of the time from the mainshock origin time and of the distance from the

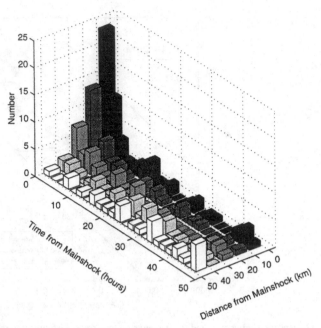

Fig. 5.24 The stacked space-time distribution of the largest foreshocks preceding individual mainshocks that had occurred in and around Japan (Maeda, 1999).

mainshock hypocenter location. Since it is often the case that only a few foreshocks precede a mainshock, he selected the largest foreshock as a representative of a sequence of foreshocks for an individual mainshock, and showed the stacked space-time distribution of the largest foreshocks preceding individual mainshocks (Figure 5.24). Figure 5.24 indicates that foreshocks occurring more immediately before the mainshock occurrence, concentrate more in the vicinity of the mainshock hypocenter. This prominent feature can be well explained by a model of the rupture nucleation put forward based on laboratory experiments.

Microseismicity is activated during the nucleation process in the brittle regime, when the fault surfaces have geometric irregularities (or roughnesses) with a fractal nature. During the nucleation process, slip failure deformation concentrates and accelerates in the localized zone of nucleation. In the nucleation zone, shear strength degrades with ongoing slip, and this ongoing slip causes the fracture of projection portions (or micro-asperities) preferentially in and around the zone on the geometrically irregular fault surfaces during the nucleation process. Therefore, the energy is exclusively expended in the nucleation zone, causing activated microseismicity during the nucleation process. In fact, laboratory experiments have demonstrated that a great number of microseismic (or acoustic emission) activities are induced during the quasi-static nucleation process, prior to the macroscopic shear failure of intact rock (Lockner *et al.*, 1991, 1992; Sammonds *et al.*, 1992), or prior to the dynamic instability of a stick-slip failure on a precut fault with surfaces having fractal roughnesses (Ohnaka *et al.*, 1994; Ohnaka, 1995b; Sammonds and Ohnaka, 1998). Such a typical example is shown in Figure 5.25 (see also Figure 7.29). We can clearly see

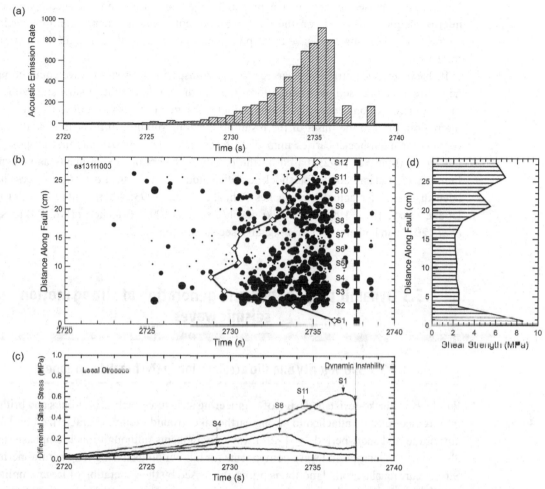

Fig. 5.25 Time series variation of acoustic emission activities induced during the quasi-static nucleation process, prior to the dynamic instability of a stick-slip rupture on a precut fault having surfaces with fractal roughnesses. (a) A plot of the number of acoustic emissions per unit time against time, (b) Spatio-temporal distribution of acoustic emission sources (black dots). The black dot sizes are different according to the maximum amplitudes of individual acoustic emissions. The white squares connected with solid lines indicate the bi-directional growth of the nucleation zone with time. (c) Time series plots of local differential shear stresses at four positions (S1, S4, S8, and S11) along the fault. (d) Precut fault strength distribution. The non-uniform distribution of local shear strength along the fault is caused by the geometric irregularity of the fault surfaces. Unpublished data from Ohnaka *et al.* (1994) are plotted.

from Figure 5.25 that microseismicity is really induced and activated during the nucleation process of stick-slip failure.

During the nucleation process of a typical large earthquake, a zone of a few kilometers or more in dimension (referred to as the nucleation zone) on the seismogenic fault necessarily slips, initially at a very slow and steady rate and subsequently at accelerating rates, over an amount of the order of 1 m, prior to the mainshock occurrence.

In general, whether or not such a preslip in the nucleation zone is accompanied by microearthquakes depends on the fault zone structure and ambient conditions such as temperature. Therefore, we have to keep in mind that some earthquakes do not carry any foreshock.

If, however, geometric irregularities (or micro-asperities) of short wavelength components are superimposed on a nonuniform fault in the brittle regime, a slow slip failure in the "asperity" region necessarily brings about the fracture of micro-asperities in and around the region, prior to the onset of the resulting dynamic rupture. In this case, the dynamic rupture (or mainshock) carries immediate foreshocks. This model was first proposed in the 1990s (Ohnaka, 1992), based on the results of laboratory experiments as shown in Figure 5.25, and immediate foreshock activities induced during the nucleation process have been discussed for actual earthquakes (Ohnaka, 1992, 1993; Abercrombie *et al.*, 1995; Dodge *et al.*, 1995, 1996; Shibazaki and Matsu'ura, 1995; Bouchon *et al.*, 2011; Kato *et al.*, 2012; and others). For more details, see subsection 7.2.2.

5.3 Dynamic propagation and generation of strong motion seismic waves

5.3.1 Slip velocity and slip acceleration in the breakdown zone

For the earthquake-resistant design of engineering structures such as buildings and bridges, importance must be attached to strong earthquake ground motion characterized by large-amplitude and short-period ($\leq$ 1–2 s) wave components, although much emphasis must also be put on long-period wave components for tall buildings having long natural periods. Strong earthquake ground motion is primarily caused by the generation of large-amplitude and short-period seismic waves at the earthquake source; however, the properties of seismic waves on the Earth's surface can in general be affected by the effect of the propagation path from the source to the site and by local site response effects, because seismic waves are attenuated by the path effect, and may be attenuated or amplified by the site effects. Near-surface geological site conditions can have a significant impact on ground shaking, and earthquake damage can be increased by the ground motion amplification, due to soft soil and unconsolidated sediments near the ground surface.

In this section, however, we will exclusively address the physical mechanism of large-amplitude and short-period (or high-frequency) seismic wave generation at the source, from the standpoint of fracture mechanics, based on significant results regarding the generating mechanism of elastic wave radiation obtained in laboratory experiments.

The generating mechanism of strong motion seismic waves during dynamic rupture propagation on a fault is related to the inhomogeneity of the faulting process. Strong motion seismic waves are radiated not from the whole area of a dynamically expanding fault but from the localized zone behind the front of a dynamically propagating rupture on the fault. The localized zone behind the rupture front is called the breakdown zone, and characterized as the zone in which stresses, slip displacement, slip acceleration, and slip velocity vary greatly, as will be shown later in Figure 5.27

The breakdown process is governed by constitutive law, and the properties of elastic waves generated during the breakdown process are strongly influenced by the physical parameters dictating the constitutive law. Such physical parameters are the breakdown displacement and the breakdown stress drop. The breakdown displacement D_c is defined as the critical slip displacement required for the shear strength to degrade to a residual friction stress level, and is directly related to the breakdown zone size X_c (see Eq. (3.28)). The breakdown stress drop $\Delta\tau_b$ is defined as the peak shear strength τ_p minus the residual friction stress τ_r in the breakdown zone. These physical quantities are intrinsically linked to physical, mechanical, and geometric properties of the breakdown zone, and are responsible for generating strong motion seismic waves, as will be shown later.

The physically reasonable constitutive relation for shear rupture must result in the non-singularity of not only stresses but also slip acceleration at or near the front of the dynamically propagating rupture. This is crucial when strong motion source parameters such as peak slip acceleration during unstable, dynamic shear-rupture propagation are discussed in terms of the underlying physics. Note, therefore, that stricter formulation of the constitutive law is crucially important. For example, a simplified slip-weakening model as shown in Figure 2.14, in which shear traction τ on the rupturing surfaces decreases linearly with ongoing slip D from its peak value τ_p at $D = 0$ to the residual friction stress level τ_r, inevitably leads to slip acceleration singularity of the type $r^{-1/2}$ at the distance r from the tip of the breakdown zone along the fault plane (Ida, 1973). The unbounded acceleration at the rupture front is physically unreasonable, and therefore such a simplified model is unsuitable when we discuss the generating mechanism of strong motion elastic waves at the source.

The driving force of rupture extension is supplied by the release of the elastic strain energy stored in the medium surrounding the breakdown zone, and the breakdown process in the zone is governed by constitutive law. Thus, if a rationally formulated constitutive relation is specifically given, then spatial distributions of not only shear stress and slip displacement but also slip velocity and slip acceleration in the breakdown zone behind the rupture front can be uniquely determined, and obtained from theoretical and numerical calculations (Ida, 1972, 1973; Ohnaka and Yamashita, 1989). Let us specifically address this below.

The slip velocity and acceleration are the first and second partial time derivatives of $D(x)$ in the coordinate system fixed to the medium. If we use the same coordinate system adopted in subsection 2.2.3, the slip velocity and acceleration are expressed as

$$\frac{\partial}{\partial t}D(x - Vt) = \frac{V}{x_{\max}}D_c\phi'\left(\frac{Vt - x}{x_{\max}}\right) \tag{5.34}$$

and

$$\frac{\partial^2}{\partial t^2}D(x - Vt) = \left(\frac{V}{x_{\max}}\right)^2 D_c\phi''\left(\frac{Vt - x}{x_{\max}}\right), \tag{5.35}$$

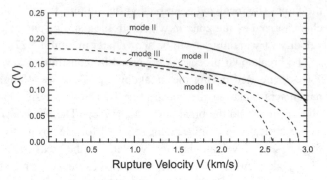

Fig. 5.26 A plot of $C(V)$ against the rupture velocity V (Ohnaka and Matsu'ura, 2002). The solid lines indicate $C(V)$ calculated under the assumption that $V_P = 6\,\mathrm{km/s}$ and $V_S = V_P/1.73$, and the broken lines indicate $C(V)$ calculated under the assumption that $V_P = 4.4\,\mathrm{km/s}$ and $V_S = 2.9\,\mathrm{km/s}$.

respectively, where V is the rupture velocity, $\phi((Vt - x)/x_{\mathrm{max}})$ is the dimensionless slip displacement defined as $\phi = D/D_c$, x_{max} is the parameter defined in subsection 2.2.3, and the prime denotes the derivative with respect to the argument.

The condition that the shear stress τ is finite at the crack-tip results in the following relational expression (Ohnaka and Yamashita, 1989):

$$x_{\mathrm{max}} = \frac{\pi^2 \mu C(V) D_c}{\Gamma \tau_p}, \tag{5.36}$$

where Γ is the dimensionless parameter defined by Eq. (2.33) in subsection 2.2.3, and $C(V)$ is a function of V, specifically expressed as (see subsection 2.2.3)

$$C(V) = \frac{2V_S^2}{\pi V^2} \left[\left(1 - V^2/V_P^2\right)^{1/2} - \frac{\left(1 - V^2/2V_S^2\right)^2}{\left(1 - V^2/V_S^2\right)^{1/2}} \right] \tag{5.37}$$

for an in-plane shear rupture (mode II), and

$$C(V) = \frac{1}{2\pi} \left(1 - \frac{V^2}{V_S^2}\right)^{1/2} \tag{5.38}$$

for an anti-plane shear rupture (mode III). Figure 5.26 shows specifically how $C(V)$ depends on the rupture velocity V. In this figure, the solid lines indicate $C(V)$ calculated under the assumption that $V_P = 6$ km/s and $V_S = V_P/1.73$, and the broken lines indicate $C(V)$ calculated under the assumption that $V_P = 4.4$ km/s and $V_S = 2.9$ km/s. From Figure 5.26, one can see to what extent $C(V)$ is affected by the rupture velocity V.

Given the relation (5.36) and that $\tau_r = 0$ has been assumed for the present calculation (see subsection 2.2.3), the slip velocity and acceleration are rewritten in terms of the well-defined physical parameters as follows (Ida, 1973; Ohnaka and Yamashita, 1989):

$$\frac{\partial}{\partial t} D(x - Vt) = \frac{\Gamma V}{\pi^2 C(V)} \frac{\Delta \tau_b}{\mu} \phi' \tag{5.39}$$

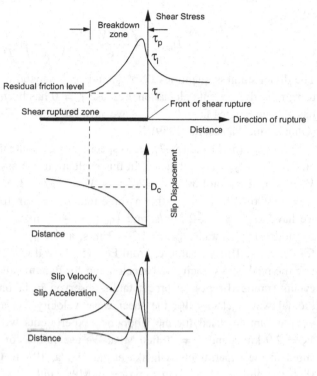

Fig. 5.27   Spatial distributions of shear stress, slip displacement, slip velocity, and slip acceleration in the breakdown zone behind the front of shear rupture (Ohnaka and Matsu'ura, 2002).

and

$$\frac{\partial^2}{\partial t^2} D(x - Vt) = \left(\frac{\Gamma V}{\pi^2 C(V)}\right)^2 \left(\frac{\Delta \tau_b}{\mu}\right)^2 \frac{1}{D_c} \phi'', \tag{5.40}$$

respectively. In the above equations, ϕ' represents the dimensionless slip velocity, and ϕ'' represents the dimensionless slip acceleration. Given that Γ and μ are invariable, we notice from Eqs. (5.39) and (5.40) that the slip velocity depends on physical quantities V and $\Delta \tau_b$, and that the slip acceleration depends on physical quantities V, $\Delta \tau_b$, and D_c.

Figure 5.27 illustrates spatial distributions of not only shear stress and slip displacement but also slip velocity and slip acceleration in the breakdown zone on a fault. It has been demonstrated that such distributions can be derived theoretically if a laboratory-derived constitutive relation as shown in Figure 3.22 is adopted. From Figure 5.27, one can see that slip velocity and slip acceleration attain their peak values within the breakdown zone. In particular, it may be worthwhile to note that it is invalid to assume that the shear stress and the slip acceleration attain their peak values at the rupture front.

From Eqs. (5.39) and (5.40), one can see that the peak slip velocity $\dot{D}_{\max}$ and the peak slip acceleration $\ddot{D}_{\max}$ are written as follows:

$$\dot{D}_{\max} = \frac{\Gamma V}{\pi^2 C(V)} \frac{\Delta \tau_b}{\mu} \phi'_{\max} \tag{5.41}$$

and

$$\ddot{D}_{max} = \left(\frac{\Gamma V}{\pi^2 C(V)}\right)^2 \left(\frac{\Delta\tau_b}{\mu}\right)^2 \frac{1}{D_c} \phi''_{max}. \tag{5.42}$$

The dimensionless quantities Γ, ϕ'_{max}, and ϕ''_{max} depend on $(\tau_i - \tau_r)/\Delta\tau_b$, which is equal to τ_i/τ_p in the present calculation, because $\tau_r = 0$ has been assumed. When $(\tau_i - \tau_r)/\Delta\tau_b$ has a value ranging from 0.5 to 0.8, we have $\Gamma\phi'_{max} = 1.37\text{–}2.75$, and $\Gamma^2\phi''_{max} = 13.7\text{–}37.2$ (Ohnaka and Yamashita, 1989).

From Eqs. (5.41) and (5.42), let us specifically evaluate the peak slip velocity and acceleration for a typical earthquake. In this evaluation, we assume typical values for V_P, V_S, V, $\Delta\tau_b$, and μ as follows: $V_P = 6$ km/s, $V_S = V_P/1.73$, $V = 0.8V_S$, $\Delta\tau_b = 10$ MPa, and $\mu = 30\,000$ MPa. If we further assume that $D_c = 1$ m for the typical large earthquake, we have $\dot{D}_{max} = 1.1\text{–}2.2$ m/s, and $\ddot{D}_{max} = 9\text{–}25$ m/s^2 = 900–2500 gal for the rupture of mode II type, and $\dot{D}_{max} = 1.3\text{–}2.7$ m/s, and $\ddot{D}_{max} = 13\text{–}36$ m/s^2 = 1300–3600 gal for the mode III type rupture. From Eqs. (5.41) and (5.42), for reference, we also calculate the peak slip velocity and acceleration for dynamically propagating stick-slip failure events (mode II type) on precut faults observed in laboratory experiments. The longitudinal wave velocity V_P, the shear wave velocity V_S, and the rigidity μ of Tsukuba granite samples used for the laboratory experiments were as follows: $V_P = 4.4$ km/s, $V_S = 2.9$ km/s, and $\mu = 20\,000$ MPa. We use values for V, $\Delta\tau_b$, and D_c actually measured in the experiments (Ohnaka et al., 1987a, 1987b; Ohnaka and Kuwahara, 1990). If, for example, $V = 1$ km/s, $\Delta\tau_b = 1$ MPa, and $D_c = 3$ μm are assumed, we have $\dot{D}_{max} = (4.1\text{–}8.3) \times 10^{-2}$ m/s, and $\ddot{D}_{max} = (4.1\text{–}11.2) \times 10^3$ m/s^2 = $(4.1\text{–}11.2) \times 10^5$ gal. If $V = 2$ km/s, $\Delta\tau_b = 1$ MPa, and $D_c = 3$ μm are assumed, then we have $\dot{D}_{max} = (11.5\text{–}23.5) \times 10^{-2}$ m/s, and $\ddot{D}_{max} = (23.2\text{–}130.4) \times 10^3$ m/s^2 = $(23.2\text{–}130.4) \times 10^5$ gal. These calculated values are consistent with the values observed in the laboratory experiments (see Figure 3.21). Thus, it is worth noting that $\ddot{D}_{max}$ for dynamically propagating stick-slip rupture events observed in the laboratory is remarkably high in comparison to $\ddot{D}_{max}$ for typical earthquakes, although $\dot{D}_{max}$ for dynamically propagating stick-slip rupture events is lower than $\dot{D}_{max}$ for typical earthquakes. This will be discussed in more detail below.

Equation (5.41) can be rewritten as

$$\frac{\dot{D}_{max}}{V} = \frac{\Gamma\phi'_{max}}{\pi^2 C(V)} \frac{\Delta\tau_b}{\mu}. \tag{5.43}$$

If we assume, for example, $V_P = 6$ km/s, $V_S = V_P/1.73$, and $V = (0.2\text{–}0.8) \times V_S$, then we have $\Gamma\phi'_{max}/[\pi^2 C(V)] = 0.7\text{–}2.9$. This indicates that $\Gamma\phi'_{max}/[\pi^2 C(V)]$ is of the order of unity. Thus, we have from Eq. (5.43)

$$\frac{\dot{D}_{max}}{V} \approx \frac{\Delta\tau_b}{\mu}. \tag{5.44}$$

This theoretical relationship indicates that the peak slip velocity normalized to the rupture velocity is of the order of the breakdown stress drop divided by the rigidity, which has been obtained empirically from laboratory experiments on dynamically propagating stick-slip ruptures on a precut fault (see Figure 3.20).

From Eqs. (5.41) and (5.42), we have

$$\ddot{D}_{\max} = \frac{h}{D_c}(\dot{D}_{\max})^2, \tag{5.45}$$

where

$$h = \phi''_{\max}/(\phi'_{\max})^2. \tag{5.46}$$

The dimensionless parameter h slightly depends on $(\tau_i - \tau_r)/\Delta\tau_b$; that is, h has a value in the range 4.9–7.2 according to the value of $(\tau_i - \tau_r)/\Delta\tau_b$ in the range 0.5–0.8 (Ohnaka and Yamashita, 1989). If such a slight change may be regarded as virtually constant, we can see from Eq. (5.45) that the peak slip acceleration $\ddot{D}_{\max}$ is a function of the peak slip velocity $\dot{D}_{\max}$ and the breakdown displacement D_c.

Figure 5.28 shows a plot of the logarithm of the peak slip acceleration $\ddot{D}_{\max}$ against the logarithm of the peak slip velocity $\dot{D}_{\max}$ (Ohnaka, 2003). Solid lines in the figure represent specific theoretical relationships between $\ddot{D}_{\max}$ and $\dot{D}_{\max}$, where $h = 5$ has been assumed, and where an amount of D_c has been assumed to be 1 μm, 10 μm, 1 mm, 10 cm, 1 m, or 10 m. The black dots in the figure denote data points obtained in the laboratory experiments on dynamically propagating stick-slip rupture events (mode II type) on precut faults, shown in Figure 3.21. From Figure 5.28, one can see that the peak slip acceleration for stick-slip rupture events of laboratory-scale is very high, ranging from 10^2 to 10^5 m/s^2, while the peak slip velocity ranges from 1 to 40 cm/s. D_c for these laboratory events is independently measurable, and has been found to be very small, ranging from 1 to 18 μm (Ohnaka et al., 1987a, 1987b; Ohnaka, 2003). If we compare theoretical Eq. (5.45) to empirical Eq. (3.10), we have $D_c = 10^{-6}h$ m $= (4.9 \sim 7.2) \times 10^{-6}$ m $= 4.9$–7.2 μm. Figure 5.28 corroborates that the theoretically derived formula (5.45) explains the experimental data on dynamically propagating stick-slip ruptures in quantitative terms.

On the other hand, near-source strong motion records for typical major-to-great earthquakes have shown that the peak slip acceleration has a value in the range of 5 to 50 m/s^2, and that the peak slip velocity has a value in the range of 1 to 10 m/s. These ranges in the peak slip acceleration and velocity for typical earthquakes are depicted as the shaded portion in Figure 5.28. From this figure, we notice that the peak slip velocity for dynamically propagating stick-slip rupture events occurring on precut faults of laboratory-scale is lower than that for typical earthquakes occurring on faults of field-scale. In contrast, we notice that the peak slip acceleration for dynamically propagating stick-slip rupture events on faults of laboratory-scale is remarkably high in comparison to that for typical earthquakes on faults of field-scale. This fact can be accounted for quantitatively only by the scale-dependence of D_c. The breakdown displacement D_c for stick-slip rupture events observed in laboratory experiments is remarkably small in comparison to that for typical earthquakes. As Eq. (5.40) or (5.42) exhibits explicitly, the slip acceleration is inversely proportional to the breakdown displacement D_c. Since D_c is scale-dependent, the slip acceleration inversely scales with the breakdown displacement D_c. For details of the scale-dependence of scale-dependent physical quantities inherent in the shear rupture, see Chapter 6.

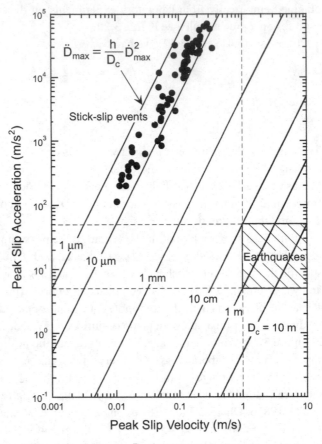

Fig. 5.28 A plot of the logarithm of the peak slip acceleration $\ddot{D}_{max}$ against the logarithm of the peak slip velocity $\dot{D}_{max}$ (Ohnaka, 2003). The black dots denote laboratory data on frictional stick-slip ruptures that propagate dynamically on precut faults. The solid lines represent theoretical relationships between $\ddot{D}_{max}$ and $\dot{D}_{max}$ under the assumption that $h = 5$, and $D_c = 1\,\mu m$, $10\,\mu m$, 1 mm, 10 cm, 1 m, or 10 m.

The limited ranges in the peak slip acceleration and velocity for typical earthquakes, shown in Figure 5.28, lead to the limited values of D_c for earthquakes, under the prerequisite of theoretical formula (5.45). It is predicted from Figure 5.28 that D_c should be larger than 10 cm for major earthquakes. Figure 5.28 also predicts that proper values of D_c for great earthquakes are of the order of one meter or greater. Otherwise, either $\ddot{D}_{max}$ or $\dot{D}_{max}$ would necessarily be forced to have physically unrealistic values. For example, if we assume that D_c for major earthquakes are as small as the values of D_c ($= 1$–$10\,\mu m$) for stick-slip rupture events observed in the laboratory, and that $\ddot{D}_{max}$ for major earthquakes are of the order of $10\,m/s^2$ ($\approx 1\,g$), then we have from formula (5.45) that $\dot{D}_{max}$ necessarily has values as small as 1 to 10 mm/s, which is unrealistic for real, major earthquakes. If we assume for major earthquakes that D_c has values in the range 1–10 μm, and that $\dot{D}_{max}$ has values in the range 1–10 m/s, then we have from formula (5.45) that $\ddot{D}_{max}$ necessarily has values as large as 10^6–$10^7\,m/s^2$, which is also unrealistic for real, major earthquakes. We can thus conclude

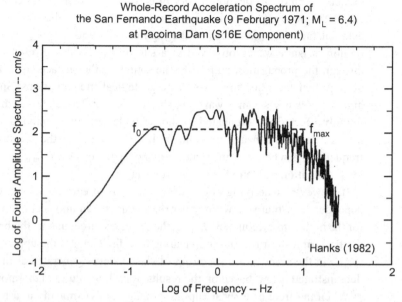

Fig. 5.29 The seismic wave acceleration amplitude spectrum of the 1971 San Fernando earthquake observed at Pacoima Dam in California, USA. Reproduced from Hanks (1982).

that D_c must be scale-dependent, and that D_c for major earthquakes must be larger than 10 cm. This is consistent with the results of D_c estimated for real earthquakes. In fact, the values of D_c estimated for major earthquakes range from several tens of centimeters to one meter (Papageorgiou and Aki, 1983b; Ide and Takeo, 1997; Mikumo *et al.*, 2003; Mikumo and Yagi, 2003; Ruiz and Madariaga, 2011). In addition, the values of D_c estimated for great earthquakes are of the order of 1 m or greater (Papageorgiou and Aki, 1983b).

Thus, D_c for large (major to great) earthquakes is 5–6 orders of slip amount greater than D_c for small-scale stick-slip rupture events in the laboratory. The scale-dependence of D_c has been confirmed in this subsection, independently of the derivation in Chapter 3.

5.3.2 The cutoff frequency f^s_{max} of the power spectral density of slip acceleration at the source

In general, the acceleration amplitude spectrum of real seismic waves observed at a site on the Earth's ground surface is characterized by band-limited flat and decreasing spectral amplitudes at frequencies lower than low cutoff frequency f_0 and at frequencies higher than high cutoff frequency f_{max}, as exemplified in Figure 5.29. The cutoff frequency f_0 is attributed to the source size $\tilde{L}$, and directly proportional to $V_S/\tilde{L}$ (e.g., Brune, 1970; Madariaga, 1976; Papageorgiou and Aki, 1983a). For example, for a circular fault model, $\tilde{L}$ represents the radius of the circular fault.

In contrast, the cause of the high cutoff frequency f_{max} has been discussed by seismologists (e.g., Hanks, 1982; Papageorgiou and Aki, 1983a). In the case of classic fault models based on dislocation theory of earthquakes, there is no causal factor for the high cutoff frequency f_{max} in the seismic source; therefore, there is no choice but to expect external factors as the cause of the high cutoff frequency f_{max}. As noted earlier in this section, seismic waves observed at a site on the Earth's ground surface are attenuated through the propagation path from the source to the surface site, and may be attenuated or amplified according to near-surface geological site conditions. Since the wavelength of higher-frequency seismic waves is shorter, higher-frequency seismic waves can be profoundly affected and attenuated by shorter-scale structures included in the propagation path medium and in the near-surface layer. If this is the case, the spectral amplitude at higher frequencies can be more attenuated, and hence the cause of f_{max} can be narrowed down to the propagation path and the near-surface layer.

If, however, underlying physical fault models of earthquake sources based on the slip-dependent constitutive law for shear rupture are taken into account, there is a causal factor for the high cutoff frequency f_{max} in the seismic source; that is, the breakdown zone behind the front of a shear rupture propagating on a fault causes the upper-limit cutoff frequency (f_{max}^s) in the acceleration spectrum of elastic waves generated at the source. This will be demonstrated below, based on the results of high-resolution laboratory experiments.

When the front of a shear rupture propagates dynamically at a high speed on a preexisting fault, how can the signals of shear stress, slip acceleration, slip velocity, and slip displacement recorded at a position in the close vicinity of the rupture front be graphically expressed? It is very difficult (or almost impossible) to observe all the signals of shear stress, slip acceleration, slip velocity, and slip displacement for a real earthquake, at one location very close to the source; indeed, these signals for a natural earthquake have not been observed concurrently at one location in the close vicinity of the fault. Hence, Figure 5.30 shows a typical example of those signals recorded at one position in the vicinity of the rupture front for a dynamically propagating stick-slip rupture (mode II type crack growth) generated in the laboratory.

Figure 5.30 specifically indicates the signals of (a) dynamic shear stress, (b) slip acceleration, (c) slip velocity, and (d) slip displacement recorded at one position, just 5 mm distant from the fault plane, for a dynamically propagating stick-slip rupture generated under the normal stress of 6.48 MPa (Ohnaka and Yamashita, 1989). The slip velocity and the slip acceleration shown in Figure 5.30 are the time derivative of the slip displacement and the time derivative of the slip velocity, respectively. In Figure 5.30, T_c indicates the *breakdown time*, and T_w denotes the half-pulse width of slip acceleration. The breakdown time T_c is defined as the time interval required for local breakdown at one position on a fault. T_c can be estimated from the shear stress versus time curve recorded at a position along the fault. The T_c estimated in this way is useful in evaluating strong motion seismic wave properties, as shown below.

The predominant period of acceleration waves near an earthquake fault will be closely related to the pulse width of slip acceleration on the fault plane, and the pulse width of slip acceleration on the fault is expected to be proportional to T_c. Figure 5.31 shows a plot of the breakdown time T_c against the half value width T_w of the slip acceleration pulse for dynamically propagating stick-slip rupture events observed in the laboratory. The experimental data points, plotted as black dots in Figure 5.31, are considerably scattered.

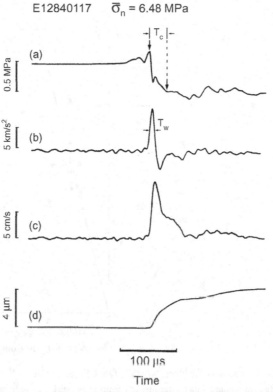

$$E12840117 \qquad \bar{\sigma}_n = 6.48 \text{ MPa}$$

Fig. 5.30 A typical example of the signals of (a) dynamic shear stress, (b) slip acceleration, (c) slip velocity, and (d) slip displacement recorded at one position, 5 mm distant from the fault, for a dynamically propagating stick-slip rupture generated under the normal stress $\sigma_n = 6.48$ MPa (Ohnaka and Yamashita, 1989). The slip velocity and slip acceleration are the time derivative of the slip displacement and the slip velocity, respectively. T_c denotes breakdown time, and T_w denotes half pulse width.

Nevertheless, taking into account that the scattered data points are due to experimental error, one can see from Figure 5.31 that the pulse width of slip acceleration on a fault is directly proportional to the local breakdown time T_c. The broken straight line in the figure denotes the theoretical relationship $T_c = 4.1T_w$ (Ohnaka and Yamashita, 1989).

Figure 5.32 shows a log–log plot of the normalized power spectral density against frequency for the slip acceleration–time record shown in Figure 5.30(b). As shown in this figure, f_{max}^s is defined as the upper-limit cutoff frequency. The cutoff frequency f_{max}^s of the power spectral density of the time series variation of slip acceleration at a position on the fault is related to the local breakdown time T_c at the same position. Figure 5.33 shows the relation between f_{max}^s and $1/T_c$ for experimental data on dynamically propagating stick-slip rupture events on precut faults in the laboratory (Ohnaka and Yamashita, 1989). For the individual stick-slip rupture events observed, T_c was estimated empirically from a local shear stress versus time record at a position along the fault, whereas f_{max}^s was estimated from the upper-limit cutoff frequency of the power spectral density calculated

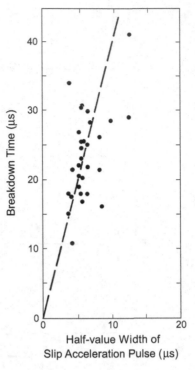

Fig. 5.31 Relationship between the breakdown time T_c and the half-value width T_w of slip acceleration pulse. The black dots denote experimental data on dynamically propagating stick-slip ruptures, and the broken straight line denotes the relationship $T_c = 4.1\,T_w$. Reproduced from Ohnaka and Yamashita (1989).

from the time series variation of local slip acceleration recorded at the same, or nearly the same, position along the fault. Therefore, the data on f_{max}^s and on T_c are mutually independent; nevertheless, there is a good correlation between f_{max}^s and $1/T_c$, as shown in Figure 5.33. We can thus conclude from Figure 5.33 that

$$f_{max}^s = 1/T_c. \tag{5.47}$$

The breakdown time T_c has a physical meaning not only of the time required for the rupture front to be broken down, but also of the time during which the rupture front proceeds by the distance of the breakdown zone length X_c. Therefore, expression (5.47) indicates that the cutoff frequency f_{max}^s of the power spectral density of slip acceleration on the fault is inversely related to the length X_c of the breakdown zone. In other words, expression (5.47) implicitly indicates that the breakdown zone length is ascribed as the cause of the cutoff frequency f_{max}^s of strong motion acceleration waves observed at or near the source.

Given that the breakdown time T_c is equal to the time required for the rupture front to proceed by the length X_c of the breakdown zone at a speed V, T_c can be expressed as

$$T_c = \frac{X_c}{V}, \tag{5.48}$$

E12840117 - CH1. 9SSM. 1 - 2048

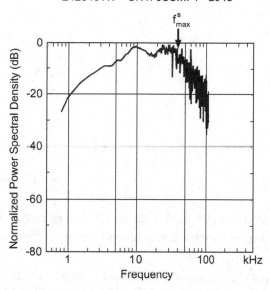

Maximum Power Spectral Density = 9.9953 x 10^{15}
Frequency Resolution = 4.8828 x 10^{2} Hz
Foldover Frequency = 5.0000 x 10^{6} Hz = Point 1024

Fig. 5.32 A log–log plot of the normalized power spectral density against frequency for the slip acceleration–time record shown in Figure 5.30(b). The upper-limit cutoff frequency is denoted by f^s_{max}. Reproduced from Ohnaka and Yamashita (1989).

which is rewritten as (see subsection 3.3.5)

$$T_c = \frac{1}{k}\frac{D_c}{V}\frac{\mu}{\Delta\tau_b} = (0.24 \sim 0.41)\pi^2 C(V)\frac{D_c}{V}\frac{\mu}{\Delta\tau_b},\qquad(5.49)$$

where D_c is the breakdown displacement, $\Delta\tau_b$ is the breakdown stress drop, μ is the rigidity, $C(V)$ is a function of V defined by Eqs. (5.37) and (5.38), and k is the dimensionless quantity defined by Eq. (3.29).

From Eqs. (5.47) and (5.48), we have

$$f^s_{max} = \frac{V}{X_c}.\qquad(5.50)$$

This relation has been postulated by Papageorgiou and Aki (1983a) in the analysis of seismological data to infer earthquake-source parameters such as the breakdown zone size and the breakdown stress drop, based on their own specific barrier model. Equation (5.50) explicitly shows that the cutoff frequency f^s_{max} is attributed to the breakdown zone length X_c.

In general, f^s_{max} is related to f_{max} (obtained from the acceleration spectrum of seismic waves observed at a site on the Earth's ground surface) by (Ohnaka and Matsu'ura, 2002)

$$f_{max} \leq f^s_{max}.\qquad(5.51)$$

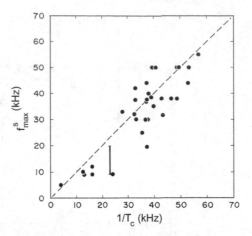

Fig. 5.33   Relationship between f_{max}^s and $1/T_c$ (Ohnaka and Yamashita, 1989). The black dots indicate the data points observed in laboratory experiments on stick-slip ruptures dynamically propagating on precut faults.

If $f_{max} < f_{max}^s$, f_{max} can be attributed to the higher frequency attenuation of seismic waves through the propagation path from the source to the observation site. In contrast to this, if $f_{max} = f_{max}^s$, the breakdown zone behind the rupture front in the source can be ascribed as the cause of f_{max}, as shown above.

Let us derive theoretical relations between $\ddot{D}_{max}$, $\dot{D}_{max}$, and f_{max}^s. The dimensionless parameter χ_c in Eq. (3.29) is originally defined by (Ohnaka and Yamashita, 1989)

$$\chi_c = X_c / x_{max}, \tag{5.52}$$

where X_c is the breakdown zone size, and x_{max} is the parameter defined in subsection 2.2.3. Substituting x_{max} from Eq. (5.52) into Eqs. (5.34) and (5.35), we have

$$\frac{\partial}{\partial t} D(x - Vt) = \chi_c D_c \frac{V}{X_c} \phi' \tag{5.53}$$

and

$$\frac{\partial^2}{\partial t^2} D(x - Vt) = \chi_c^2 D_c \left(\frac{V}{X_c} \right)^2 \phi''. \tag{5.54}$$

Given that $V/X_c = 1/T_c = f_{max}^s$, the peak slip velocity $\dot{D}_{max}$ and the peak slip acceleration $\ddot{D}_{max}$ can be expressed from Eqs. (5.53) and (5.54) as

$$\dot{D}_{max} = \chi_c \phi'_{max} \frac{D_c}{T_c} = \chi_c \phi'_{max} D_c f_{max}^s \tag{5.55}$$

and

$$\ddot{D}_{max} = \chi_c^2 \phi''_{max} \frac{D_c}{T_c^2} = \chi_c^2 \phi''_{max} D_c (f_{max}^s)^2, \tag{5.56}$$

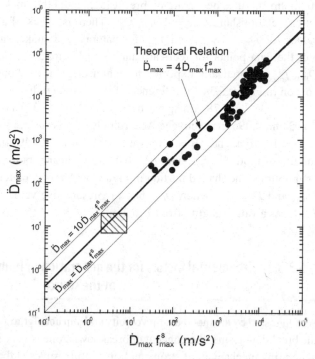

Fig. 5.34 Relationship between $\ddot{D}_{max}$ and $\dot{D}_{max} f^s_{max}$. The theoretical relation is compared with experimental data, denoted by black dots in the figure, on dynamically propagating stick-slip ruptures generated on precut faults. The shaded portion in the figure indicates the ranges in $\ddot{D}_{max}$ and in $\dot{D}_{max} f^s_{max}$ expected for typical earthquakes (see text). Data from Ohnaka and Yamashita (1989).

respectively. From Eqs. (5.55) and (5.56), we finally have

$$\ddot{D}_{max} = \chi_c \left(\frac{\phi''_{max}}{\phi'_{max}} \right) \dot{D}_{max} f^s_{max}. \tag{5.57}$$

Theoretically, the dimensionless quantity $\chi_c(\phi''_{max}/\phi'_{max})$ depends slightly on the parameter $(\tau_i - \tau_r)/\Delta\tau_b$; that is, $\chi_c(\phi''_{max}/\phi'_{max}) = 3.6$–$4.4$ when $(\tau_i - \tau_r)/\Delta\tau_b$ has a value ranging from 0.5 to 0.8 (Ohnaka and Yamashita, 1989). Practically, however, the quantity can be regarded as virtually constant. Similarly, $\chi_c\phi'_{max}$ has a value ranging from 0.56 to 0.91 when $(\tau_i - \tau_r)/\Delta\tau_b = 0.5$–$0.8$. Thus, Eqs. (5.55) and (5.57) are reduced to

$$\dot{D}_{max} = (0.56 \sim 0.91) D_c f^s_{max} \tag{5.58}$$

and

$$\ddot{D}_{max} = (3.6 \sim 4.4) \dot{D}_{max} f^s_{max}, \tag{5.59}$$

respectively.

Figure 5.34 shows a log–log plot of $\ddot{D}_{max}$ against $\dot{D}_{max} f^s_{max}$. Solid circles in this figure denote data points obtained in laboratory experiments on dynamically propagating stick-slip ruptures occurring on precut faults, and the thick solid line represents the theoretical relationship $\ddot{D}_{\max} = 4\dot{D}_{\max} f^s_{\max}$. There is a lack of a reliable data set of individual values for $\ddot{D}_{\max}$, $\dot{D}_{\max}$, and $f^s_{\max}$ for natural earthquakes, and hence it is not possible to plot the data points for real earthquakes in Figure 5.34. Instead, the ranges in $\ddot{D}_{\max}$ and $\dot{D}_{\max} f^s_{\max}$ theoretically expected for major or great earthquakes are depicted as a shaded portion in Figure 5.34 for reference. If it is assumed that $V_P = 6$ km/s, $V_S = V_P/1.73$, and $V = 0.8\ V_S = 2.78$ km/s, we have from Eqs. (5.37) and (5.38) that $C(2.78\text{km/s}) = 0.115$ and 0.096, respectively. Accordingly, if it is further assumed that $\mu = 30\ 000$ MPa, $\Delta\tau_b = 10$ MPa, and $D_c = 1$ m, we have from Eq. (5.49) that $T_c = 0.30$–0.50 s for shear rupture of mode II type, and 0.24–0.42 s for shear rupture of mode III type. Under these assumptions, the shaded portion in Figure 5.34 was depicted as representing the ranges in $\ddot{D}_{\max}$ and $\dot{D}_{\max} f^s_{\max}$ when $\ddot{D}_{\max}$ has a representative value ranging from 7 to 20 m/s^2 and $\dot{D}_{\max}$ has a value ranging from 1 to 2 m/s.

5.3.3 Environmental factors for the generation of high-frequency strong motion at the source

As suggested by the fact that slip velocity and slip acceleration attain their peak values within the breakdown zone, the concept of breakdown zone is very important in understanding the generating mechanism of strong motion seismic waves at the source, within the framework of fracture mechanics. Indeed, the peak slip velocity depends on the breakdown stress drop $\Delta\tau_b$, and the peak slip acceleration depends on the breakdown stress drop $\Delta\tau_b$ and the breakdown displacement D_c (see subsection 5.3.1). Both $\Delta\tau_b$ and D_c are the physical quantities related to the breakdown zone (or its size X_c). In addition, the upper-limit cutoff frequency $f^s_{\max}$ of the power spectral density of the slip acceleration on a fault is also attributed to the breakdown zone size X_c, as demonstrated in subsection 5.3.2.

The size (or length) X_c of the breakdown zone is directly related to D_c by Eq. (3.28). Let us specifically evaluate X_c from Eqs. (3.28) and (3.29), assuming the following typical values for the longitudinal wave velocity V_P, the shear wave velocity V_S, and the rupture velocity V: $V_P = 6$ km/s, $V_S = V_P/1.73$, and $V = 0.8V_S$. The parameters Γ and χ_c in Eq. (3.29) are dimensionless quantities, and can be regarded as virtually constant. Here, we assume that $\Gamma/\chi_c = 3.3$ (see subsection 3.3.5). Under these assumptions, we have from Eq. (3.29) that $k = 2.9$ for an in-plane shear rupture (mode II), and $k = 3.5$ for an anti-plane shear rupture (mode III). Since we here intend to roughly estimate the size of the breakdown zone, we simply assume $k = 3$ without any distinction of in-plane and anti-plane shear modes. In this case, Eq. (3.28) is reduced to

$$X_c \approx \frac{1}{3}\frac{\mu}{\Delta\tau_b}D_c. \qquad (5.60)$$

As shown in Figure 5.20, $\Delta\tau_b$ for many earthquakes has a value in the range 1–100 MPa, irrespective of their magnitudes (or seismic moments), and the average value of $\Delta\tau_b$ is estimated to be 10 MPa for these earthquakes. If, therefore, it is assumed that $\Delta\tau_b = 10$ MPa for typical earthquakes, and that the rigidity μ of the rock mass surrounding the fault is 30 000 MPa, Eq. (5.60) is further reduced to

$$X_c \approx 10^3 D_c. \tag{5.61}$$

It has been shown previously (Section 3.3, and Section 5.2 and subsection 5.3.1) that D_c is scale-dependent. As shown in Figure 5.21, D_c for small earthquakes has a value of a few centimeters, whereas D_c for large earthquakes has a value of a few meters. Thus, it can be roughly estimated from Eq. (5.61) that X_c is several tens of meters for small earthquakes, and a few kilometers for large earthquakes.

As noted previously, real faults embedded in the seismogenic layer are inherently hetero-geneous, and the faults are composed of local strong areas highly resistant to rupture growth (called "asperities"), with the rest of the faults having low (or little) resistance to rupture growth. Such a local strong "asperity" area highly resistant to rupture growth on a fault is capable of storing a greater amount of the elastic strain energy in the surrounding medium than the amount of the elastic strain energy that can be stored in the medium surrounding the rest of the fault having low (or little) resistance to rupture growth. The driving force for strong earthquake motion can be supplied by the release of the stored strain energy only when such local strong "asperity" areas are broken down.

The breakdown stress drop $\Delta\tau_b$ in local strong "asperity" areas on a fault is greater than that in the rest of the fault area. Accordingly, the peak slip velocity $\dot{D}_{max}$ and the peak slip acceleration $\ddot{D}_{max}$ are expected to be higher in the local strong "asperity" areas than in the rest of the fault area. This has already been confirmed through the analysis of seismological data on actual earthquakes, showing that the amplitude of the slip velocity time function is much larger for the "asperity" area than for the rest of the fault area (Miyake et al., 2003).

Let us examine what causes the generation of short-period strong motion at the source, and under what environmental conditions such strong motion is generated at the source. High slip velocity and acceleration on the fault are required for generating strong motion. Equations (5.41) and (5.42) indicate that a high rupture velocity V and/or large breakdown stress drop $\Delta\tau_b$ are required to produce a high slip velocity $\dot{D}$ and high slip acceleration $\ddot{D}$ on the fault. In addition, Eq. (5.42) indicates that a small breakdown displacement D_c is also effective for high slip acceleration $\ddot{D}$ on the fault. On the other hand, Eq. (5.49) indicates that a high rupture velocity V, a large breakdown stress drop $\Delta\tau_b$, and/or small breakdown displacement D_c are required for generating short-period seismic motion. These consistently lead to the conclusion that a high rupture velocity V, large breakdown stress drop $\Delta\tau_b$, and small breakdown displacement D_c effectively generate short-period strong seismic motion. This means that the breakdown of a strong but small-sized "asperity" area generates short-period strong motion.

Let us direct our attention to the environmental conditions under which high slip velocity $\dot{D}$ and high slip acceleration $\ddot{D}$ are produced on a heterogeneous fault. As noted previously in Chapter 1, the process of a real earthquake rupture occurring on a heterogeneous fault at shallow crustal depths is not a simple process of frictional slip failure on a uniformly precut weak fault, but a more complex process, including the fracture of initially intact rock at local strong areas on the fault. The breakdown of a strong area highly resistant to rupture growth (or "asperity") on a fault releases a larger amount of the stored strain energy than does the

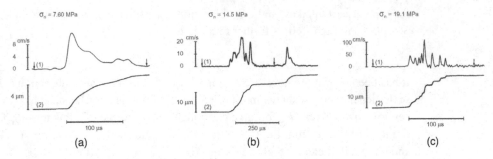

Fig. 5.35 Plots of (1) slip velocity and (2) slip displacement against time, observed locally at one position along a precut fault for stick-slip ruptures dynamically propagating at a steady speed, under the condition that the normal stress: (a) $\sigma_n = 7.6$ MPa, (b) 14.5 MPa, and (c) 19.1 MPa. Reproduced from Ohnaka *et al.* (1986).

rest of the fault having low (or little) resistance to rupture growth; consequently, strong motion is necessarily generated at and around such a strong area highly resistant to rupture growth (or "asperity"), as noted above. The resistance to rupture growth is approximately given by the product of $\Delta\tau_b$ and D_c, and therefore a large amount of elastic strain energy can be accumulated in the medium surrounding an asperity area where both $\Delta\tau_b$ and D_c are large. Accordingly, a large earthquake will be generated at the site of an "asperity" area where both $\Delta\tau_b$ and D_c are large. It would be easy to understand that high-frequency strong motion is generated at the site of an "asperity" area having large $\Delta\tau_b$ and small D_c on a fault.

In the case of a precut fault whose surfaces are geometrically irregular, facing fault-surfaces do not make contact with each other over the entire area, but make contact only at a number of asperities, under an applied compressive normal stress across the fault in the brittle regime. These contact junction areas act as strong areas highly resistant to rupture growth on the precut fault, and become stronger at a higher compressive normal stress. The breakdown stress drop $\Delta\tau_b$ in the contact junction areas becomes greater under a higher normal stress. This suggests that strong motion at the source is generated at a high normal stress. In addition, at a higher compressive normal stress, greater deformation of the asperities that are in contact with the opposing fault-surface causes a greater number of asperities to make contact with the opposing surface. This narrows the spacing between neighboring asperity-contact sites on the fault, which results in the generation of high-frequency strong motion, when a frictional slip (stick-slip) rupture propagates dynamically at a steady speed on the fault. These have been demonstrated in laboratory experiments (Ohnaka *et al.*, 1986), as shown below.

Figure 5.35 shows plots of (1) slip velocity and (2) slip displacement against time, observed locally at one position along a precut fault for frictional stick-slip ruptures dynamically propagating at a steady speed (about a few km/s) on the fault in the laboratory (Ohnaka *et al.*, 1986). These time-series data plotted in Figures 5.35(a), (b), and (c), were obtained under normal stress conditions $\sigma_n = 7.6$, 14.5, and 19.1 MPa, respectively. From Figure 5.35, one can clearly see that the plotted forms of the observed time-series records depend upon the applied normal stress σ_n. The slip displacement–time function

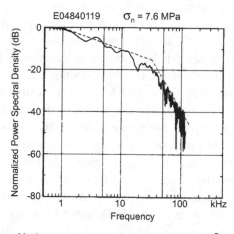

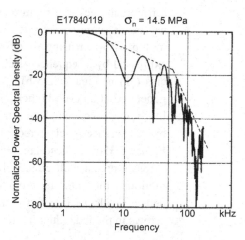

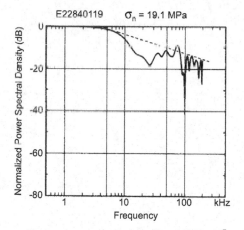

Fig. 5.36 Plots of the normalized power spectral densities calculated from the slip velocity–time curves over the time intervals indicated by arrows in the individual figures for the stick-slip rupture events shown in Figures 5.35(a), (b), and (c), respectively (Ohnaka *et al.*, 1986).

for a frictional stick-slip rupture event generated at a low compressive normal stress ($\sigma_n = 7.6$ MPa) is simple and smooth, and can be represented by a single ramp functional form. Since the slip velocity is the time derivative of the slip displacement–time function, the slip velocity–time curve for the event generated at $\sigma_n = 7.6$ MPa is expressed by a gentle pulse form ((1) in Figure 5.35(a)). In contrast, the slip displacement–time function for a frictional stick-slip rupture event generated at a higher normal stress ($\sigma_n = 14.5$ MPa) exhibits a more complicated ramp functional form, with an irregular stepwise-increase in slip with increasing time. Consequently, the slip velocity–time curve for this event is expressed by a sequence of clustered sharp pulses containing significantly high frequencies ((1) in Figure 5.35(b)). The slip displacement–time function for a stick-slip event generated at an even higher normal stress ($\sigma_n = 19.1$ MPa) more clearly exhibits a step-like ramp functional form, and consequently the slip velocity–time curve is expressed by a sequence of clustered sharper pulses containing much higher frequencies ((1) in Figure 5.35(c)). These laboratory data indicate that a higher compressive normal stress across a precut fault effectively contributes to the generation of higher-frequency strong motion on the fault. This is confirmed by Figures 5.36(a)–(c), which show the power spectral densities calculated from the slip velocity–time curves over the time intervals indicated by arrows in the individual figures for the events shown in Figures 5.35(a)–(c), respectively.

Physical scale-dependence

6.1 Introduction

As described in Chapter 1, fracture (or failure) phenomena are observed over a very broad range of size scales, from atomistic-scale to microscopic-scale to macroscopic-scale fractures. A shear failure (or rupture) of laboratory-scale encompassed by continuum mechanics would be roughly of the order of 10^{-3} to 1 m. In contrast, shear rupture phenomena occurring in the Earth's interior, including microearthquakes and huge earthquakes, encompass a much broader range of size scales from 10^{-1} to 10^6 m. Rupture phenomena over such a broad scale range covering both laboratory-scale and field-scale encompassed by continuum mechanics, are characterized by scale-dependent physical quantities inherent in the rupture.

In general, physical quantities inherent in rupture phenomena can be categorized into two groups: scale-dependent physical quantities, and scale-independent physical quantities. As shown in previous chapters, the scale-dependent quantities include the breakdown zone length X_c, the breakdown time T_c (or its reciprocal f^a_{max}), the nucleation zone length L_c, and the fault-slip acceleration $\ddot{D}$. Thus, it is an unavoidable fact that rupture phenomena are scale-dependent. As noted earlier, therefore, it is essential that the constitutive law must be formulated in such a way that the scaling property inherent in the rupture breakdown is incorporated into the law; otherwise, scale-dependent physical quantities inherent in the rupture over a broad scale range cannot be treated consistently and quantitatively in a unified manner in terms of a single constitutive law.

In this chapter, we focus on how specifically the scaling property is incorporated into the constitutive law (Section 6.2), and what the root cause of the scale-dependence is (Section 6.3). In Chapter 5, it has already been demonstrated that physical scaling relationships hold between the seismic moment M_0 of an earthquake and its nucleation zone length L_c, and between M_0 and D_c. In Section 6.4 of this chapter, we will confirm the theoretically derived scaling relationship between D_c and L_c or X_c, based on field data on real earthquakes and laboratory data on shear rupture (subsection 6.4.1). In addition, we will deal with other scale-dependent physical quantities: the duration time of shear rupture nucleation (subsection 6.4.2), and the apparent shear rupture energy or the resistance to shear rupture growth G_c defined by Eq. (2.44) (subsection 6.4.3). Throughout this chapter, it will be corroborated that the scale-dependence of the scale-dependent quantities is attributed to the scale-dependent breakdown displacement D_c or the characteristic length λ_c.

In the last section of this chapter (Section 6.5), it will be clarified that the b-value of the Gutenberg–Richter frequency–magnitude relation is directly related to fault heterogeneity with a fractal nature in the seismogenic layer.

6.2 Scaling property incorporated into the slip-dependent constitutive law

As noted previously (Section 3.1), the shear rupture of rock in the brittle regime is an inhomogeneous and nonlinear process during which inelastic shear deformation concentrates in a thin zone, resulting in bond-shearing and the release of the shear stress along the shear-rupturing surfaces with ongoing slip displacement. The fault zone (or shear zone) may thus be defined as a thin zone in which the concentration of shear deformation is highly localized, and in which the macroscopic rupture surfaces are eventually formed.

The preexisting fault zone may contain asperities on the fault surfaces, gouge fragments, and/or highly damaged (or high crack density) thin layers consisting of subsidiary, minute cracks developed in the vicinity of the fault surfaces. Interstitial pores in the fault zone may possibly be filled with water, which has the potential to enhance the inelasticity of the fault zone and to lower the fault zone strength. No matter how thin it may be, an actual fault zone formed not only in a heterogeneous rock material but also in the seismogenic layer has its own thickness. The effective fault zone thickness may be defined as the thickness of a highly damaged zone characterized by inelastic deformation, and the outside of the fault zone is primarily characterized by elastic deformation (Ohnaka, 2003).

It is therefore important to note that the constitutive law for shear rupture is formulated in terms of the shear stress acting on both walls of the fault zone thickness and the relative displacement between both walls of the fault zone thickness (see Figure 3.4). As described in Chapter 3, the slip-dependent constitutive law for shear failure can be prescribed in terms of the following parameters: τ_i, τ_p, $\Delta\tau_b$, D_a, and D_c (or D_{wc}). Of these parameters, displacement-related parameters D_a and D_c (or D_{wc}) are scale-dependent, as demonstrated in Section 3.3. In contrast, stress or strength-related parameters τ_i, τ_p, and $\Delta\tau_b$ are scale-independent.

As experimentally and theoretically shown in previous chapters, scale-dependent physical quantities inherent in the rupture are intrinsically related to the breakdown displacement D_c, and the scale-dependence of the scale-dependent physical quantities is directly attributed to the scale-dependence of D_c. A typical example of the scale-dependent physical quantities is the critical length $2L_c$ (L_c, half length) of the nucleation zone, which is expressed in terms of the constitutive law parameters $\Delta\tau_b$ and D_c as (see Eq. (5.8))

$$2L_c = \frac{2}{k}\frac{\mu}{\Delta\tau_b}D_c,$$

which indicates that $2L_c$ directly scales with D_c. Another typical example of the scale-dependent physical quantities would be the slip acceleration $\ddot{D}$ in the breakdown zone on a fault, and $\ddot{D}$ is expressed in terms of $\Delta\tau_b$ and D_c as (see Eq. (5.40))

$$\ddot{D} = \left(\frac{\Gamma V}{\pi^2 C(V)}\right)^2 \left(\frac{\Delta\tau_b}{\mu}\right)^2 \frac{\phi''}{D_c},$$

which indicates that $\ddot{D}$ inversely scales with D_c. In Chapter 5, the scale-dependence of these quantities has been examined in detail from a physical viewpoint, and it has been demonstrated that the scale-dependence of those quantities is attributed to the scale-dependence of D_c.

The fact that scale-dependent physical quantities inherent in the rupture are directly related to D_c is a natural consequence of the physical property of the breakdown process, which is the basic concept needed to account for the transitional zone of strength degradation behind the front of a propagating rupture in terms of fracture mechanics (see also Section 2.2). The breakdown displacement D_c, defined as the critical slip displacement required for the breakdown of the rupture front, is the slip displacement at the end of the breakdown process, and is directly related to the geometric length of the coherent zone of the rupture breakdown (see Eq. (3.28)). The parameter D_c or a parameter directly related to D_c is necessarily incorporated into the constitutive law when the law is formulated as a slip-dependent constitutive law.

It is obvious that D_c is the most appropriate scaling parameter for the physical scaling of scale-dependent physical quantities inherent in the rupture, and that the slip-dependent constitutive formulation is best suited for the shear rupture, in order to scale the scale-dependent physical quantities inherent in the rupture over a broad scale range consistently and quantitatively in a unified manner, as noted earlier in Chapter 4.

In order to avoid the misunderstanding that strength-related parameters τ_p and $\Delta\tau_b$ are scale-dependent, we here refer to the issue of whether the strength is scale-dependent or scale-independent. It is often said that the strength of an inhomogeneous body exhibits scale-dependence. This is true; however, the strength in this case does not represent local shear strength in the vicinity of the front of shear rupture, but represents the overall body strength. An inhomogeneous larger body stochastically contains more numerous and higher stress-concentration sources such as minute preexisting cracks. Accordingly, the probability that the overall body strength will degrade becomes higher, as the body size becomes larger. In this sense, the overall body strength is scale-dependent. Note, however, that the aforementioned strength-related constitutive law parameters τ_p and $\Delta\tau_b$ are defined in terms of local shear strength in the vicinity of the shear rupture front, and that the parameters thus defined are scale-independent.

In fact, as shown in Figure 3.23 and Figure 5.20, the values of $\Delta\tau_b$ estimated for earthquakes are in a range from roughly 1 to 100 MPa, and fall in between the values of $\Delta\tau_b$ for the shear fracture of intact rock samples and those for frictional stick-slip failure on a precut rock interface (see Figure 3.23). The sizes of these samples used in the laboratory experiments are very small in comparison to the size of earthquake sources. This indicates that $\Delta\tau_b$ is scale-independent. More specifically, Figure 3.23 shows that $\Delta\tau_b$ for earthquakes is of the same order of, or less than, the magnitude of $\Delta\tau_b$ for the shear fracture of intact rock tested in the laboratory. This fact meets the requisite condition that $\Delta\tau_b$ for earthquakes cannot exceed the magnitude of $\Delta\tau_b$ for the shear fracture of intact rock at seismogenic crustal conditions; in other words, $\Delta\tau_b$ for the shear fracture of intact rock is the upper limit of $\Delta\tau_b$ for earthquake rupture.

By contrast, Figure 3.23 also shows that the amount of D_c for earthquakes is much larger than the amount of D_c for the shear fracture of intact rock and for frictional stick-slip failure on a precut rock interface tested in the laboratory. Indeed, the earthquake data points in Figure 3.23 do not fall on or around the trend line of the data points on the shear fracture of intact rock and on the frictional stick-slip failure on a precut rock interface, but systematically deviate rightward from the trend line. From Figure 3.23, one can see that D_c for earthquakes falls within a range of 4 mm to 3 m, whereas D_c for the laboratory-scale frictional stick-slip failure is of the order of 1 to 10 μm, and D_c for the laboratory-scale shear fracture of intact rock is of the order of 1 mm. In particular, D_c for large earthquakes falls within the range 1–3 m, which is roughly 10^3–10^6 times greater than D_c for the laboratory-scale shear fracture and frictional stick-slip failure. This is contrasted with the aforementioned result of $\Delta\tau_b$. Thus, both sets of data obtained from laboratory experiments (see Chapter 3) and from seismological observations lead to the consistent conclusion that D_c is scale-dependent.

As noted in Chapter 3, the fundamental cause of the scaling property of D_c lies in the characteristic length λ_c representing the geometric irregularity of the shear-rupturing surfaces. This suggests that the scale-dependence of D_c is attributable to the fault zone structure. This will be explained in more detail in the next section, based on solid facts obtained from high-resolution laboratory experiments, and will be considered in concrete terms by adopting a simplified model of the fault zone, serving as an example.

6.3 Root cause of scale-dependence

A large number of preexisting faults of various scales are embedded in the seismogenic layer. Hence, the seismogenic layer is inherently inhomogeneous. Individual faults embedded therein also exhibit structural inhomogeneities of various scales; however, they cannot be self-similar at all scales, but exhibit self-similarities over limited bandwidths. For example, real fault surfaces are nonplanar, and exhibit geometric irregularity with band-limited self-similarities (e.g., Aviles *et al.*, 1987; Okubo and Aki, 1987). In general, the geometric irregularity of the fault surfaces cannot be self-similar beyond the depth of the seismogenic layer and the size of the fault segment, as noted in subsection 5.2.1. The self-similarity can also be limited by the fact that the shear rupture process smoothes away the geometric irregularity of the shear-rupturing surfaces.

When a shear-rupture surface has multiple band-limited self-similarities, a different fractal dimension can be calculated for each band, and a characteristic length can be defined as the corner wavelength that separates the neighboring two bands with different fractal dimensions, as noted previously (Chapter 3 and Section 5.1). The characteristic length defined as such represents a predominant wavelength component of the geometric

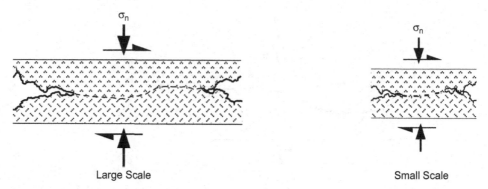

Large Scale   Small Scale

Fig. 6.1   Shearing failure of an adhesively bonded (or alternatively, cohesive) area on a fault. A great amount of displacement is required for the breakdown of a large-scale area, whereas only a small amount of displacement is required for the breakdown of a small-scale area.

irregularity of the rupture surface. If the amplitude and wavelength of the predominant wavelength component are particularly large and long in the fault zone, it may be regarded as what is called an "asperity" or "barrier" in the field of earthquake seismology. In this case, the geometric size of the "asperity" or "barrier" would be a representative characteristic length in the fault zone.

Specifically, structural inhomogeneities of a fault include not only asperities on the fault surfaces but also fault bend and fault-segment stepover. Resistance to rupture growth is high at the sites of such fault bend and fault-segment stepover as well as at interlocking asperity areas on the fault surfaces. Adhering junction (or cohesive) areas on a fault are also highly resistant to rupture growth. In general, a large slip is required for a large-scale area of high resistance to rupture growth to break down (Figure 6.1). Accordingly, it is natural that the breakdown displacement D_c is scale-dependent. The sizes of such local sites or areas highly resistant to rupture growth on a fault can be characteristic length scales, departing from the self-similarity on the fault. If this is the case, natural faults contain a wide range of characteristic length scales departing from the self-similarity (see also Aki, 1979, 1984, 1996).

Let us consider a situation where two asperities are interlocked in the fault zone (Figure 6.2). In order for the fault to be laterally sheared in this situation, it is necessary for these interlocking asperities to fracture when the compressive normal stress σ_n across the fault is sufficiently high, or alternatively, to slide over each other when the normal stress σ_n is sufficiently low. When a geometrically larger asperity is fractured in such a situation, a relatively longer predominant wavelength is necessarily contained in the irregular rupturing surfaces of the asperity, because of the fractal nature involved when the other conditions are equal, and hence the effective characteristic length λ_c, defined in this case as the predominant wavelength, is longer. On the other hand, when two interlocking asperities slide over each other to cause frictional slip failure, the effective characteristic length λ_c in this case would be determined from the slip distance required for one asperity to slide over the other, and hence λ_c is also longer when the interlocking asperities are larger.

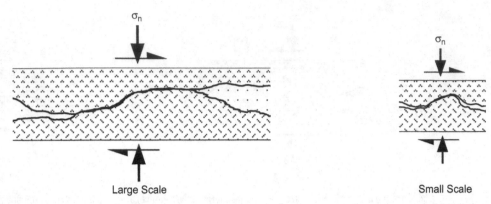

Fig. 6.2 Shearing failure of interlocking asperities in a fault zone. A great amount of displacement is required for the breakdown of a large asperity, whereas only a small amount of displacement is required for the breakdown of a small asperity.

In either case, therefore, a large slip is required for a large asperity to fracture or to slide over the other, and hence the critical amount of slip (termed breakdown displacement) D_c is large for a large asperity to fracture or to slide over the other. With this in mind, the term "breakdown" has been comprehensively used in this book, including not only the case of the shear fracture of intact rock but also the case of the frictional stick-slip failure on a precut rock interface (see Chapter 3). Indeed, the constitutive scaling relation (Eq. (3.15) in Section 3.3)

$$D_c = m(\Delta\tau_b/\tau_p) \times \lambda_c$$

has been derived empirically from laboratory data on both shear fracture and frictional stick-slip failure (see Section 3.3). In other words, the above formula (or constitutive scaling relation, Eq. (3.15)) is a specific formulation of the fact that the critical amount of slip (or the breakdown displacement) D_c required for a larger asperity to break down is necessarily larger, irrespective of the shear fracture of intact rock or the frictional stick-slip failure on a precut rock interface.

The above formula physically means that the root cause of the scaling property of D_c lies in the characteristic length λ_c representing a predominant length departing from the self-similarity on the fault, and hence the formula is critically important as the constitutive scaling expression that directly relates D_c to λ_c. As noted earlier (Section 3.3), the parameter m in the above formula is a monotonically increasing function of $\Delta\tau_b/\tau_p$, and has the maximum value of 0.66 at $\Delta\tau_b/\tau_p = 1$ (see Figure 3.26). Therefore, the constitutive scaling relation between D_c and λ_c is affected by the parameter $\Delta\tau_b/\tau_p$. For example, the formula leads to the following specific scaling relations: $D_c = 0.66\lambda_c$ if $\Delta\tau_b/\tau_p = 1$; $D_c = 0.10\lambda_c$ if $\Delta\tau_b/\tau_p = 0.1$; and $D_c = 0.014\lambda_c$ if $\Delta\tau_b/\tau_p = 0.01$. From this, one can see that D_c always has a value smaller than λ_c, even if $\Delta\tau_b$ has the maximum value ($= \tau_p$).

In general, in a statistical sense, a large fault tends to include geometrically large, local tough areas highly resistant to rupture growth. The irregular rupture surfaces of such geo-metrically large local areas contain a long predominant wavelength component λ_c because the rupture surfaces exhibit a fractal nature. If a number of strong areas (or "asperities") of different sizes having high resistance to rupture growth are inhomogeneously distributed on a fault, the critical amount of slip required for each asperity to break down differs from one asperity to another on the fault. Of a number of such asperity areas on a fault, however, the largest asperity area is most noteworthy, because the largest asperity area highly resistant to rupture growth on the fault is capable of accumulating an adequate amount of the elastic strain energy as the driving force to bring about an earthquake having the largest magnitude on the fault.

Accordingly, it is highly probable that a large earthquake results from the breakdown of a large asperity area capable of sustaining a large amount of the elastic strain energy in the surrounding medium, whereas a small earthquake results from the breakdown of a small asperity area capable of sustaining a small amount of the elastic strain energy. Indeed, this inference is consistent with seismological observations. Analysis results of observational data on real earthquakes commonly indicate that the largest asperity area on a fault scales with the seismic moment (Somerville et al., 1999; Miyake et al., 2003). In particular, Miyake et al. (2003) confirmed that a strong motion generation area coincides with the asperity area, and showed that a strong motion generation area self-similarly scales with the seismic moment (or magnitude).

In their analyses, the asperity area is defined as a region on the fault rupture surface that experiences a large slip in comparison to the slip averaged over the entire fault. Note that the asperity area defined as such corresponds to a tough area highly resistant to rupture growth. A large area highly resistant to rupture growth is capable of accumulating a large amount of the elastic strain energy in the surrounding medium. Once the area breaks down, a large amount of the elastic strain energy stored in the surrounding medium is released, and the released energy acts as the driving force to create a large earthquake, leading to a large slip. (Note that the seismic moment is practically determined by the product of the amount of slip and the fault area.)

A consistent result has also been achieved by a completely different approach. Mikumo et al. (2003) estimated D_c for the 2000 Tottori earthquake (the moment magnitude $M_w 6.6$) and for the 1995 Kobe earthquake ($M_w 7.2$), from the slip velocity functions obtained from the inversion of strong motion records. Their results show that the values of D_c for subfaults of those earthquake faults, estimated with the use of the data in the frequency band between 0.05 and 0.5 Hz, were in a range from 40 to 90 cm, and that these estimated values for D_c are heterogeneously distributed on the faults. Based on these results, they found that D_c tends to linearly scale with the local maximum slip on individual subfaults.

Since D_c scales with λ_c according to the above formula (that is, Eq. (3.15)), the characteristic length λ_c plays a crucial role in scaling scale-dependent physical quantities inherent in the rupture. This poses a question about how large the effective characteristic length λ_c for real earthquakes is. The effective characteristic length λ_c for earthquakes can be inferred under the assumption that the laboratory-derived constitutive scaling law (that is, Eq. (3.15)) can be extrapolated to a field-scale earthquake rupture. The constitutive scaling law predicts: $\lambda_c \cong 1.5 D_c$ if $\Delta\tau_b/\tau_p = 1$; $\lambda_c \cong 10 D_c$ if $\Delta\tau_b/\tau_p = 0.1$; $\lambda_c \cong 70 D_c$ if $\Delta\tau_b/\tau_p = 0.01$; and $\lambda_c \cong 480 D_c$ if $\Delta\tau_b/\tau_p = 0.001$. It can thus be concluded that λ_c is of

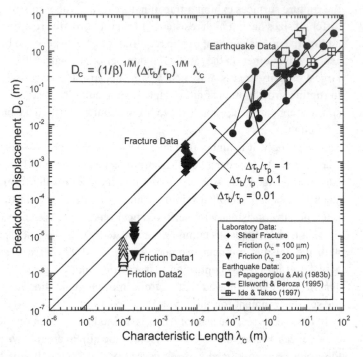

$$D_c = (1/\beta)^{1/M}(\Delta\tau_b/\tau_p)^{1/M}\lambda_c$$

Fig. 6.3 Scaling relation between the breakdown displacement D_c and the characteristic length λ_c. The solid lines in the figure represent theoretical scaling relations between D_c and λ_c under the assumption of $\Delta\tau_b/\tau_p = 0.01, 0.1,$ or 1. Field data on earthquakes and laboratory data on shear fracture and on frictional stick-slip rupture can be consistently accounted for in a unifying manner in terms of the theoretical scaling relation $D_c = (1/\beta)^{1/M}(\Delta\tau_b/\tau_p)^{1/M}\lambda_c$. Reproduced from Ohnaka (2003).

the same order of D_c if $\Delta\tau_b/\tau_p = 1$ (complete stress drop); however, λ_c is between one and two orders of magnitude greater than D_c if $\Delta\tau_b/\tau_p < 0.1$. For real earthquakes, it is likely that $\Delta\tau_b/\tau_p \geq 0.01$; however, it may be unrealistic to assume that $\Delta\tau_b/\tau_p < 0.01$. It would be much more unrealistic to assume that $\tau_p > 1000$ MPa at depths in the brittle seismogenic layer, because τ_p for intact rock tested at crustal conditions simulated in the laboratory does not exceed 1000 MPa.

The effective characteristic length λ_c can specifically be inferred for earthquakes for which the constitutive law parameters $\Delta\tau_b$ and D_c have been estimated, if τ_p can be appropriately assumed. For example, if $\tau_p = 100$ MPa is assumed for California earthquakes analyzed by Papageorgiou and Aki (1983b), and Ellsworth and Beroza (1995), then λ_c/D_c has a value ranging from 1.5 to 150, depending on a specific value for $\Delta\tau_b/\tau_p$. Figure 6.3 shows a plot of D_c against λ_c for laboratory data on the shear fracture of intact rock and on the frictional stick-slip rupture on a precut rock interface, and for field data on earthquake rupture compiled by Ohnaka (2000). In this plot, the peak shear strength τ_p for earthquakes has been assumed to be 100 MPa. If τ_p for these earthquakes is higher than 100 MPa, data points for the earthquakes shift rightward in Figure 6.3, and if τ_p for those earthquakes is lower than 100 MPa, the data points shift leftward in Figure 6.3, with the constraint that

$\Delta \tau_b / \tau_p \leq 1$. The solid straight lines in the figure represent constitutive scaling relations between D_c and λ_c for the three cases where $\Delta \tau_b / \tau_p$ has been assumed to be 0.01, 0.1, or 1. From Figure 6.3, it is clearly seen that both D_c and λ_c for field data on earthquakes are large in comparison to those for laboratory data on the shear fracture of intact rock samples and the frictional stick-slip failure on a precut interface of rock samples.

Let us consider a specific case where $D_c = 1$ m, $\Delta \tau_b = 10$ MPa, and $\tau_p = 100$ MPa, which may be regarded as a representative example for large earthquakes. In this case, we know from Eqs. (3.15) and (3.16) that the effective characteristic length λ_c has a value of 10 m. If it is further assumed that $k = 3$ and $\mu = 30\ 000$ MPa, we have from Eqs. (3.15), (3.16), and (3.28)

$$
\begin{aligned}
X_c &= \frac{1}{k} \left(\frac{1}{\beta} \right)^{1/M} \frac{\mu}{\Delta \tau_b} \left(\frac{\Delta \tau_b}{\tau_p} \right)^{1/M} \lambda_c \\
&= 0.221 \times \frac{3 \times 10^4}{\Delta \tau_b} \left(\frac{\Delta \tau_b}{\tau_p} \right)^{0.833} \lambda_c \\
&= 0.221 \times \frac{3 \times 10^4}{10} \left(\frac{10}{100} \right)^{0.833} \times 10\,\text{m} \\
&= 0.97 \times 10^3\,\text{m},
\end{aligned}
\tag{6.1}
$$

From this calculation, we know that the breakdown zone size X_c is approximately 1 km. This means that a predominant wavelength component of 10 m is contained in the geometric irregularity of the rupture surfaces over the breakdown zone length of 1 km. This may be paraphrased as follows: under the stress condition that $\Delta \tau_b / \tau_p = 0.1$, one meter for D_c is required to break down a cohesive zone of 1 km, and the newly created rupture surfaces have a geometric irregularity characterized by a predominant wavelength component of 10 m. It would not be unreasonable to postulate that an effective characteristic length of 10 m is contained in the geometric irregularity of the rupture surfaces over a distance of 1 km.

6.4 Physical scaling of scale-dependent physical quantities

6.4.1 Scaling relationships between X_c and D_c, and between L_c and D_c

From Eq. (3.28) and Eq. (5.8), we know

$$
L_c = X_c = \frac{1}{k} \frac{\mu}{\Delta \tau_b} D_c.
\tag{6.2}
$$

This relationship theoretically predicts that X_c and L_c both linearly scale with D_c. Let us specifically verify this theoretical scaling relationship by using not only field data on real earthquakes but also laboratory data on the shear failure of intact rock samples and on the frictional stick-slip failure on a precut rock interface.

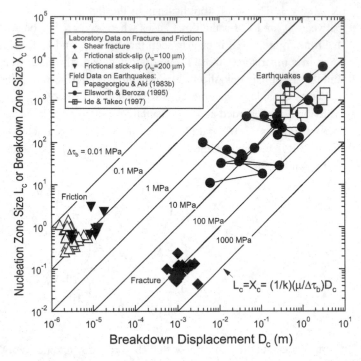

Fig. 6.4 A plot of L_c or X_c against D_c for field data on earthquakes compiled by Ohnaka (2000) and for laboratory data on the shear failure of intact rock samples and on the frictional stick-slip rupture on precut rock interfaces (Ohnaka, 2003). The solid lines in the figure represent theoretical scaling relations between L_c or X_c and D_c under the assumption of $\Delta \tau_b = 0.01, 0.1, 1, 10, 100,$ or 1000 MPa.

Figure 6.4 shows a plot of L_c or X_c against D_c for field data on earthquakes compiled by Ohnaka (2000) and for laboratory data on the shear failure of intact rock samples and on the frictional stick-slip failure on precut rock interfaces (Ohnaka, 2003, 2004a). For laboratory data on the shear failure and on the frictional stick-slip failure, L_c corresponding to the critical size of the earthquake nucleation phase can be estimated from Eq. (6.2) under the same assumption ($k = 3$ and $\mu = 30\,000$ MPa) made for the field data on earthquakes. For comparison, the values for L_c thus estimated for laboratory data on shear fracture and on frictional stick-slip rupture have also been plotted against D_c in Figure 6.4.

Equation (6.2) theoretically predicts that both L_c and X_c are directly proportional to D_c if $\Delta \tau_b$ is constant. With this in mind, the proportional relationships between L_c or X_c and D_c derived from Eq. (6.2) under the assumption that $\Delta \tau_b = 0.01, 0.1, 1, 10, 100,$ or 1000 MPa, are also shown as solid straight lines in Figure 6.4 for reference. For real earthquakes, $\Delta \tau_b$ may take any value in a wide range of approximately 1 to 100 MPa, depending on the seismogenic environment, and hence, in reality, the scaling relationship between L_c or X_c and D_c is severely affected by the magnitude of $\Delta \tau_b$, as shown in Figure 6.4.

Laboratory data plotted in Figure 6.4 show that $\Delta \tau_b$ has a very low value in a range of 0.01 to 0.2 MPa for frictional stick-slip failure on a precut rock interface, and has a very high value in a range of 50 to 700 MPa for the shear failure of intact rock (Ohnaka, 2003). Therefore, the effect of $\Delta\tau_b$ needs to be equalized, when the scaling relationship between L_c (or X_c) and D_c for earthquakes is compared to the corresponding relations for the shear failure of intact rock samples and for the frictional stick-slip rupture on a precut interface of rock samples in the laboratory.

It is estimated from Eq. (6.2) that L_c or X_c has a value of roughly 7 cm for the shear failure of intact rock because of a very high value for $\Delta\tau_b$, and has a value between 5 cm and 1 m for frictional stick-slip rupture because of a very low value for $\Delta\tau_b$, in spite of a small amount of D_c. On the other hand, L_c (or X_c) for earthquakes has been estimated to be between 10 m and 10 km, according to the amount of D_c and the magnitude of $\Delta\tau_b$ (refer to Table 1 of Ohnaka (2000), and see also Figure 6.4). Thus, given the effect of $\Delta\tau_b$, one can see from Figure 6.4 that formula (6.2) provides a unified comprehension for both field data on earthquakes and laboratory data on the shear failure of intact rock samples and on the frictional stick-slip rupture on a precut interface of rock samples. This justifies formula (6.2) as a scaling law that consistently accounts in quantitative terms for the scale-dependence of L_c and X_c on D_c.

6.4.2 Physical scaling of the duration time of shear rupture nucleation

As described in Chapter 5, the nucleation process of a shear rupture consists of an initial, stable and quasi-static phase (phase I), and the subsequent, spontaneous and accelerating phase (phase II). The rupture growth length L during the nucleation process is given by (see Eqs. (5.12) and (5.15))

$$L(t) = \begin{cases} L_0 + V_{st}t & (0 < t < t_{sc}) \quad \text{during phase I} \\ L_c \left(\dfrac{t_a - t_c}{t_a - t}\right)^{1/(n-1)} & (t_{sc} \leq t \leq t_c) \quad \text{during phase II} \end{cases}, \qquad (6.3)$$

where L_0 is the initial length, V_{st} is the steady, rupture growth rate controlled by an applied loading rate, t_{sc} is the critical time at which the behavior of the rupture growth changes from a steady, quasi-static phase (phase I) controlled by the loading rate to a self-driven, accelerating phase controlled by inertia, t_c is the other critical time at which $L = L_c$ is attained, and t_a is defined by Eq. (5.16).

The duration time during phase II is given by Eq. (5.14); that is,

$$t_c - t_{sc} = \int_{L_{sc}}^{L_c} \frac{dL}{V(L)} = \frac{\lambda_c^n}{\alpha V_S(1 - n)}\left(L_c^{-n+1} - L_{sc}^{-n+1}\right). \qquad (6.4)$$

As indicated by Eq. (5.5), the rupture growth rate (or velocity) V during phase II is a monotonically increasing function of the rupture growth length L. Hence, the duration time $t_c - t_{sc}$ during phase II can be expressed as an explicit function of V in place of L. Using Eq. (5.5), Eq. (6.4) can be rewritten as

$$t_c - t_{sc} = \frac{\lambda_c}{V_S} f\left(\frac{V_{sc}}{V_S}, \frac{V_c}{V_S}\right), \qquad (6.5)$$

where $f(V_{sc}/V_S, V_c/V_S)$, which is a function of V_{sc}/V_S and V_c/V_S, is expressed as

$$f\left(\frac{V_{sc}}{V_S}, \frac{V_c}{V_S}\right) = \frac{\alpha^{-1/n}}{(n-1)} \left(\frac{V_c}{V_S}\right)^{(-n+1)/n} \left[\left(\frac{V_{sc}/V_S}{V_c/V_S}\right)^{(-n+1)/n} - 1\right]. \tag{6.6}$$

In the above equations, V_{sc} denotes the rupture growth velocity at $t = t_{sc}$ (or $L = L_{sc}$), V_c denotes the rupture growth velocity at $t = t_c$ (or $L = L_c$), and V_S denotes the shear wave velocity. Since V_S and $f(V_{sc}/V_S, V_c/V_S)$ are scale-independent, one can see from Eqs. (6.5) and (6.6) that the duration time $t_c - t_{sc}$ directly scales with the characteristic length λ_c of the shear-rupturing surface roughness. Equation (6.5) indicates that the rougher the rupture surfaces, the longer the duration time of shear rupture nucleation.

The duration time from an initial length $L_0(< L_{sc})$ at $t = 0$ to the rupture growth length L_{sc} at $t = t_{sc}$ during an initial, stable and quasi-static phase (phase I) is given by

$$t_{sc} = \int_{L_0}^{L_{sc}} \frac{dL}{V_{st}} = \frac{L_{sc} - L_0}{V_{st}}. \tag{6.7}$$

Thus, the entire duration time t_c of a shear rupture nucleation from its initial length $L_0(< L_{sc})$ to the critical length L_c is given by

$$t_c = \frac{L_{sc} - L_0}{V_{st}} + \frac{\lambda_c}{V_S} f\left(\frac{V_{st}}{V_S}, \frac{V_c}{V_S}\right), \tag{6.8}$$

where the relation $V_{sc} = V_{st}$ at $t = t_{sc}$ has been used. If V_{st} and $L_{sc} - L_0$ are normalized to the shear wave velocity V_S and the characteristic length λ_c, respectively, then V_{st} can be expressed as $V_{st} = c_1 V_S$, and $L_{sc} - L_0$ as $L_{sc} - L_0 = c_2 \lambda_c$ (c_1 and c_2 being numerical constants). Thus, Eq. (6.8) can be rewritten as

$$t_c = \frac{\lambda_c}{V_S} \left[\frac{c_2}{c_1} + f\left(c_1 + \frac{V_c}{V_S}\right)\right]. \tag{6.9}$$

This theoretical scaling relation also shows that t_c directly scales with λ_c, and that the rougher the rupture surfaces, the longer the duration time of shear rupture nucleation.

Let us closely examine how long the duration time of shear rupture nucleation is. Figure 6.5 shows to what extent the following function $f(V/V_S, V_c/V_S)$ depends on V/V_S and V_c/V_S in phase II ($V_{st} \leq V < V_c$). In this plot, the experimentally derived values of $\alpha = 8.87 \times 10^{-29}$ and $n = 7.31$ have been employed. Since V_c is a steady, high rupture speed close to the shear wave velocity V_S, it is unlikely that V_c/V_S has a value less than 0.1. With this in mind, we find from Figure 6.5 that $f(V/V_S, V_c/V_S)$ does not depend on V_c/V_S, but depends only on V/V_S in the range of $V/V_S < 10^{-2}$. This means that the duration time t_c from a stage of any value for V/V_S in phase II to the critical stage V_c/V_S is virtually not affected by the critical value for V_c. In this case, therefore, t_c is predictable from the following relation (Ohnaka and Shen, 1999):

$$t_c = \frac{\lambda_c}{V_S} f\left(\frac{V}{V_S}, \frac{V_c}{V_S}\right). \tag{6.10}$$

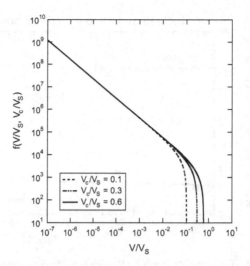

Fig. 6.5 A log–log plot of $f(V/V_S, V_c/V_S)$ against V/V_S (Ohnaka and Shen, 1999).

For example, for frictional stick-slip rupture nucleation on a precut smooth interface ($\lambda_c = 46\ \mu m$) of granite samples with $V_S = 2.9\ km/s$ tested in the laboratory, t_c from a stage of $V/V_S = 10^{-5}$ in phase II to the critical stage V_c/V_S is given by

$$t_c = \frac{46\ \mu m}{2.9\ km/s} \times f\left(10^{-5}, \frac{V_c}{V_S}\right) \cong 0.36\ s,$$

where $f(10^{-5}, V_c/V_S) = 2.26 \times 10^7$ has been used. For frictional stick-slip rupture nucleation on a precut rough interface ($\lambda_c = 200\ \mu m$) of granite samples with $V_S = 2.9 km/s$ tested in the laboratory, t_c from the stage of $V/V_S = 10^{-5}$ in phase II to the critical stage V_c/V_S is given by

$$t_c = \frac{200\ \mu m}{2.9\ km/s} \times f\left(10^{-5}, \frac{V_c}{V_S}\right) \cong 1.6\ s.$$

In contrast, for major or great earthquakes for which $\lambda_c = 1 \sim 10\ m$ has been inferred (see Figure 6.3), t_c from the stage of $V/V_S = 10^{-5}$ in phase II to the critical stage V_c/V_S is given by

$$t_c = \frac{1 \sim 10\ m}{(6/1.73)\ km/s} \times f\left(10^{-5}, \frac{V_c}{V_S}\right) \cong 1.8 \sim 18\ hours,$$

where $V_P = 6\ km/s$ and $V_S = V_P/1.73$ have been assumed.

The above values for t_c estimated theoretically from Eq. (6.10) can be verified by using laboratory data on frictional stick-slip rupture nucleation. A typical example of such verification is shown in Figure 6.6. This figure shows a plot of the rupture growth length $L(t)$ against time t for data in the accelerating phase (phase II) of frictional stick-slip rupture nucleation that occurred on a precut smooth fault with its surface roughness characterized by $\lambda_c = 46\ \mu m$. In this figure, the black dots denote the data points, the solid curve denotes a graph expression of Eq. (6.3) fitted to the data points, with $L_c = 27\ cm$,

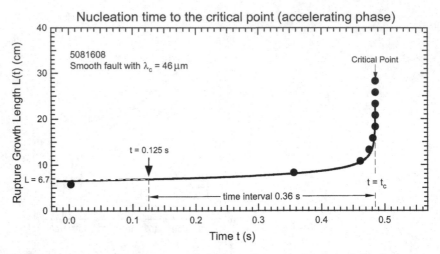

Nucleation time to the critical point (accelerating phase)

5081608
Smooth fault with $\lambda_c = 46\,\mu$m

Critical Point

$t = 0.125$ s

$L = 6.7$

$t = t_c$

time interval 0.36 s

Time t (s)

Fig. 6.6 A plot of the rupture growth length $L(t)$ against time t for data in the accelerating phase of stick-slip rupture nucleation that occurred on a precut smooth fault with its characteristic length $\lambda_c = 46\,\mu$m. The black dots denote data points, and the thick curve denotes the graph expression of Eq. (6.3) in the text, fitted to the data points with $L_c = 27$ cm, $t_a - t_c = 6.329 \times 10^{-5}$ s, $t_a = 0.48554329$ s, and $n = 7.31$. The plotted data are from Ohnaka (2004a, 2004b).

$t_a - t_c = 6.329 \times 10^{-5}$ s, $t_a = 0.48554329$ s, and $n = 7.31$, and the origin of time is such that $t - t_{sc} = 0$. The critical point attains at $t = t_c$, at which $V = V_c$ (or $L = L_c$). From Figure 6.6, one can see how well Eq. (6.3) accounts for laboratory data on frictional stick-slip rupture nucleation in quantitative terms. This corroborates the theoretical result that the nucleation process leading up to the critical point obeys the power law of Eq. (6.3).

When $V/V_S = 10^{-5}$ and $\lambda_c = 46\,\mu$m, we have from formula (5.5) that $L = 6.6$ cm. On the other hand, Figure 6.6 indicates that $L \cong 6.7$ at $t = 0.125$ s, and that the time interval from $t = 0.125$ s to $t = t_c (= 0.485$ s) is 0.36 s. Thus, $t_c \cong 0.36$ s from the stage of $V = 10^{-5}V_S$ in phase II to the critical point, predicted from theoretical relation (6.10), is verified by experimental data on frictional stick-slip rupture nucleation on a precut smooth fault with $\lambda_c = 46\,\mu$m, shown in Figure 6.6.

In Figure 6.7, experimental data on the nucleation process leading up to the critical point tested on precut faults with three different surface roughnesses are compared. The fault surface roughnesses used have been shown in Figure 5.1 for comparison in Chapter 5. The rough fault surface is characterized by the predominant wavelength of $\lambda_c = 200\,\mu$m, the smooth fault surface is characterized by $\lambda_c = 46\,\mu$m, and the extremely smooth fault surface is characterized by $\lambda_c = 10\,\mu$m (see Section 5.1).

It is clear from Figure 6.7 that the rupture growth length and the duration time to the critical point are both very long for the nucleation that proceeds on a rough fault with $\lambda_c = 200\,\mu$m, if compared to those for the nucleation that proceeds on a smooth fault with $\lambda_c = 46\,\mu$m, or on an extremely smooth fault with $\lambda_c = 10\,\mu$m (see the inset of Figure 6.7). This indicates that the effect of the geometric irregularity of fault surfaces on shear rupture nucleation is very significant.

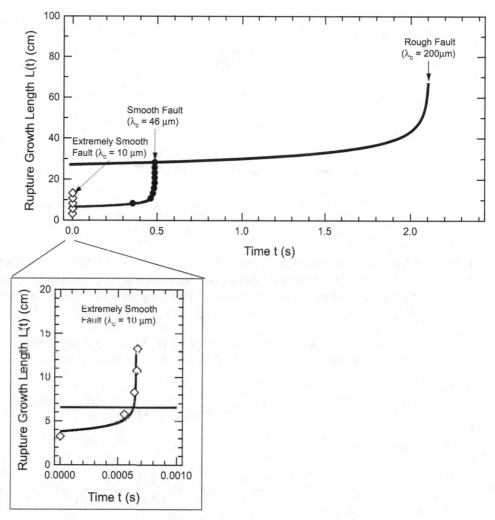

Fig. 6.7 Plots of the rupture growth length $L(t)$ against time t during the accelerating phase of the nucleation up to the critical point, for experimental data on stick-slip rupture nucleation processes that occurred on precut faults with three different surface roughnesses. The plot of $L(t)$ against t for the nucleation on an extremely smooth fault with $\lambda_c = 10\,\mu$m is enlarged in the inset. Reproduced from Ohnaka (2004a).

As shown in Figure 6.7, the rupture growth length L and the duration time $|t - t_c|$ to the critical point during the nucleation process both depend on the characteristic wavelength λ_c representing the geometric irregularity of fault surfaces, and both L and $|t - t_c|$ increase with an increase in λ_c. This means that the rupture growth length and the duration time to the critical point are both scale-dependent, and that λ_c plays a crucial role in scaling the nucleation process.

The power law expressed by Eq. (6.3) for the accelerating phase (phase II) of shear rupture nucleation accounts well in quantitative terms for experimental data obtained in the laboratory (see Figures 6.6 and 6.7). It is clear from Figure 6.7, however, that Eq. (6.3)

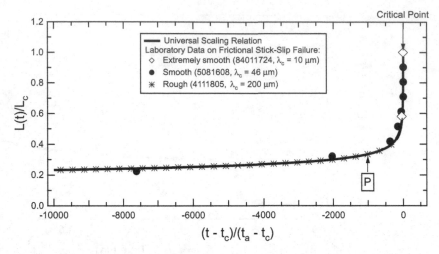

Scale-independent relationship between $L(t)/L_c$ and $(t - t_c)/(t_a - t_c)$. $L(t)/L_c$ is plotted against $(t - t_c)/(t_a - t_c)$ for the data on frictional stick-slip rupture nucleation shown in Figure 6.7. The theoretical relation (6.11) in the text is also over-plotted for comparison with the experimental data. Reproduced from Ohnaka (2004a).

is a scale-dependent mathematical expression. To derive a scale-independent universal expression, we rewrite Eq. (6.3) as follows (Ohnaka, 2004a, 2004b):

$$\frac{L(t)}{L_c} = \left(\frac{1}{1 - (t - t_c)/(t_a - t_c)} \right)^{1/(n-1)} \qquad (t \leq t_c), \qquad (6.11)$$

where

$$\frac{t - t_c}{t_a - t_c} = \alpha(n - 1) \left(\frac{L_c}{\lambda_c} \right)^{n-1} \frac{t - t_c}{(\lambda_c/V_S)}. \qquad (6.12)$$

Expression (6.11) has a mathematical form that the relationship between $L(t)/L_c$ and $(t - t_c)/(t_a - t_c)$ is scale-invariant, if n is scale-invariant. In fact, the laboratory data on frictional stick-slip rupture nucleation shown in Figure 6.7 can be unified completely in quantitative terms by this expression (see Figure 6.8).

Figure 6.8 shows a plot of $L(t)/L_c$ against $(t - t_c)/(t_a - t_c)$ for the data on frictional stick-slip rupture nucleation shown in Figure 6.7. The theoretical relation (6.11) has also been over-plotted in Figure 6.8 for comparison with the experimental data. From Figure 6.8, one can see that different sets of experimental data on frictional stick-slip rupture nucleation tested on faults with different λ_c are unified completely in quantitative terms by expression (6.11). Since $t - t_c$ scales with λ_c/V_S (see Eq. (6.12)), it is obvious from Figure 6.8 that not only the rupture growth length but also the duration time to the critical point during the nucleation process scale with the characteristic length λ_c. It has thus been confirmed that the characteristic length λ_c plays a crucial role in scaling both the nucleation zone length and the nucleation time to the critical point.

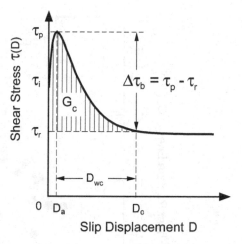

Fig. 6.9   A slip-dependent constitutive relation $\tau(D)$ for shear rupture. See the text for explanation of the following
parameters: τ_i, τ_p, τ_r, $\Delta\tau_b$, D_a, D_c, D_{wc}, and G_c.

6.4.3 Scale-dependence of apparent shear rupture energy

In this subsection, we will demonstrate that the shear rupture energy or the resistance to
rupture growth, G_c defined by Eq. (2.44) is scale-dependent.

Here we will reintroduce a diagram of the constitutive relation $\tau(D)$ for shear rupture
in Figure 6.9. In this figure, τ_i denotes the initial stress at the onset of slip, τ_p denotes
the peak shear strength, τ_r denotes the residual friction stress, D_a denotes the critical
slip displacement at which the peak shear strength is attained, D_c denotes the breakdown displacement defined as the critical amount of slip required for the shear traction to
degrade to the residual friction stress level τ_r, $\Delta\tau_b$ is the breakdown stress drop defined as
$\Delta\tau_b = \tau_p - \tau_r$, and D_{wc} is the slip-weakening displacement defined as $D_{wc} = D_c - D_a$.

When a constitutive relation $\tau(D)$ as shown in Figure 6.9 is specifically given, G_c can
be evaluated directly by integrating Eq. (2.44). The G_c thus evaluated depends upon the
values of $\Delta\tau_b$ and D_c. The dependence of G_c on $\Delta\tau_b$ and D_c would be easily recognized
from Figure 6.9, given that the G_c defined by Eq. (2.44) is equal to the vertical-lined area
in Figure 6.9. As demonstrated in previous chapters, $\Delta\tau_b$ is scale-independent. However,
D_c is inevitably scale-dependent. This fact necessarily leads to the conclusion that the G_c
defined by Eq. (2.44) is also scale-dependent. In other words, the G_c defined by Eq. (2.44)
not only depends on the material property and ambient environmental conditions but also
depends on the size scale.

As described in Chapter 2, G_c is defined as the critical energy per unit area of fracture-
surface, and thus the dimensions of G_c are [energy/area] $= [\text{J/m}^2]$. Therefore, if the fracture
surface is a completely flat plane, G_c does not depend on the size scale (i.e., is scale-
independent). However, the fracture surface of a heterogeneous body such as rock is not
a flat plane but exhibits geometric irregularity (or roughness) with a fractal nature. In this
case, the real area of the fracture surface is distinctly different from, and specifically larger

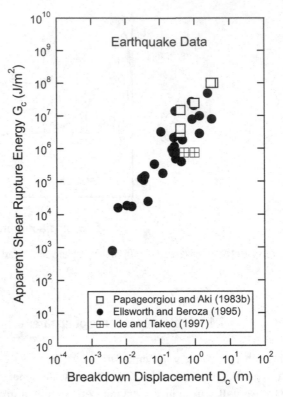

Fig. 6.10 A plot of apparent shear rupture energy G_c against breakdown displacement D_c for earthquakes. Data compiled by Ohnaka (2000, 2003) are plotted.

than, the apparent area (or nominal area), under the tacit assumption that the fracture surface is a flat plane, as noted in Section 2.2. The rougher the fracture surface, the larger the real surface area becomes. However, this fact has been overlooked and is not taken into account when G_c for earthquakes is estimated, despite the fact that G_c is, in a macroscopic sense, defined by Eq. (2.44). Given that a real fracture surface has geometric irregularity with a fractal nature, it is natural that the G_c defined in a macroscopic sense by Eq. (2.44) is scale-dependent. The G_c defined by Eq. (2.44) should therefore be referred to as *apparent shear rupture energy*.

Figure 6.10 shows a plot of the G_c defined by Eq. (2.44) against D_c for earthquakes. In this figure, white squares denote data points for G_c estimated for earthquakes by Papageorgiou and Aki (1983b), black circles denote those for earthquakes analyzed by Ellsworth and Beroza (1995), and white squares with crosses denote those for the 1995 Kobe earthquake analyzed by Ide and Takeo (1997). Since Ellsworth and Beroza (1995) and Ide and Takeo (1997) did not estimate G_c for earthquakes, G_c for these earthquakes has been calculated by the author in an earlier paper (Ohnaka, 2003) from Eq. (2.45) under the assumption that $\Gamma = 1$. To calculate G_c for an earthquake from Eq. (2.45), we need to know $\Delta\tau_b$ and D_c. However, Ellsworth and Beroza (1995) did not estimate D_c for earthquakes, and hence D_c for those earthquakes has been calculated from Eq. (5.8), under the same assumptions made by Ellsworth and Beroza (see subsection 5.2.3).

From Figure 6.10, one can see that G_c increases with an increase in D_c. This is a direct manifestation of the fact that G_c and D_c are both scale-dependent. From this figure, one can also see that G_c for large earthquakes is of the order of 10^7 to 10^8 J/m^2, which are 2 to 3 orders of magnitude greater than G_c for the shear fracture of intact rock samples tested at seismogenic crustal conditions simulated in the laboratory (see Chapter 3). The cohesive shear strength of intact rock is the upper limit of frictional strength on a precut rock interface. Hence, the cohesive shear strength of intact rock is higher than, or equal to, any preexisting fault strength, if the other conditions are equal. If, therefore, G_c were real scale-independent fracture energy, G_c for any earthquake could not exceed the critical energy (per unit area) required for the shear fracture of intact rock at seismogenic crustal conditions. As mentioned above, however, actual values for G_c estimated for large earthquakes are 2 to 3 orders of magnitude greater than G_c for the shear fracture of intact rock samples tested at seismogenic crustal conditions simulated in the laboratory. Note that D_c for large earthquakes is 2 to 3 orders of slip amount greater than that for the shear fracture of intact rock samples tested in the laboratory (see Figure 3.23). Thus, it is true that the G_c defined by Eq. (2.44) is scale-dependent.

Once it is accepted that G_c is scale-dependent, the observation that G_c for large earthquakes is greater than G_c for small earthquakes can be rationally accounted for in terms of the scale-dependent D_c. Scale-dependent G_c is consistent and physically reasonable, if G_c is construed as the resistance to rupture growth.

6.5 Fault heterogeneity and the Gutenberg–Richter frequency–magnitude relation

We have presented an inhomogeneous fault model in which the nucleation zone size and the final size of the resulting earthquake are both prescribed by a common sizable "asperity" area highly resistant to rupture growth on the fault. In this section, we show that the b-value of the Gutenberg–Richter frequency–magnitude relation is directly related to fault heterogeneities with a fractal nature in the seismogenic layer.

If $n(M)dM$ denotes the number of earthquakes with magnitudes ranging from M to $M + dM$, generated in a certain region for a fixed time period, the Gutenberg–Richter frequency–magnitude relation is expressed as

$$\log n(M) = a - bM, \tag{6.13}$$

where a and b are constants. This relation is one of the well-known Gutenberg–Richter formulas. If $N(M)$ denotes the total number of earthquakes with magnitudes equal to or greater than M, the following relation holds:

$$N(M) = \int_M^\infty n(M)dM. \tag{6.14}$$

Thus, the Gutenberg–Richter frequency–magnitude relation is also expressed in terms of $N(M)$ and M as follows:

$$\log N(M) = A - bM, \tag{6.15}$$

where

$$A = a - \log(b \ln 10). \tag{6.16}$$

The coefficient b is called the *b-value*, and has a numerical value close to unity.

Equation (6.15) indicates that the cumulative number of earthquakes with magnitudes equal to or greater than M is proportional to 10^{-bM}; that is,

$$N(M) \propto 10^{-bM}. \tag{6.17}$$

This is an essential property of the Gutenberg–Richter frequency–magnitude relation, and can be derived from fault inhomogeneities with a fractal nature.

The shape of an "asperity" area highly resistant to rupture growth on an inhomogeneous fault will be, in general, complicated. Despite this, it is possible to define the characteristic length representing the geometric size of such an "asperity" area on the fault. The simplest way may be to obtain the "asperity" area S_A, and then to define the characteristic length R by $R = \sqrt{S_A/\pi}$. In any case, let R represent the geometric size of an "asperity" area highly resistant to rupture growth, and let $N(R)$ represent the total number of such "asperities" having characteristic lengths equal to or greater than R on inhomogeneous faults in the seismogenic layer. If a power-law relation holds for the "asperity" size distribution (Aki, 1981), the relation between $N(R)$ and R is expressed as

$$N(R) \propto R^{-d}, \tag{6.18}$$

where d denotes the fractal dimension. The power-law relation with $d \approx 2$ has commonly been observed for crater size distribution, and fragment size distribution resulting from the crushing of a heterogeneous body.

Since the breakdown displacement D_c is, by definition, the critical amount of the slip displacement required for the breakdown of an "asperity" with the characteristic length R in the present context, a proportional relationship holds between D_c and R; that is,

$$D_c \propto R. \tag{6.19}$$

On the other hand, the seismic moment M_0 is related to the magnitude M by (Kanamori, 1977; Hanks and Kanamori, 1979)

$$M_0 \propto 10^{1.5M}. \tag{6.20}$$

From Eqs. (6.18), (6.19), and (6.20), together with Eq. (5.30) or (5.32), we have

$$N(M) \propto 10^{-(d/2)M}. \tag{6.21}$$

This relation is equivalent to Eq. (6.17). Thus, we have from Eqs. (6.17) and (6.21)

$$b = \frac{d}{2}. \tag{6.22}$$

This relation agrees with the relation derived by Aki (1981). When $b \cong 1$, we have $d \cong 2$ for the "asperity" size distribution from Eq. (6.22). The present model leads to the conclusion that the b-value is a manifestation of the fractal dimension for the "asperity" size distribution arising from fault heterogeneities in the seismogenic layer. Accordingly, it will be understood that the b-value is directly related to structural heterogeneity with a fractal nature in the seismogenic layer.

7 Large earthquake generation cycles and accompanying seismic activity

7.1 The cyclical process of typical large earthquakes on a fault

When a large earthquake occurs in a region, a large amount of the elastic strain energy stored in the region is released and expended as the driving force for the earthquake generation. Consequently, the residual energy in the region is lowered to a subcritical level immediately after the event. The released energy in the region cannot be restored to its critical level without the contribution of tectonic loading due to perpetual slow and steady plate motion. Accordingly, it is reasonable to assume that large earthquakes occur repeatedly at quasi-periodic intervals on a fault, as described in Chapter 5 (see subsection 5.2.1). In fact, it has been proven statistically that a series of large earthquakes along a tectonic plate interface occur repeatedly on a quasi-periodic basis. If it is well-documented over a long period of time that large earthquakes have occurred repeatedly along a tectonic plate interface in a region, we can statistically examine whether or not these events have indeed occurred at quasi-periodic intervals, as described below.

Whether temporal distribution of earthquake occurrences on a specific fault is clustered or random or quasi-periodic can be examined statistically under the assumption that the sequence of identical events is independently distributed (e.g., Utsu, 1999, 2002). Let t be the time that has elapsed since the last occurrence of an earthquake on the fault, and let $\mu(t)dt$ denote the probability that the next earthquake will occur on the fault in a short time interval from t to $t + dt$. If it is further assumed that individual time intervals (denoted by T) of earthquake occurrences on the fault are distributed around their average value (denoted by T_{av}), and that the probability density function $w(T)$ of the time interval distribution is represented by the Weibull distribution

$$w(T) = \alpha\beta T^{\beta-1} \exp(-\alpha T^{\beta}), \tag{7.1}$$

where α and β are positive parameters, then $\mu(t)$ is expressed as (e.g., Utsu, 1999, 2002)

$$\mu(t) = \frac{w(t)}{\displaystyle\int_{t}^{\infty} w(t)dt} = \alpha\beta t^{\beta-1}. \tag{7.2}$$

The probability $p(u|t)$ that the next earthquake will occur on the fault in a time interval from t to $t + u$ is expressed in terms of $\mu(t)dt$ as (e.g., Utsu, 1999, 2002)

$$p(u|t) = 1 - \exp\left\{-\int\limits_{t}^{t+u} \mu(t)dt\right\}. \tag{7.3}$$

Using Eq. (7.2), Eq. (7.3) can be integrated and expressed as

$$p(u|t) = 1 - \exp\{-\alpha[(t + u)^{\beta} - t^{\beta}]\}. \tag{7.4}$$

The mean time interval $E[T]$ $(= T_{\text{av}})$ and its variance $V[T]$ are given by (Utsu, 1999)

$$E[T] = \alpha^{-1/\beta}\Gamma(1/\beta + 1) \tag{7.5}$$

and

$$V[T] = \alpha^{-2/\beta}\{\Gamma(2/\beta + 1) - [\Gamma(1/\beta + 1)]^2\}, \tag{7.6}$$

respectively, where $\Gamma()$ denotes the Gamma function. The Gamma function $\Gamma(x)$ for $x > 0$ is defined as $\Gamma(x) = \int_0^{\infty} \tau^{x-1}e^{-\tau}d\tau$. The time series of event occurrences on a single fault can be classified in terms of the parameter β into the following four cases (Utsu, 1998): (1) $0 < \beta < 1$ if events are clustered; (2) $\beta = 1$ if events are random; (3) $\beta > 1$ if events are intermittent; and (4) $\beta \to \infty$ if events are strictly periodic. Utsu (1998) showed in his simulation that events virtually occur periodically when $\beta > 10$, and quasi-periodically even when $\beta = 3 - 6$.

As described in subsection 5.2.1, paleoseismic, historical, and seismic instrumentation records indicate that large earthquakes along plate boundaries have occurred repeatedly, not in clusters or at random but quasi-periodically, on a single fault, and that average recurrence time intervals are well defined for those events. In particular, large earthquakes along the tectonic plate interface in the Nankai region in the southwest of Japan are well documented. Specifically, it is fully documented that a series of large earthquakes along the tectonic plate interface in the Nankai region occurred in the years 1361, 1498, 1605, 1707, 1854, and 1946 (e.g., Ishibashi and Satake, 1998; Utsu, 1998). Therefore, this series of earthquakes along the tectonic plate interface in the Nankai region is the best to use for statistically proving that they have occurred repeatedly at quasi-periodic intervals. For these events, indeed, it has been proven by using the aforementioned statistical method that they have occurred at quasi-periodic intervals ($\beta = 6.09$), and the average recurrence time interval with its standard deviation has been calculated to be 117.1 ± 21.2 years (Utsu, 1998). In this manner, average recurrence time intervals for large-earthquake occurrences can be well defined.

The elastic strain energy builds up to the critical level much faster in the elastic medium along tectonic plate boundaries than in regions away from plate boundaries. Hence, large earthquakes occur intermittently more often on plate boundary faults than on faults in regions away from plate boundaries, and the recurrence time interval for inter-plate earthquakes occurring on plate boundary faults is much shorter than that for intra-plate earthquakes occurring on faults in regions away from plate boundaries. The recurrence time interval is of the order of 10^2 years for large earthquakes that occur on a plate boundary fault between two tectonic plates whose relative motion has a rate of a few cm/year. Such a typical example is the aforementioned series of large historical earthquakes along the plate interface in the Nankai region in the southwest of Japan.

Let us envision a simple case where two tectonic plates move sideways against each other except at the plate boundary where the interface is stuck under the compressive normal load. If the rate of relative plate motion is a few cm/year, the relative displacement between a position in each of these two plates, distant from the plate interface, becomes a few meters in 100 years, and corresponding amounts of shear deformation and strain energy accumulate in the elastic medium surrounding the plate boundary. When a great earthquake rupture occurs along this plate boundary, the elastic rebound of both sides of the plate boundary fault causes slip displacement on the fault plane, and releases a great amount of elastic strain energy. It is known that the amount of slip displacement for such a great earthquake is in general a few meters, which corresponds to the amount of the deformation built up by the relative plate motion during the preceding 100 years. This would explain in simple terms why the recurrence time interval for a great inter-plate earthquake is of the order of 10^2 years.

In contrast, it has been inferred for intra-plate paleoearthquakes that the recurrence time interval is of the order of 10^3 to 10^4 years or longer, depending on the rate of tectonic strain buildup (e.g., Kumamoto, 1998; Matsuda, 1998). With slower loading rates, the length of the recurrence time interval can be affected more by factors other than the loading rate. Nevertheless, observations indicate that the recurrence time interval for intra-plate paleoearthquakes primarily depends on the tectonic loading rate. Specifically, for example, the recurrence time interval is of the order of 10^3 years for events on an active fault cumulatively displaced at an average rate of 1–10 mm/year along the fault, and of the order of 10^4 years for events on an active fault cumulatively displaced at an average rate of 0.1–1 mm/year along the fault. From these examples, we can understand that a difference in the recurrence time interval is primarily attributed to a difference in crustal deformation rate (or tectonic loading rate). We have to keep in mind, however, that recurrence time intervals T_i ($i = 1, 2, 3, \ldots, N$) for intra-plate seismic events occurring on an active fault in a region away from plate boundaries are widely distributed around their average value T_{av} ($= \sum_{i=1}^{N} T_i / N$). This is because the length of the recurrence time interval can be affected more by factors other than the tectonic loading rate, with slower loading rates, as pointed out above.

Let us consider why the length of the recurrence time interval for earthquakes occurring repeatedly on a single fault fluctuates. The seismogenic layer and individual faults therein are inherently heterogeneous, and therefore it is quite unlikely that nearly equal sized earthquakes occur repeatedly at a periodic interval on a single fault in such a way that frictional stick-slip events periodically repeat on a uniformly precut interface, observed in the laboratory experiments. There can be several causal factors for the recurrence time interval fluctuation. A stress perturbation produced by a dynamic rupture (or earthquake) locally occurring at a site transfers elastically through the brittle seismogenic layer to induce an earthquake rupture at another site. A stress perturbation produced by such a dynamic rupture in the seismogenic layer may also transfer through the viscoelastic layer below the base of the seismogenic layer, and in this case, a delayed influence can be exerted on other sites in the seismogenic layer, creating the possibility of a delayed rupture. At the stage where the tectonic stress is in close proximity to the critical level at which an earthquake faulting occurs, the probability of an earthquake occurrence is enhanced by the triggering effect of a stress perturbation due to an external factor such as tides. If the pressure of interstitial pore water present in a seismogenic fault zone changes for any cause, then the effective normal stress changes according to the change in pore water pressure. This causes a change in the fault strength because the fault strength depends on the effective normal stress. Thus, a change in pore water pressure in a fault zone can exert an influence on the occurrence time of an imminent earthquake on the fault. The chemical reaction between silica (SiO_2), a main component of the Earth's crust, and pore water (H_2O) is promoted at stress concentrations in the fault zone to produce weak reaction products (fault strength degradation). This may also influence the occurrence time of an earthquake.

As noted above, there are several factors having the potential to cause the recurrence time interval fluctuation. Specifically, we will see here that the recurrence time interval for dynamic shear ruptures occurring on a single heterogeneous fault is influenced by dynamic stress transfer due to local fractures on the fault, and that the recurrence time interval fluctuates based on the fault heterogeneity. Senatorski (2002) conducted a computer simulation of seismic activity occurring repeatedly on a deterministic, single fault model. In his simulation, a heterogeneous fault of asperity type was assumed as shown in Figure 7.1, in which local fault strengths are distributed randomly according to a Gaussian distribution with a mean value of 8.4 MPa and a standard deviation of 1 MPa. Senatorski's fault model is characterized by two strongest local asperity areas having the highest strength of 14.4 MPa and one weakest local area having the lowest strength of 3 MPa. This model fault was assumed to be a pure strike-slip fault of rectangular shape ($50\,\text{km} \times 30\text{km}$), and vertically embedded in a three-dimensional, elastic half-space.

In the simulation, the slip-dependent constitutive relation expressed as

$$\tau(D) = \frac{\tau_p}{c}(D + \alpha c) \exp\left[\frac{1}{2} - \frac{(D + \alpha c)^2}{2c^2}\right] \tag{7.7}$$

was assumed for the fault slip failure (Senatorski, 2002). In the above equation, τ_p denotes the peak shear strength, and α and c are constants. Specifically, it was assumed that the strength τ_p is heterogeneously distributed as shown in Figure 7.1. It was also assumed that $\alpha = 0.6$, $c = 0.3$ m, and the residual friction stress $\tau_r = 0$. Since τ_r has been assumed to be zero, τ_p equals the breakdown stress drop $\Delta\tau_b$ in this simulation. As boundary conditions, $\tau_p = 0$ was assumed along the lower edge of the fault and at greater depths, and along both lateral edges of the fault and beyond, to avoid sudden energy releases when a rupture reaches these fault edges.

The relation between shear traction τ and slip rate (or velocity) $\dot{D}$ at a local position on the fault is in general represented as shown in Figure 4.1(b). Thus, a local position on the fault is re-strengthened as $\dot{D}$ at the position approaches zero after the occurrence of an earthquake rupture. This fault re-strengthening process was treated in the simulation by introducing the function H as follows (Senatorski, 2002):

$$H = \begin{cases} 1 & \text{when the driving stress is greater than a threshold stress level.} \\ 0 & \text{otherwise} \end{cases}$$

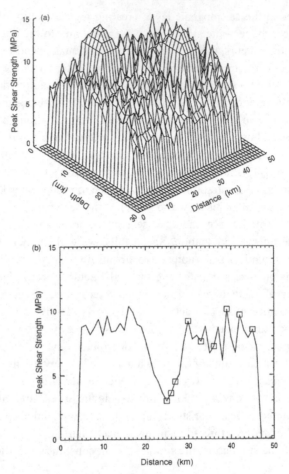

Fig. 7.1   A heterogeneous fault model of asperity type (Senatorski, 2002).

That is, the local fault strength is instantaneously restored immediately after the local slip movement stops. Although the slip rate dependence of the strength at slow slip rates has not been assimilated into the simulation, this slip rate dependence is a secondary effect on fault re-strengthening, and therefore there is no problem in examining how recurrence time intervals for dynamic ruptures occurring on a single fault are influenced by stress perturbations produced by the fault strength heterogeneity.

Figure 7.2 shows the result obtained from the computer simulation of seismic activity over 600 years on the heterogeneous fault model shown in Figure 7.1, under the assumption that the relative plate motion has a rate of 10 cm/year (Senatorski, 2002). Figure 7.2(a) indicates plots of the cumulative seismic moment released and the cumulative energy radiated against time, and Figure 7.2(b) indicates plots of the seismic energy release rate and time series variation of tectonic stress against time. From Figure 7.2(a), we can see that the trends in the cumulative seismic moment released and the cumulative seismic energy radiated both steadily slope upward with time over 600 years. If, however, we focus our

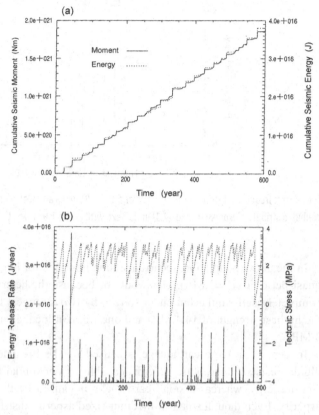

Fig. 7.2 The result obtained from the computer simulation of seismic activity over 600 years on the heterogeneous fault model shown in Figure 7.1 (Senatorski, 2002). (a) Plots of the cumulative seismic moment (solid line) and the cumulative radiated energy (dotted line) against time. (b) Plots of the seismic energy release rate (solid line) and time series variation of tectonic stress (dotted line) against time.

attention on large events having seismic energy release rates greater than 8×10^{15} J/year, we notice from Figure 7.2(b) that those large events have occurred repeatedly, not steadily or at random, but quasi-periodically on a 30 to 40 year time-interval basis.

From Figure 7.2(b), one can read individual time intervals T between adjacent large events having seismic energy release rates greater than 8×10^{15} J/year, calculate the average value T_{av} (34.4 years) and the standard deviation (± 9.7 years), and plot the relative frequency distribution of the ratio of T to T_{av} in Figure 7.3. If this relative frequency distribution is approximated by the Weibull distribution expressed as Eq. (7.1), then the parameters α and β in Eq. (7.1) can be determined by the method of maximum likelihood as follows: $\alpha = 1.941 \times 10^{-6}$, and $\beta = 3.616$ (Ohnaka and Matsu'ura, 2002). The dotted curve shown in Figure 7.3 denotes the Weibull distribution with $\alpha = 1.941 \times 10^{-6}$ and $\beta = 3.616$. From this result, we can see that the fluctuation of recurrence time intervals for large events derives from inhomogeneous distribution of fault strength.

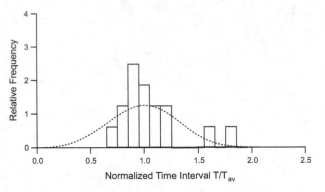

Fig. 7.3 The relative frequency distribution of the ratio of T to T_{av} (Ohnaka and Matsu'ura, 2002). The dotted curve denotes the Weibull distribution expressed as Eq. (7.1) in the text, with $\alpha = 1.941 \times 10^{-6}$ and $\beta = 3.616$.

In the above simulation, recurrence time intervals for large events were not random but quasi-periodic ($\beta = 3.616$). This would be because the heterogeneous fault model was not completely self-similar but characterized by two strongest local areas (asperities) having the highest strength of 14.4 MPa and one weakest local area having the lowest strength of 3 MPa.

In general, a large earthquake is caused by the breakdown of a large-scale asperity highly resistant to rupture growth, capable of accumulating a large amount of elastic strain energy, whereas a small earthquake is caused by the breakdown of a small-scale asperity. Even though small-to-medium-sized asperity distribution is self-similar, the size of a large-scale asperity provides a length scale departing from the self-similarity, or alternatively, the upper limit of the self-similarity. Thus, the length scale provided by such a large-scale asperity is a characteristic length representing the size (or magnitude) of a large earthquake. When such a large-scale asperity providing a characteristic length scale exists on a fault, large earthquakes can occur repeatedly at a quasi-periodic interval on the fault.

Thus, it would be understood that large earthquakes occur repeatedly on a single fault not randomly but quasi-periodically, although the recurrence time interval can fluctuate because of various secondary causal factors. This is an important fact to be recognized.

The crustal region just after a large earthquake is in a subcritical state, and crustal deformation in the region proceeds toward the critical state (at which the next earthquake occurs) under tectonic loading. The process from a subcritical state to the critical state is repeated intermittently on a single preexisting fault in regions under perpetual tectonic loading. Let us envision a brittle, inhomogeneous crustal region in which is embedded a master fault that has the potential to cause a large earthquake, as described in subsection 5.2.1. If we specifically consider the deformation process of a crustal region including a master fault, from a subcritical state toward the critical state under tectonic loading, the entire process of one cycle for a typical, large earthquake may commonly be divided into five phases, as shown in Figure 7.4 (Ohnaka, 1998, 2004a). Just after the arrest of a large-earthquake

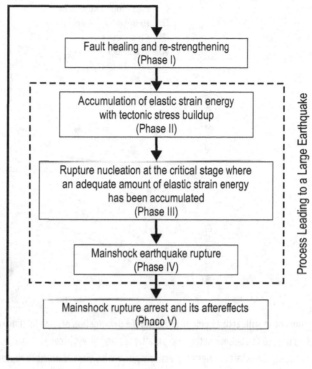

Fault healing and re-strengthening
(Phase I)

Accumulation of elastic strain energy
with tectonic stress buildup
(Phase II)

Rupture nucleation at the critical stage where
an adequate amount of elastic strain energy
has been accumulated
(Phase III)

Mainshock earthquake rupture
(Phase IV)

Process Leading to a Large Earthquake

Mainshock rupture arrest and its aftereffects
(Phase V)

Fig. 7.4   One cycle for a typical large earthquake, divided into five phases.

rupture, the preexisting master fault begins to heal and to be re-strengthened (phase I in Figure 7.4). The fault re-strengthening can be attained by an increase in the sum of the real areas of asperity contact, such as asperity-interlocking and asperity-ploughing, on the fault surfaces with progressive deformation along the fault under tectonic loading, and hence the elastic strain energy can again accumulate in the region surrounding the fault (phase II in Figure 7.4), as tectonic stress builds up. The fault thus regains the potential to cause the next large earthquake.

At an early stage of phase II, the tectonic stress is far below the critical level, and the amount of the elastic strain energy accumulated in the region is inadequate. This stage is therefore characterized by quiescent seismicity (i.e., background seismicity; see Figure 7.5), and may be recognized as a quiescent period of seismicity when attention is paid to its time domain, and as a seismic gap when attention is paid to its spatial domain. Indeed, the well-known concept of seismic gaps (Imamura, 1928/29; Fedotov, 1965; Mogi, 1968; Sykes and Nishenko, 1984) was proposed several decades ago on the basis of seismicity studies.

Figure 7.5 shows a model of seismic activity enhanced during the crustal deformation process leading up to a large earthquake. Seismic activity occurs after tectonic stress exceeds a threshold level (point a in the figure), and thereafter seismicity becomes progressively active with time.

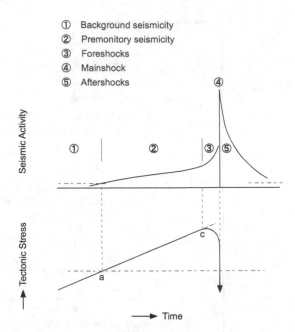

① Background seismicity
② Premonitory seismicity
③ Foreshocks
④ Mainshock
⑤ Aftershocks

Fig. 7.5 A model of seismic activity enhanced during the deformation process leading up to a large earthquake (Ohnaka and Matsu'ura, 2002). Seismic activity occurs after tectonic stress exceeds a threshold level (point a in the figure), and thereafter seismicity becomes progressively active with time. Mainshock rupture nucleation begins to proceed at point c in the figure.

At later stages of phase II, the tectonic stress reaches higher levels, and gradually approaches the critical level at which an earthquake occurs. Accordingly, the crustal region begins to behave in-elastically, and seismicity in the region surrounding the master fault becomes progressively active with time, as shown in Figure 7.5. This is because the Earth's crust is inherently heterogeneous and includes numerous faults of small-to-moderate sizes. In particular, at the stage where the tectonic stress is in close proximity to the critical level, seismic activity is enhanced by fluid-rock interaction, and by the triggering effect of stress transfer. Thus, the later stages of phase II can be characterized by activation of seismicity. Activated seismicity at later stages of phase II may be recognized as premonitory phenomena for an ensuing large earthquake that will take place on the master fault (see Figure 7.5).

Eventually, crustal deformation tends to concentrate locally along the master fault, and a rupture nucleation begins to occur at a location where the resistance to rupture growth is lowest on the fault (phase III in Figure 7.4), when the tectonic stress has attained the critical level, and when an adequate amount of elastic strain energy has been accumulated. The nucleation process necessarily leads to the ensuing mainshock rupture on the fault (phase IV in Figure 7.4), accompanied by a rapid stress drop and the dissipation of a great amount of the elastic strain energy, resulting in the radiation of seismic waves. The arrest of the mainshock rupture results in its aftereffects that include re-distribution of local stresses on and around the fault area, leading to aftershock activity (phase V in Figure 7.4).

Aftershock activity tends to migrate from the mainshock hypocentral region toward the periphery, and further outward, with time. The aftereffects of the mainshock arrest can be characterized by time-dependent stress relaxation and stress transfer.

If geometric irregularities (or micro-asperities) of short wavelength components are superimposed on sizable asperity areas on a master fault, slow slip failure at such an asperity area will necessarily bring about the fracture of micro-asperities superimposed on the asperity area in the brittle regime. If this is the case, the fracture of micro-asperities occurs during the nucleation process despite the fact that the overall shear stress decreases (because of slip-weakening) in the nucleation zone. Accordingly, the nucleation process possibly carries micro-earthquakes (i.e., immediate foreshocks; see Figure 7.5). Immediate foreshock activities induced during the mainshock nucleation process will be described in the next section.

What has been presented is a model of the cycle process for a typical, large earthquake that takes place on a master fault in the brittle seismogenic layer characterized by heterogeneities. In this model, phases I to V come in a cycle. In particular, phases II and III are integral parts of the process leading up to a typical, large earthquake that takes place in the brittle seismogenic layer. Thus, we will focus on seismicity activated in phases II and III in the next section.

7.2 The process leading up to a large earthquake and seismic activity

7.2.1 Seismic activity at later stages of the recurrence interval

The Kanto earthquake that occurred along the Sagami Trough in Sagami Bay (e.g., Kanamori and Ando, 1973; Ando, 1974) on September 1, 1923 was one of the most disastrous earthquakes in the history of Japan. The total number of dead and missing exceeded 140 000, the number of partially or completely destroyed houses exceeded 250 000, and more than 440 000 houses burned down. The earthquake was later estimated to have had a magnitude (M) of 7.9 (on the JMA magnitude scale) or 8.2 (on the surface-wave magnitude scale). The Kanto area (Figure 7.6) includes Metropolitan Tokyo and Yokohama, both of which are densely populated, and hence it has been, and still is, of great concern from the viewpoints of earthquake forecasting and of disaster prevention, to know whether or not the 1923 Kanto earthquake was preceded by anomalous seismic activity. So, first let us cast a spotlight on the sequence of seismic activity in the Kanto area before and after the 1923 Kanto earthquake.

Figure 7.7 shows the spatial distribution of epicenters of shallow focus earthquakes with $M \geq 5.8$ and focal depth < 80 km in the Kanto area during the period 1885–1982 (Ohnaka, 1984/85). The area shown in Figure 7.7 is specified by the range 34.4 °N– 37.0 °N latitude and 138.5 °E–141.6 °E longitude, and tentatively called the Kanto area. In plotting the location of the earthquake epicenters on the map shown in Figure 7.7,

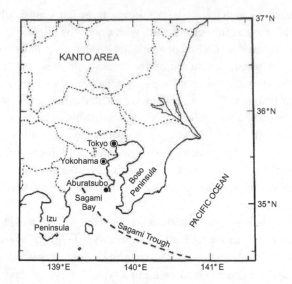

Fig. 7.6 Kanto area, including Metropolitan Tokyo and Yokohama which are densely populated (Ohnaka, 1984/85).

earthquakes with $M \geq 5.8$ were selected from reliable catalogues of earthquake data, in order to avoid apparent change in seismic activity with time. Although the most reliable catalogues, published by Utsu (1979, 1981, 1982a, 1982b) and the Japan Meteorological Agency (JMA) (1982), were used, we had no choice but to select earthquakes with $M \geq 5.8$ from the catalogues. This is because the observation capability for smaller earthquakes was improved with time, and because more small earthquakes with $M < 5.8$ were observed as measurement capabilities were improved during the period 1885–1982. If, therefore, we had included such smaller earthquakes, a real change in seismic activity with time would have been masked by observational deficiencies.

In Figure 7.7, the broken-line ellipse indicates the focal region of the 1923 Kanto earthquake, and the individual circles indicate the epicentral locations of earthquakes that took place during the period January 1885–August 1982. In particular, Figure 7.7(c) shows the epicentral locations of the Kanto earthquake (mainshock, largest black circle in the figure), and the subsequent aftershocks, plus other earthquakes that occurred during the 2-year period from September 1, 1923 to August 31, 1925. Figures 7.7(a), (b), (d), (e), and (f) show the epicentral locations of earthquakes that occurred during every 19-year interval from January 1885 to August 1904, from September 1904 to August 1923, from September 1925 to August 1944, from September 1944 to August 1963, and from September 1963 to August 1982, respectively. From these figures, one may notice that seismic activity in the Kanto area had been relatively high during the 20 years preceding the Kanto earthquake, and continued at elevated levels for more than 20 years immediately after it.

In order to clarify the temporal transition of seismic activity before and after the Kanto earthquake, the cumulative number of earthquakes generated in the Kanto area was plotted against time in Figure 7.8. A broken straight line in the figure indicates a background seismicity level accumulation with time. From this figure, one can see clearly that seismicity had been active for the nearly 20 years before the Kanto earthquake, and that active

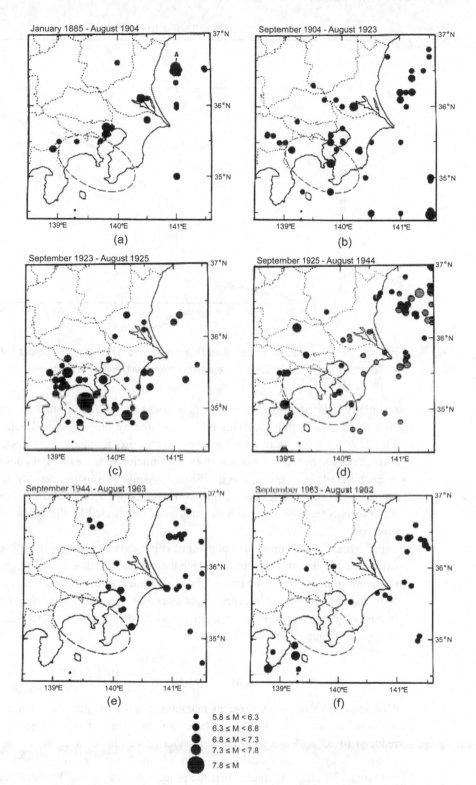

Fig. 7.7 Spatial distribution of epicenters of shallow focus earthquakes in the Kanto area from January 1885 to August 1982 (Ohnaka, 1984/85). The broken-lined ellipse indicates the focal region of the 1923 Kanto earthquake, and the black circles indicate the epicentral locations of earthquakes with $M \geq 5.8$. The epicentral location distributions cover the following dates: (a) January 1885–August 1904; (b) September 1904–August 1923; (c) September 1923–August 1925; (d) September 1925–August 1944; (e) September 1944–August 1963; and (f) September 1963–August 1982.

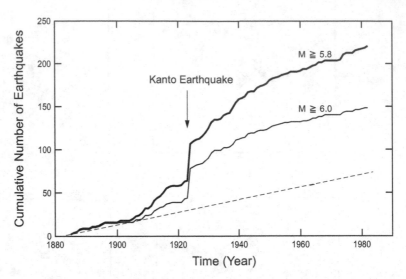

Fig. 7.8 Time variation of the cumulative number of earthquakes generated in the Kanto area (Ohnaka and Matsu'ura, 2002). The broken line in the figure indicates a background seismicity level accumulation with time.

seismicity reduced transitionally toward a steady level for more than 20 years immediately after the Kanto earthquake. This is corroborated by more qualified data of earthquakes with $M \geq 6.0$ (see Figure 7.8). The seismic activation for nearly 20 years before the Kanto earthquake can be recognized as activated seismicity at later stages of phase II as presented in Section 7.1. In contrast, the transitional change of seismic activity to a steady level immediately after the Kanto earthquake may be understood as aftershocks generated in the course of the relaxation of local stresses re-distributed by the arrest of the mainshock rupture.

In Figure 7.8, the cumulative number of earthquakes of $M \geq 6.0$ may seem remarkably small compared with the cumulative number of earthquakes of $M \geq 5.8$, despite only a slight difference of 0.2 in the lower-limit magnitude. Note, however, that such a remarkable decrease in the number of larger earthquakes comes from a natural property of the seismicity frequency–magnitude distribution. Specifically, from the Gutenberg–Richter frequency–magnitude relation (Eq. (6.15)), we have

$$\frac{N(M+1)}{N(M)} = \frac{10^{-b(M+1)}}{10^{-bM}} = 10^{-b} \cong 10^{-1}. \qquad (7.8)$$

This indicates that an increase in magnitude of 1.0 in general reduces the number of earthquakes to a tenth. Likewise, the cumulative number of earthquakes of $M \geq 6.0$ is reduced to 63% of the cumulative number of earthquakes of $M \geq 5.8$, as reflected in Figure 7.8, by the specific difference of 0.2 in the lower-limit magnitude.

Figures 7.7 and 7.8 suggest that the process leading up to the 1923 Kanto earthquake is a case where activity in moderate-sized earthquakes progressively increases with time over several decades until a great earthquake occurs. Such a phenomenon of activated seismicity during the process leading up to a great or major earthquake has been widely

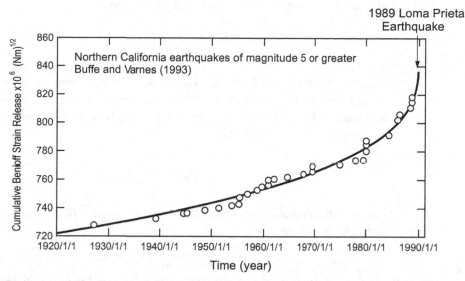

Fig. 7.9 A plot of the cumulative Benioff strain release against time. The open circles indicate data points of northern California earthquakes with $M \geq 5.0$ during the period 1927–1988, and the thick curve represents the best-fit solution for n and t_f in Eq. (7.10). Reproduced from Bufe and Varnes (1993).

observed in other regions as well (e.g., Sykes and Jaume, 1990; Bufe and Varnes, 1993; Bufe et al., 1994; Brehm and Braile, 1998; Jaume and Sykes, 1999), as exemplified below (see Figure 7.9).

Thus, let us focus on how such seismic activity can be accounted for in quantitative terms in the context of the seismic cycle model presented in Section 7.1. Phase II defined in Section 7.1 can be regarded as the preparation process for a large earthquake, in the sense that the elastic strain energy as the driving force to bring about the next large earthquake builds up in this phase, and also in the sense that the crustal deformation proceeds toward catastrophic failure with tectonic loading. In view of this property of phase II, models for phenomena observable at the intermediate-term stage of the process leading up to a large earthquake have been proposed.

A typical model for this is the time-to-failure function model (e.g., Bufe and Varnes, 1993; Bufe et al., 1994; Jaume and Sykes, 1999). The model focuses on accelerating seismic activity observed during the deformation process of the inhomogeneous crust leading up to a large earthquake with tectonic loading, and its theory is based on the laboratory observation that the deformation process of an inhomogeneous body leading to catastrophic failure obeys a power law of the form (Saito, 1969; Varnes, 1989)

$$\frac{d\Omega(t)}{dt} = \frac{k}{(t_f - t)^n}. \qquad (7.9)$$

In the above equation, $\Omega(t)$ denotes strain or any measurable quantities, such as event count, Benioff strain, or seismic moment, as a function of time t, t_f denotes the time of failure, and k and n are constants. Integration of Eq. (7.9) leads to

$$\Omega(t) = \Omega_f - \frac{k}{1-n}(t_f - t)^{1-n} \qquad (n \neq 1),$$ (7.10)

where $\Omega_f \equiv \Omega(t_f)$.

Figure 7.9, reproduced from Bufe and Varnes (1993), exemplifies the relation between $\Omega(t)$ and time t for the period 1927–1988 for northern California earthquakes of magnitude 5 or greater. In Figure 7.9, $\Omega(t)$ is defined as the cumulative Benioff strain release, which is given by (Bufe and Varnes, 1993)

$$\Omega(t) = \sum_{i=1}^{N(t)} \omega_i,$$ (7.11)

where ω_i is the Benioff strain release defined by $\omega_i = \sqrt{E_{si}}$ (E_{si}, seismic energy) for each event, calculated from

$$\log \omega_i = 0.75M_i + d.$$ (7.12)

Here, ω_i and M_i represent the Benioff strain release and the magnitude, respectively, of the ith event in the sequence of N earthquakes that occurred in a certain region, and d is a constant.

In Figure 7.9, open circles indicate data points of earthquakes with $M \geq 5.0$, and the thick curve graphically represents the best-fit solution for n and t_f in Eq. (7.10). The Loma Prieta earthquake with $M7.1$ occurred on October 18, 1989 (UTC, Coordinated Universal Time). Figure 7.9 indicates that activity in moderate-sized earthquakes in the region accelerated progressively towards the 1989 Loma Prieta event, and that the accelerating seismicity can be well accounted for by the power law of Eq. (7.10).

The observation of accelerating seismicity in a region, as presented above, provides indirect evidence that the regional fault network system has approached criticality, and in this case the largest event is produced finally at the critical point. Using data on earthquakes covering a relatively wide range of magnitudes from 4.8 to 8.6, it has been found empirically that the magnitude of the final event in a region scales with the logarithm of the critical size of the region (Bowman et al., 1998), as shown in Figure 7.10. The empirical scaling relation shown in Figure 7.10 can be rewritten in terms of the critical region radius R_c and the ruptured fault area S of the final event, as follows (Rundle et al., 2000):

$$R_c = 10S^{1/2}.$$ (7.13)

This relation indicates that the preparation process for a larger earthquake develops in a broader region, and that accelerating seismic activity during the crustal deformation process leading up to a great or major event occurs in a region having a critical radius R_c of ten times larger than the characteristic fault length $S^{1/2}$.

Let us consider another model. A large earthquake occurs on a subduction zone plate boundary fault segment along an ocean trench or trough. In this case, the plate boundary fault between an overlying (continental) plate and a subducting (oceanic) plate (called a "slab") is locked during the interseismic period, because of frictional resistance on the fault plane. Thus, the contact edge of the overlying plate is dragged down by the subducting slab.

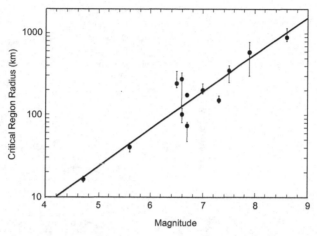

Fig. 7.10   Empirical relationship between the magnitude of the final seismic event in a region and the critical size of the region. Reproduced from Bowman *et al.* (1998).

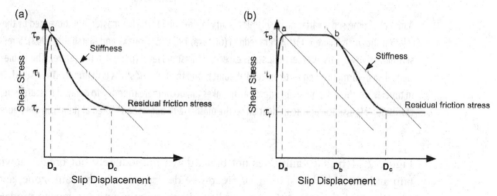

Fig. 7.11   Two different types of fault constitutive relations (Ohnaka and Matsu'ura, 2002). (a) A typical fault constitutive relation. In this type, the fault immediately breaks down when tectonic shear stress attains the critical level τ_p of the fault strength. (b) A fault constitutive relation accompanied by a certain amount of stable sliding or creep-like deformation. In this type, the fault does not break down until slip displacement reaches the critical value D_b, despite the fact that the tectonic shear stress has attained the critical level τ_p of the fault strength.

Consequently, the overlying plate is forced to deform in order for its contact edge to keep up with the subducting slab. In this situation, tectonic shear stress along the plate boundary fault plane builds up and elastic strain energy also accumulates owing to the elastic shear deformation of the overlying plate during the interseismic period.

Eventually, the tectonic shear stress attains the critical level of the fault strength, and immediately the fault breaks down when the constitutive relation for the fault is expressed in the form shown in Figure 7.11(a). In this case, time-series variations of tectonic stress and resultant seismic activity are modeled as already shown schematically in Figure 7.5. However, when the constitutive relation is expressed in such a form as shown in

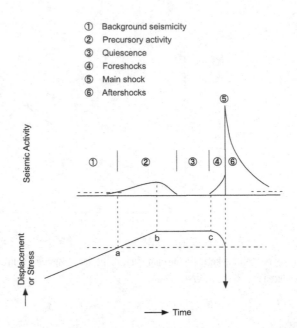

Fig. 7.12 A model of time series variation of seismic activity in a region in which a master fault governed by the constitutive relation shown in Figure 7.11(b) is embedded (Ohnaka, 1984/85; Ohnaka and Matsu'ura, 2002). Stress-induced seismic activity occurs after tectonic stress exceeds a threshold level (point a in the figure). The stable sliding or creep-like deformation begins to proceed at point b, and the tectonic stress is maintained at a steady level after point b in time. Nevertheless, time-dependent earthquakes can occur thereafter due to delayed fracture (stress-aided corrosion). Mainshock rupture nucleation on the master fault begins to proceed at point c in the figure.

Figure 7.11(b), the fault does not break down immediately but breaks down after a certain amount of stable sliding, or creep-like deformation in the fault zone, and time-series variations of the tectonic stress and resultant seismic activity can be modeled as shown schematically in Figure 7.12. A significant feature of this model is that the tectonic stress is maintained at a critical level prior to the occurrence of the fault breakdown resulting in an earthquake. In this case, the fault does not break down until slip displacement D reaches the critical value D_b, despite the fact that the tectonic shear stress τ has attained the critical level τ_p of the fault strength. In other words, the onset of the fault breakdown is governed not by the fault strength τ_p but by the critical displacement D_b. In this respect, the properties of the fault governed by the constitutive relation expressed in the form shown in Figure 7.11(b) differ distinctly from those of the fault governed by the constitutive relation expressed in the form shown in Figure 7.11(a).

There is a clear difference in time-series variation of seismic activity between faults whose breakdown is governed by the constitutive relation shown in Figure 7.11(a) and faults preceded by stable sliding or creep-like deformation and governed by the constitutive relation shown in Figure 7.11(b). Let us examine the latter case in more detail.

In Figure 7.12, activity in moderate-size earthquakes begins to occur at stress level a in a relatively wide region, and progressively increases with an increase in the tectonic stress while the subducting plate boundary fault is locked. Stable sliding or creep-like deformation begins to proceed at point b in time, and the tectonic stress is maintained at a steady level after point b in time. Seismic activity induced by stress-increase ceases while the tectonic stress is held constant. This does not mean, however, that seismic activity is not observable after point b in time. Moderate-size earthquakes can occur due to delayed fracture even after the tectonic stress is held constant.

In general, a rock material to which a constant load is applied can fracture after the elapse of a certain period of time, despite the fact that the applied load is less than the strength of the material. Such a fracture is often called delayed fracture. Chemical reaction between rock-forming minerals and corrosive agents is promoted at locations of stress concentration such as crack-tips to produce weakened products. If the strength of the products of the reaction is weaker than the strength of the original material, the material necessarily fractures under the applied constant load. Specifically, strong Si–O bonds of silica minerals at some locations of stress concentration can react with water molecules (H_2O) operating as a corrosive agent, to produce very weak hydrogen-bonded hydroxyl groups (e.g., Michalske and Freiman, 1982; Atkinson and Meredith, 1987; see also subsection 3.4.2), which can easily be broken at low stresses. Indeed, it has been shown in laboratory experiments that rock strength can be weakened by stress corrosion (e.g., Scholz, 1972; Atkinson, 1982, 1984). It has also been shown in laboratory experiments that the cumulative number $N(t)$ of acoustic emission events generated in a brittle rock specimen to which a constant compressive load is applied is proportional to the logarithm of time t; that is (Ohnaka, 1983),

$$N(t) = A \log(t + 1) + N_0, \tag{7.14}$$

where A and N_0 are constants, and t is the time measured from the point in time when the applied load is held constant. Thus, the rate $n(t)$ of the occurrence of acoustic emission events caused by delayed fracture lessens with time t, in accordance with

$$n(t) = \frac{dN(t)}{dt} = \frac{A}{(t + 1)}. \tag{7.15}$$

This empirical equation can be derived theoretically (Scholz, 1968; Ohnaka, 1983; Ohnaka and Matsu'ura, 2002), as shown below.

Let us consider the situation where a constant compressive stress σ_c is applied to a heterogeneous brittle rock in a corrosive environment at time $t = 0$. Diverse cracks and pores preexisting in the rock fluctuate the stress field, and the stress corrosion accelerates at locations of stress concentration such as crack-tips where tensile stresses are locally high. Therefore, acoustic emissions are generated by microcracking at locations of stress concentration in such a heterogeneous rock body in the brittle regime, even under a constant compressive load. We assume that the heterogeneous body can be divided into a large number of small regions that contain only one source of stress concentration. Let σ be the maximum local stress around the location of stress concentration in each small region, and let σ_s be the instantaneous fracture strength of a small region. A small region in the body locally fails when σ becomes equal to σ_s in the region. Let $f(\sigma, \sigma_c, t)d\sigma$ be the probability that the maximum local stress σ in a small region is in a range from σ to $\sigma + d\sigma$ at time t, where $f(\sigma, \sigma_c, t)$ denotes the probability density function of σ. When each small region undergoes static fatigue independently, the mean fracture time $\bar{t}$ can be expressed by the following equation (Yokobori, 1955; Wiederhorn *et al.*, 1980):

$$1/\bar{t} = a \exp[-(E - b\sigma)/RT], \tag{7.16}$$

where a and b are constants, E is the activation energy of the corrosion reaction, R is the gas constant, and T is absolute temperature. The probability $g(\sigma)dt$ that a small region at stress σ will fracture in a short time interval from t to $t + dt$ is related to the mean fracture time $\bar{t}$ by (Yokobori, 1955; Scholz, 1968)

$$g(\sigma)dt = dt/\bar{t}. \tag{7.17}$$

Thus, we have from Eqs. (7.16) and (7.17)

$$g(\sigma)dt = a \exp[-(E - b\sigma)/RT]dt. \tag{7.18}$$

The rate $n(t)(= dN(t)/dt)$ of acoustic emissions generated by microfractures within the body is given by

$$n(t) = \frac{dN(t)}{dt} = N^* \int_0^{\sigma_s} f(\sigma, \sigma_c, t)g(\sigma)d\sigma, \tag{7.19}$$

where N^* is a constant. If each region fails only once, the number of available small regions in a stress range $\sigma \sim \sigma + d\sigma$ decreases at a rate expressed in the following equation:

$$\frac{\partial}{\partial t} f(\sigma, \sigma_c, t) = -f(\sigma, \sigma_c, t)g(\sigma). \tag{7.20}$$

From this equation, we have

$$f(\sigma, \sigma_c, t) = f(\sigma, \sigma_c, 0) \exp[-g(\sigma)(t + t_0)], \tag{7.21}$$

where t_0 is a constant. From Eqs. (7.18) and (7.21), Eq. (7.19) can be rewritten as follows:

$$n(t) = \frac{dN(t)}{dt} = \frac{N^*RT}{b} \int_{g(0)}^{g(\sigma_s)} f(\sigma, \sigma_c, 0) \exp[-g(\sigma)(t + t_0)]dg(\sigma). \tag{7.22}$$

Under the condition that $f(\sigma, \sigma_c, 0)$ is continuous for $0 \leq \sigma \leq \sigma_s$ and does not as steeply vary with σ as $\exp[-g(\sigma)(t + t_0)]$ does, Eq. (7.22) can be simply expressed as (Scholz, 1968; Ohnaka, 1983; Ohnaka and Matsu'ura, 2002)

$$n(t) = \frac{dN(t)}{dt} \cong \frac{N^*RT}{b} \frac{S(\sigma_c)}{t + t_0}, \tag{7.23}$$

where $S(\sigma_c)$ is a function of σ_c. Thus, the form of the empirical equation (7.15) has been derived theoretically. Comparing Eq. (7.23) with empirically derived Eq. (7.15), we have for A and t_0

$$\left. \begin{array}{l} A = \dfrac{N^*RTS(\sigma_c)}{b} \\[2mm] t_0 = 1 \end{array} \right\}. \tag{7.24}$$

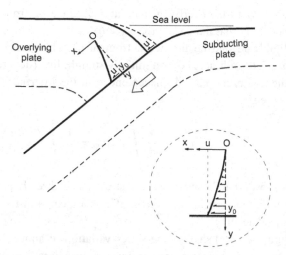

Fig. 7.13   Diagram showing the cross section of a subducting plate boundary zone.

With these results in mind, it would be easily understandable that the strengths of minor asperities on the plate boundary interface and/or minor preexisting faults in a region surrounding the plate boundary fault can be weakened by stress corrosion. Furthermore, it can be expected that seismic activity occurring after the tectonic stress has been held constant (after point b in time in Figure 7.12) lessens with time in accordance with Eq. (7.15).

A pre-slip (or stable sliding) in the localized but growing zone of a mainshock rupture nucleation on the fault begins to accelerate at point c in time in Figure 7.12, and collaterally causes local fractures at minor asperities or projections distributed in and around the zone. This activates seismicity (immediate foreshocks, ④ in Figure 7.12) in the vicinity of the mainshock hypocenter.

Stress corrosion can have a significant effect on seismic activity at depths to several tens of km in a subduction zone. Water will permeate the Earth's crust to some depth from the ground surface (or sea bottom). Water at greater depths may be conveyed by the subducting plate of the oceanic lithosphere from the sea bottom and released by the decomposition reaction of hydrous minerals in the subducting plate. The decomposition reaction can occur at different depths to release water in stages (Raleigh and Lee, 1969; Murrell and Ismail, 1976), resulting in marked weakening and embrittlement of rock containing hydrous minerals (Raleigh and Paterson, 1965; Murrell and Ismail, 1976). In addition, subcritical crack growth in rock caused by stress corrosion is enhanced by increased temperature (Meredith and Atkinson, 1985). All of these favor the assumption that stress corrosion can have a significant effect on activity in moderate-size earthquakes at depths to several tens of kilometers in a subduction zone.

Figure 7.13 is a diagram showing the cross section of a subducting plate boundary zone, where an oceanic plate (or slab) is subducted underneath a continental plate. Let the xz plane be parallel to the subducting plate boundary plane, and let the y-axis be normal to the subducting plate boundary plane. In addition, let the origin of the coordinate system be point O in the overlying plate at a distance y_0 from the plate boundary fault plane, chosen such that the shear stress τ_{yx} ($= \mu \partial u / \partial y$, where μ is the modulus of rigidity) is negligible beyond the distance y_0 from the plate boundary fault plane. A curved solid line in Figure 7.13 indicates how an initially straight line (denoted by a broken line) is deformed by the overlying plate displacement u on the interface. The average shear stress $\overline{\tau}_{yx}$ is expressed as

$$\overline{\tau}_{yx} = \frac{1}{y_0} \int_0^{y_0} \tau_{yx} dy = \frac{1}{y_0} \int_0^{y_0} \left(\mu \frac{\partial u}{\partial y} \right) dy = \frac{\mu}{y_0} u. \tag{7.25}$$

This shows that the average shear stress $\overline{\tau}_{yx}$ is directly proportional to the displacement u, of which the vertical component affects a change in the altitude at a spot on land above mean sea level.

Figure 7.14(a) shows time-series variation of annual mean sea level, measured with a tide gauge at Aburatsubo on the coast of the Miura Peninsula, Kanagawa Prefecture, Japan (Ohnaka, 1984/85; data from Coastal Movements Data Center, 1976). The 1923 Kanto earthquake occurred just beneath Aburatsubo (see Figure 7.6). During the interseismic period, a slow increase in the mean sea level resulted from the subsidence of the overlying plate on which Aburatsubo is located. A rapid backlash change in the mean sea level was caused by the elastic rebound of the overlying plate when the Kanto earthquake occurred. The Kanto earthquake resulted in a coseismic uplift of 140 cm at Aburatsubo (Ohnaka, 1984/85).

Relatively short period variations in sea level, such as daily or seasonal variations due to meteorological conditions, oceanic conditions or tidal effects, have been eliminated from the annual mean sea level data. However, sea level variations over a period longer than one year, due to oceanic conditions such as the motion of a cold water mass, are contained in the data. Hence, low-pass digital filtering (30 year cutoff period and 24 dB/octave attenuation slope) was conducted for the annual mean sea level data to eliminate such long-period sea level variations resulting from oceanic conditions (Ohnaka, 1984/85). Before filtering, the coseismic offset of 140 cm associated with the occurrence of the Kanto earthquake was removed from the original data, and the offset was restored after filtering. The low-pass filtered data are plotted against time in Figure 7.14(b).

Figures 7.14(a) and (b) indicate that there is a systematic deviation of the data points from an extrapolation of the linear upward trend over about 10 years prior to the Kanto earthquake. Such a systematic deviation can be seen as a precursory crustal deformation resulting from pre-slip (stable sliding) or creep-like deformation in the plate boundary fault zone prior to the occurrence of the Kanto earthquake. Figure 7.15 shows time-series variation of the low-pass filtered data on the annual mean sea level, in comparison to the sequence of seismic activity prior to the 1923 Kanto earthquake. In Figure 7.15(b), the low-pass filtered annual mean sea level data are displayed as a thick solid line, and the result of a simple running average of five data points of the annual mean sea level is displayed as the black dots connected by thin lines.

The annual number of earthquakes with $M \geq 5.8$ that occurred in the Kanto area during the period from 1904 to 1926 is plotted against time in Figure 7.15(a). Figure 7.15(a)

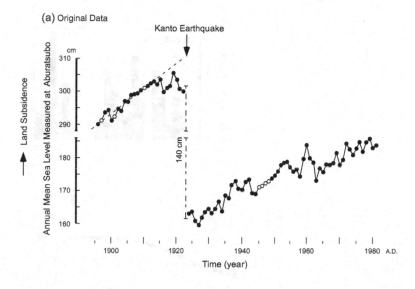

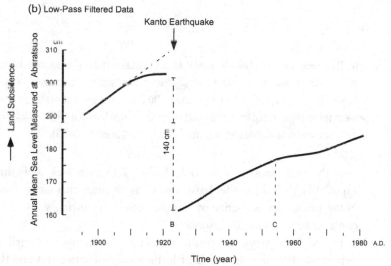

Fig. 7.14 Time-series variation of annual mean sea level, measured with a tide gauge at Aburatsubo on the coast of the Miura Peninsula, Kanagawa Prefecture, Japan (Ohnaka, 1984/85; original data from Coastal Movements Data Center, 1976). The solid smooth curves in the bottom figure denote the low-pass filtered data plotted against time (for details, see text).

indicates that the 1923 Kanto earthquake was preceded by a sequence of precursory seismic activity, quiescence, and immediate foreshocks. In Figure 7.15, the initiation time of the arrest of land subsidence is coincident with the time at the peak of the precursory seismic activity, marked A. The aforementioned model shown schematically in Figure 7.12 can thus be reasonably applied to precursory phenomena prior to the 1923 Kanto earthquake, shown in Figure 7.15. If it is confirmed that the cumulative number of earthquakes occurring

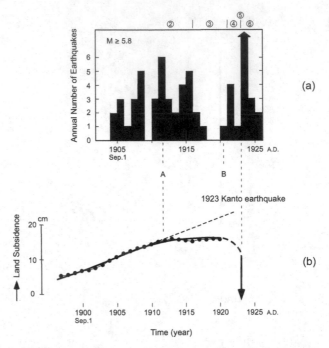

Fig. 7.15 (a) Time-series variation of annual number of earthquakes with $M \geq 5.8$ that occurred in the Kanto area from 1904 to 1926. Five seismic episodes are shown: ② seismic activity prior to quiescence, ③ quiescence, ④ immediate foreshocks, ⑤ mainshock, and ⑥ aftershocks. (b) Time-series variation of the low-pass filtered data on the annual mean sea level (thick solid line). Black dots connected by thin lines denote five-point moving averages of the annual mean sea level data. Reproduced with modification from Ohnaka (1984/85).

over the time interval A to B in Figure 7.15 increases with time in accordance with Eq. (7.14), then the model based on the assumption that stress corrosion plays a key role in the precursory sequence of seismic activity occurring while the tectonic stress is held constant becomes more reasonable.

Figure 7.16 shows a plot of the cumulative number of earthquakes occurring from September 1911 to August 1921 (the time interval between A and B in Figure 7.15) against the logarithm of time, where the origin of time has been chosen such that $t = 0$ at point A in time in Figure 7.15. From Figure 7.16, it can be confirmed that the cumulative number of earthquakes occurring over the period of time from A to B shown in Figure 7.15 basically obeys the logarithmic law expressed in Eq. (7.14).

In this subsection, we have seen that not all of a sequence of seismic activity observed during the process leading up to a great or major earthquake can be accounted for in terms of the accelerating seismic activity model, as depicted in Figure 7.5. In this model, the fault breaks down immediately after the tectonic shear stress reaches the critical level of the fault strength. However, we have also seen another case in which the fault does not break down immediately after the tectonic shear stress has reached the critical level of the fault strength, but breaks down after a certain amount of stable sliding or creep-like deformation in the fault zone. In this case, a sequence of seismic activity during the process leading up

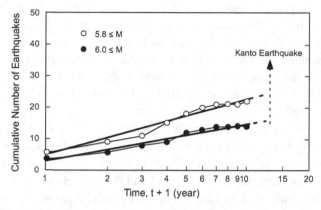

Fig. 7.16 A plot of the cumulative number of earthquakes that occurred in the Kanto area from September 1911 to August 1921 (the time interval between A and B in Figure 7.15(a)) against the logarithm of time. Reproduced with a slight modification from Ohnaka (1984/85).

to a great or major earthquake can be modeled as depicted in Figure 7.12. A difference in time-series variation of seismic activity between these two models is attributed to a difference in fault constitutive relations as shown in Figures 7.11(a) and (b).

7.2.2 Seismic activity immediately before a mainshock earthquake

As noted previously (see Chapter 5), an earthquake rupture (or unstable, dynamic shear rupture) on a preexisting fault characterized by inhomogeneities cannot begin to propagate abruptly at a high speed close to elastic wave velocities, immediately after the onset of the rupture growth, but is necessarily preceded by the transition process (called the nucleation process) consisting of two phases: an initial phase of stable and quasi-static rupture growth, and the subsequent phase of accelerating rupture extension. A physical model of the nucleation process for typical earthquakes has been shown schematically in Figure 5.17, based on the results elucidated by well-prepared, high-resolution laboratory experiments on frictional slip rupture (see Chapter 5). Given this, the nucleation process termed as phase III in Figure 7.4 will be understood as the preparation process toward an imminent earthquake rupture.

Thus, as described in Section 5.2, prior to a mainshock rupture, a localized zone on the fault slips initially at a slow and steady rate, and subsequently at accelerating rates, during the nucleation process, and the zone grows as slip progresses. Such a pre-slip within the localized but growing zone of a mainshock rupture nucleation collaterally causes local fractures at minor projections (or asperities) distributed in and around the zone. This leads to an activation of seismicity in the vicinity of the mainshock hypocenter, which rationally accounts for the decades-long observation that immediate foreshock activities concentrate in the vicinity of the epicenter of the mainshock. In other words, immediate foreshock activities are seismic activities induced as a secondary effect during the nucleation process of a mainshock rupture.

In general, however, whether or not an earthquake carries foreshock activity during the nucleation process depends upon the structural heterogeneity of the fault zone and the ambient conditions such as temperature, as noted in Chapter 5. Therefore, not all earthquakes carry foreshock activities. For example, let us think about an earthquake that begins to rupture at the base of the brittle seismogenic layer on a transform plate boundary fault. At a transform fault boundary, slips occur intermittently by seismic faulting at shallow crustal depths (in the brittle seismogenic layer), while aseismic slip or ductile shear deformation progresses continuously at greater depths, driven by horizontal steady creep-flow motion in the lower crustal asthenosphere (Savage and Burford, 1973; Turcotte and Spence, 1974; Li and Rice, 1983a). Within the brittle seismogenic layer, the fault is locked in the interseismic period, and therefore elastic strain energy can be accumulated in the elastic medium surrounding the locked part of the fault as the driving force for the next earthquake. At greater depths below the locked part of the fault, however, aseismic slip or ductile shear deformation progresses even in the interseismic period, and therefore elastic strain energy cannot be accumulated.

In this situation, earthquake rupture nucleation is liable to progress stably and quasi-statically below the base of the seismogenic layer, and the tip of the nucleation zone progresses upward uni-directionally according to a steady buildup of tectonic stress (Dmowska and Li, 1982; Li and Rice, 1983b; Yamashita and Ohnaka, 1992). This stable and quasi-static upward progression of the nucleation zone tip develops into dynamic instability, shortly after the tip of the nucleation zone has penetrated a sizable patch area highly resistant to rupture growth at or near the base of the seismogenic layer. If an adequate amount of elastic strain energy is released by the breakdown of the sizable patch area highly resistant to rupture growth, the dynamic instability gives rise to an overall seismic rupture of the locked part of the fault zone. In the course of the nucleation that grows below the base of the brittle seismogenic layer, aseismic slip or ductile deformation progresses, and therefore foreshocks are not induced.

Thus, a common characteristic of large earthquakes that begin to nucleate below the base of the brittle seismogenic layer along a transform plate boundary would be that there would be no immediate foreshocks. A typical example of this may be the aforementioned 1989 Loma Prieta earthquake, which began suddenly at a depth of 17.6 km on the San Andreas transform plate boundary fault (US Geological Survey Staff, 1990), near the San Francisco Bay area (Figure 7.17). The Loma Prieta earthquake rupture spread uni-directionally upward and bilaterally along the strike, with nearly equal components of right-lateral strike-slip and reverse slip, and indeed this earthquake carried no immediate foreshock activity (see US Geological Survey Staff, 1990). Almost all aftershocks of the Loma Prieta earthquake were located at depths shallower than the depth of the mainshock hypocenter (US Geological Survey Staff, 1990). This suggests that the base of the seismogenic layer in this particular region is at the depth of the mainshock hypocenter.

One may argue that the June 27, 1988 ($M5.0$) and August 8, 1989 ($M5.2$) Lake Elsman earthquakes were foreshocks of the Loma Prieta earthquake. However, these earthquakes would not be regarded as immediate foreshocks occurring in the nucleation zone, because they occurred more than two months prior to the Loma Prieta earthquake.

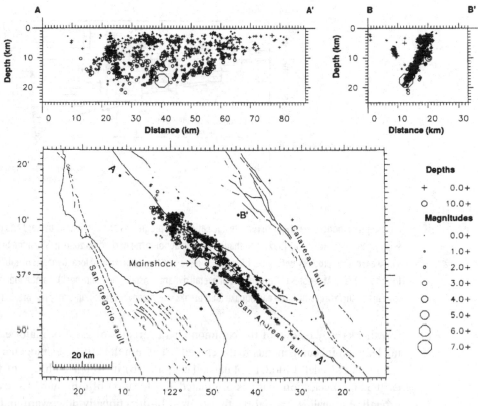

Fig. 7.17 Source locations of the 1989 Loma Prieta earthquake with M_S 7.1 and aftershocks along the San Andreas transform plate boundary fault (US Geological Survey Staff, 1990).

In contrast, when the nucleation of an earthquake grows on an inhomogeneous fault embedded within the brittle seismogenic layer, immediate foreshocks are liable to be induced during the nucleation process, even in the case where the dynamic rupture begins at or near the base of the brittle seismogenic layer. The nucleation process in the brittle seismogenic layer can be greatly influenced by the existence of local patch areas highly resistant to rupture growth in the fault zone. These patch areas of various dimensions may be produced by geological and geometric settings in the fault zone, and/or local fluctuations of pore water pressure in the fault zone. For example, if the fault surfaces are geometrically irregular, interlocking asperity areas are liable to be created in the fault zone, and the interlocking asperity areas are joint areas highly resistant to rupture growth on the fault. Accordingly, local-to-small-scale highly resistant patch areas prevail in the fault zone embedded in the brittle seismogenic layer, and minor patch areas of high resistance to rupture growth are liable to be broken down to generate minor earthquakes during the nucleation process of a great or major earthquake. This has been demonstrated in high-resolution laboratory experiments on frictional slip rupture nucleation on precut faults with fractal rough surfaces (see Figure 5.25, or Figure 7.30 in the next section). Hence, immediate

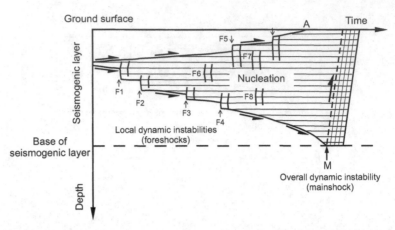

Fig. 7.18 A conceptual model of earthquake rupture nucleation that progresses bi-directionally upward and downward in the seismogenic layer (Ohnaka, 1992). The shaded portion denotes the nucleation zone in which the breakdown (or slip-weakening) proceeds with time. F1, F2, . . . represent the occurrence of local dynamic instabilities (immediate foreshocks), and M denotes the occurrence of overall dynamic instability (mainshock). Point A indicates the location on the Earth's surface at which the earthquake fault breaks out; however, some faults may not break the Earth's surface.

foreshocks may be one of the common characteristics of great or major earthquakes that nucleate on inhomogeneous faults embedded within the brittle seismogenic layer.

Let us consider a model (Ohnaka, 1992, 1995b) in which both tips of the zone of an earthquake nucleation that has originated at a location of minimum resistance to rupture growth at a shallow crustal depth, progress bi-directionally downward and upward along three-dimensionally curved paths of somewhat higher resistance to rupture growth within the brittle seismogenic layer (for this conceptual model, see Figure 7.18). We assume that the resistance to rupture growth on an inhomogeneous fault in the brittle seismogenic layer is distributed in a patchy fashion, with a non-uniform distribution of not only sizable patch areas highly resistant to rupture growth but also superimposed minor asperities somewhat highly resistant to rupture growth. Once the downward progressing tip of the nucleation zone penetrates a sizable patch area highly resistant to rupture growth at or near the base of the seismogenic layer, the sizable patch area will break down in due course. If an adequate amount of elastic strain energy is released by the breakdown of the sizable patch area, the progressing rupture nucleation develops into dynamic instability, giving rise to dynamic rupture toward the ground surface within the brittle seismogenic layer, as shown in Figure 7.18. During this nucleation process, foreshocks (F1, F2, . . . , in Figure 7.18) can be generated by the failure of minor projections (or asperities) distributed in and around the nucleation zone. In this case, the hypocentral locations of foreshocks should necessarily be restricted to within a localized region shallower than the depth of the mainshock hypocenter, located near the base of the seismogenic layer.

Such a typical example is the Mikawa earthquake of $M\,6.8$, which occurred on January 13, 1945 (JST, 9 hours ahead of UTC) in Mikawa Bay, 50 km south of Nagoya, in central Japan (for the epicentral location, see Figure 7.19). This earthquake was an intra-plate earthquake. Hamada (1986, 1987) carefully reexamined and redetermined the hypocentral locations of

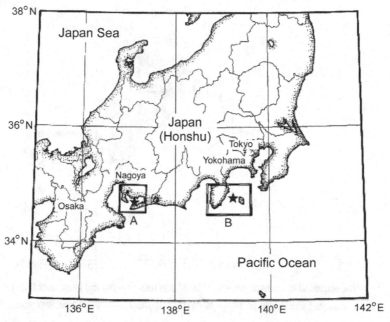

Fig. 7.19 Central Japan map indicating epicentral locations of: A, the 1945 Mikawa earthquake with $M6.8$; and B, the 1978 Izu Oshima Kinkai earthquake with $M7.0$. The asterisks indicate their epicentral locations. Reproduced from Ohnaka (1995b).

foreshocks, mainshock, and aftershocks for the Mikawa earthquake. Ten foreshocks were listed in his catalogue (Hamada, 1986), and these foreshocks began to occur 6 days before the mainshock, but most foreshocks were concentrated in the period of 27–41 hours prior to the mainshock. According to Hamada's catalogue, no foreshocks took place in the period of about 27 hours immediately before the mainshock. Hence, there was likely a lull in foreshock activity immediately before the Mikawa earthquake. The magnitudes for eight of the ten foreshocks listed in his catalogue were determined, and ranged from 4.1 to 5.7. Hamada (1987) found that the hypocentral depths of the foreshocks relocated were all shallower than the depth of the mainshock hypocenter relocated at a depth of 12.4 km. This is a significant feature expected from the model shown schematically in Figure 7.18.

Figure 7.20 shows the epicentral locations of immediate foreshocks (black circles) and mainshock (black star) for the 1945 Mikawa earthquake. From this figure, we notice that the epicentral locations of the immediate foreshocks are concentrated in the vicinity of the mainshock epicenter. The aftershocks were roughly located in the vertical-lined area in Figure 7.20 (Hamada, 1986, 1987). Most aftershocks of the Mikawa earthquake were located at depths shallower than 14 km (Hamada, 1986, 1987). This suggests that the base of the brittle seismogenic layer in this particular region is at a depth of around 14 km, and that the mainshock hypocenter is near the base of the seismogenic layer. The depth distribution of the hypocenters of these foreshocks and mainshock is shown in Figure 7.21. The vertical-lined area in Figure 7.21 indicates the mainshock fault area roughly inferred from the distribution of the aftershock hypocenter locations. The seismicity confined within

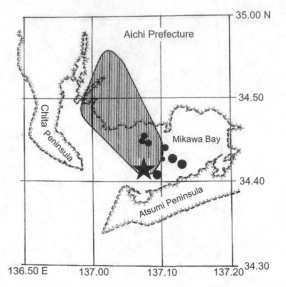

Fig. 7.20   The epicentral locations of immediate foreshocks (black circles) and mainshock (black star) for the 1945 Mikawa earthquake. The shaded portion indicates the aftershock distribution area. Reproduced from Ohnaka (1992).

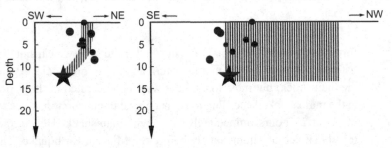

Fig. 7.21   Depth distribution of the hypocenters of immediate foreshocks and mainshock for the 1945 Mikawa earthquake. The shaded portion indicates the fault area of the mainshock roughly inferred from aftershock distribution. Reproduced from Ohnaka (1992).

a 14 km depth is compatible with depth profiles of the shear resistance τ_p, the breakdown stress drop $\Delta\tau_b$, and the breakdown displacement D_c, if the temperature gradient in this region is lower than 30 °C/km (Ohnaka, 1992, 1995b). Note that the product of $\Delta\tau_b$ and D_c is indicative of the resistance to rupture growth.

In the case where the resistance to rupture growth on a preexisting fault is higher in the downward direction than in the horizontal direction, the rupture that nucleates at a very shallow depth on the fault has no alternative but to grow along the fault strike in the brittle seismogenic layer. A typical example of this case is the Izu Oshima Kinkai earthquake of $M7.0$, which occurred on January 14, 1978 (JST) in the area between the

Izu Peninsula and Oshima Island, Japan (for the epicentral location, see the map shown in Figure 7.19). This earthquake, a typical intra-plate earthquake that nucleated within the brittle seismogenic layer, is well known for conspicuous immediate foreshock activity (Tsumura *et al.*, 1978). Indeed, the earthquake was preceded by swarm-like immediate foreshock activity, and the hypocentral depths of these foreshocks were shallower than 10 km. The mainshock hypocenter was located at a depth of about 4 km (Shimazaki and Somerville, 1979; Matsu'ura *et al.*, 1988) in the middle of these foreshock hypocenters (Figure 7.23). Its aftershock activities were also shallower than 10 km (Ohnaka, 1992). These foreshock-mainshock-aftershock hypocenter distributions confined within a 10 km depth suggest that the base of the brittle seismogenic layer in this particular region is at a depth of 10 km.

Analyses of the foreshock activity indicate that the series of immediate foreshocks with $M \geq 2.0$ began to occur at about 17:40 on January 13, 1978 (JST) in the area near the hypocenter of the mainshock (Ohnaka, 1992, 1993, 1995b). The entire period from this time to the time of the mainshock occurrence may be divided into the following three phases (see Figures 7.22 and 7.23): phase I (a 15 hour and 20 minute time-interval from 17:40 on January 13) in which foreshocks were concentrated within a very small, limited area, phase II (the subsequent 1 hour and 10 minute time interval) in which foreshocks began to migrate toward the environs, and phase III (the subsequent 2 hour and 14 minute time interval) in which foreshocks continued to migrate along the strike of the mainshock fault until the mainshock took place. The mainshock dynamic rupture propagated uni-directionally in a westward direction from the hypocenter (Shimazaki and Somerville, 1979; Kikuchi and Sudo, 1984). Why uni-directionally in a westward direction? This would be because the resistance to rupture growth was lower in the western part of the locked fault, and/or because less elastic strain energy was stored in the eastern part, so that the rupture would have no alternative but to propagate westward (note that Oshima Island is an active volcano).

As discussed so far, immediate foreshocks occur in and around the nucleation zone of an imminent mainshock. This suggests that the mainshock nucleation zone size may possibly be estimated from the distributional area of immediate foreshock activity. Based on this empirical model, the first attempt to estimate the mainshock nucleation zone size from immediate foreshock activity distribution was made for the aforementioned 1978 Izu Oshima Kinkai earthquake (Ohnaka, 1993). Figure 7.24 shows how the distributional area of a sequence of immediate foreshock activities evolved with time before the occurrence of the 1978 Izu Oshima Kinkai earthquake. It is clearly seen in Figure 7.24 that the distributional area of the foreshock activities evolved with time, as expected from the model. A consistent pattern of the evolving distributional area of immediate foreshock activities has been observed for other earthquakes (e.g., Dodge *et al.*, 1995, 1996; Kato *et al.*, 2012).

Based on the observational result shown in Figure 7.24, the critical nucleation zone size was inferred to be roughly 10 km for the 1978 Izu Oshima Kinkai earthquake. However, the foreshock sequence occurred offshore, and hence there may be a possibility that the deployment of seismometers may not have been optimal to accurately determine

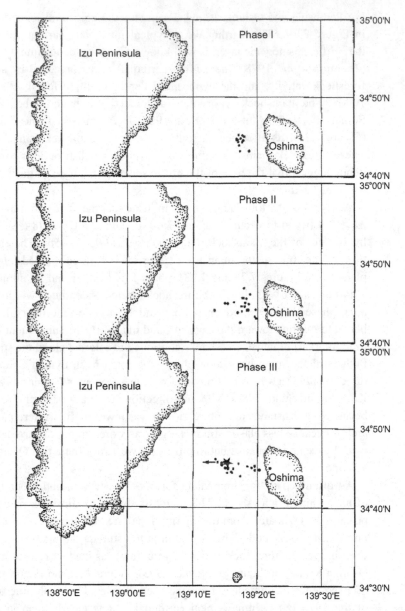

Fig. 7.22 Epicentral locations of immediate foreshocks (dots) in successive phases I, II, and III, and the mainshock (star) for the 1978 Izu Oshima Kinkai earthquake (Ohnaka, 1993). The arrow indicates the direction of the mainshock rupture propagation (Shimazaki and Somerville, 1979).

the foreshock locations. As a result, a nucleation zone size estimate of 10 km may have been exaggerated. To overcome this problem, Dodge *et al.* (1995, 1996) accurately determined the locations of foreshocks, and examined the evolution of foreshock sequences in detail for the following six moderate-to-major earthquakes that occurred in California, USA: the 1986 Mount Lewis earthquake of $M_L 5.7$, the 1986 Chalfant earthquake

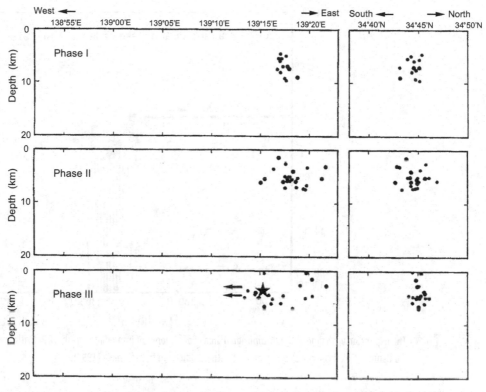

Fig. 7.23 Depth distribution of hypocenters of immediate foreshocks (dots) in successive phases I, II, and III, and the mainshock (star) for the 1978 Izu Oshima Kinkai earthquake (Ohnaka, 1993). The arrows indicate the direction of the mainshock rupture propagation (Shimazaki and Somerville, 1979).

of $M_L6.4$, the 1986 Stone Canyon earthquake of $M_L4.7$, the 1990 Upland earthquake of $M_L5.2$, the 1992 Joshua Tree earthquake of $M_W6.1$, and the 1992 Landers earthquake of $M_W7.3$ (M_L, local magnitude scale; M_W, moment magnitude scale). Dodge *et al.* first relocated the foreshock hypocenters using waveform cross-correlation analysis to accurately determine the locations of the foreshocks. Consequently, the relative locations of the foreshock sequences were successfully determined with relatively high accuracy (relative uncertainties of less than 100 m horizontally and 200 m vertically). This enabled them to identify small-scale fault zone structures that could influence nucleation, and to find that the number of immediate foreshocks increases with an increase in mainshock fault zone width.

When a fault zone is heterogeneous, the zone width thickens. With this in mind, the aforementioned expression can be paraphrased as follows: the number of foreshocks is controlled by the fault zone heterogeneity, and the more highly heterogeneous the fault zone, the greater the number of foreshocks (Dodge *et al.*, 1996). This indicates that mainshock fault zone heterogeneity is indispensable for immediate foreshocks to occur, as noted earlier in this subsection. In addition, the relatively high accuracy of the immediate foreshock locations also led to the finding that the size of the mainshock nucleation zone estimated

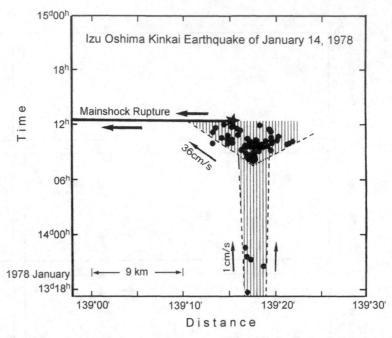

The expansion with time of the distributional area of a sequence of immediate foreshock activities prior to the 1978 Izu Oshima Kinkai earthquake. Reproduced with modification from Ohnaka (1993).

from the extent of the immediate foreshock sequences, scales with the mainshock seismic moment (Dodge *et al.*, 1996). This scaling relation is consistent with the scaling relation found between the critical length $2L_c$ of the mainshock nucleation zone determined from the seismic nucleation phase and the mainshock seismic moment M_0 (for the scaling relation between $2L_c$ and M_0, see Figure 5.22).

7.3 Predictability of large earthquakes

7.3.1 Introduction

For a period of time in the past, it had been strongly claimed by some researchers that earthquakes are unpredictable catastrophes. This was because of the pervasive hypothesis that the Earth's crust is in a state of perpetual self-organized criticality in which any small earthquake may cascade into a large event. Accordingly, the view of earthquake unpredictability seemed to have swept the world.

As described in Section 5.2, however, it has been substantiated by observational evidence that the above hypothesis is far from the truth, at least in regard to large earthquakes. A few specific pieces of observational evidence indicating the predictability of large earthquakes have already been presented. The essential points are that the spatial and temporal scales of the preparation process leading up to a larger earthquake are both greater, and that the spatial and temporal scales of the preceding phenomena of large earthquakes are distinctly different from those of small earthquakes (see Figure 7.10). It has also been shown specifically that the size (i.e., magnitude or seismic moment) of an earthquake scales with the critical size of its nucleation zone (see Figure 5.22). This suggests that whether the ongoing nucleation process develops into a large earthquake or not is in principle forecastable from preceding physical phenomena associated with the nucleation process, if the preceding phenomena are really observable under high resolution.

Our primary concern is about the predictability of large (i.e., great or major) earthquakes causing serious damage to humanity and society, and therefore let us focus attention on the predictability of large, damaging earthquakes, which usually occur at shallow depths. As presented in Chapter 5, the hypothesis of perpetual self-organized criticality is neither reasonable nor acceptable in regard to large earthquakes. In this respect, large earthquakes are distinctively different from small earthquakes. This is a significant point to be emphasized.

The ultimate aim of earthquake prediction or forecasting is to successfully achieve a short-term prediction or forecast. In order to scientifically make a short-term forecast of a large earthquake from observed reliable precursory phenomena, what is needed is a forecast scientific scenario (or predictive model), indicating in chronological order how individual precursory phenomena are associated with the preparation process leading up to an earthquake rupture (Ohnaka and Matsu'ura, 2002). We cannot make any sure prediction without knowing scientifically in advance how precursory phenomena are associated specifically with the earthquake preparation process. Therefore, a forecast scientific scenario is indispensable.

In addition, in order to bring about a short-term earthquake forecast, we have to keep in mind that the generation process of a large earthquake is affected by the geological, tectonic structure, fault-surface geometric irregularity, and thermal properties of the seismogenic fault zone (and its surrounding region), and that such properties may differ according to the individual seismogenic fault zone and its surrounding region, and may vary somewhat with the passage of time. Observable precursory phenomena may also differ according to the specific properties (geological, tectonic structure, fault-surface geometric irregularity, thermal, and electromagnetic) of an individual seismogenic fault zone and its surrounding region. If this is the case, the forecast scenario (or model) must be modified according to regional differences in the properties and/or according to changes in some specific properties with the passage of time. These make it difficult to forecast real earthquakes in the short term.

In establishing an achievable method for making a short-term forecast of a large earthquake, the fundamentally important things are: (1) physical modeling of the process leading up to a typical large earthquake, (2) figuring out the regional characteristics of the properties of the specific fault zone where an earthquake may imminently occur, and its surrounding region, and (3) constructing a forecasting system capable of making sequential forecasts/verifications iteratively by adding the latest of the observed data frequently updated on a real-time basis (Ohnaka and Matsu'ura, 2002). With this in mind, the first step in establishing the achievable method is to combine (1) with (2) to build a forecasting model into which characteristic properties of a seismogenic fault zone and its surrounding region are incorporated.

As described earlier (Chapter 5 and Sections 7.1 and 7.2 in this chapter), physical modeling of the common process leading up to a typical large earthquake has been achieved. Accordingly, we can incorporate the characteristic properties of the seismogenic fault zone and its surrounding region into the physical model of the common process leading up to a typical large earthquake. However, specific information and further detailed research are needed to construct the aforementioned forecasting system capable of making sequential forecasts/verifications iteratively. In addition, we need to develop innovative means for observing the ongoing nucleation process itself, or physical phenomena associated with the nucleation process. Accordingly, the actualization of short-term earthquake forecasts will remain a challenge. In this section, let us narrow our focus on specific evidence indicating that typical large earthquakes are in principle forecastable, and describe a possible methodological framework to forecast typical large earthquakes.

7.3.2 Long-term forecasting

As described previously, one cycle of large-earthquake generation on a fault consists of the following two processes: the buildup process of elastic strain energy owing to perpetual slow and steady tectonic loading, and the release process of the stored elastic strain energy by the occurrence of a dynamic shear rupture (earthquake). After the elastic strain energy has been released by the occurrence of a large earthquake, the elastic energy residual in the source region is lowered to a subcritical level, and therefore the next large earthquake cannot recur in the same region for a long time until an adequate amount of elastic strain energy is built up by perpetual slow tectonic loading.

Large shallow-focus earthquakes cannot occur everywhere in the Earth's crust. The occurrence of such earthquakes is restricted only to the regions where an adequate amount of elastic strain energy is accumulated. More specifically, such earthquake ruptures occur selectively along mechanically weak junction planes (or zones), such as tectonic plate boundary (inter-plate) faults and preexisting active (intra-plate) faults embedded within tectonic plates, in the regions where an adequate amount of elastic strain energy has been accumulated. Thus, it is primarily important from the standpoint of a long-term forecast to identify their locations, and to specify the geometric shapes. Since large earthquakes recur intermittently, mostly at quasi-periodic intervals, at almost the same location, there is no doubt that an identified location is a potential source region where a large earthquake will occur in the future. Thus, the primary role of the long-term forecast is in identifying the locations of plate boundary faults and active faults embedded within tectonic plates.

Large shallow-focus earthquakes occur one after another along a tectonic plate boundary, as exemplified in Figure 7.25. If we view the map on which are plotted the source areas of large shallow-focus earthquakes that have occurred along a tectonic plate boundary for an

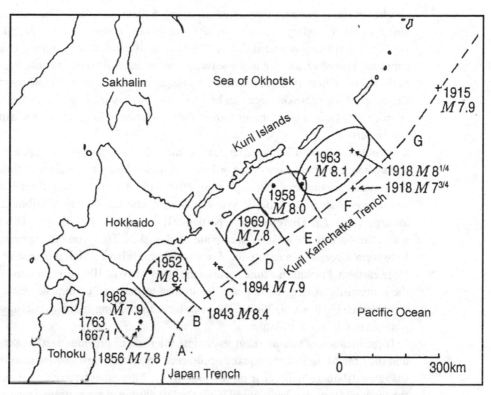

Fig. 7.25 An example identified as a seismic gap (region C) in the seismic zone almost entirely covered with the source areas of large earthquakes along the subduction zone plate boundary at the Kuril-Kamchatka Trench, southeast of the Nemuro Peninsula of Hokkaido, Japan (Utsu, 1972).

adequately long period of time, we find that the seismic zone along the plate boundary is covered with the source areas of such large earthquakes. If, however, we view the map on which are plotted the source areas of large shallow-focus earthquakes that have occurred along the same plate boundary for a certain period of time, we can find a seismic gap (region C in Figure 7.25) in the seismic zone almost covered with the source areas of large earthquakes.

About four decades ago, region C shown in Figure 7.25 was identified as a seismic gap along the subduction zone plate boundary at the Kuril-Kamchatka Trench, southeast of the Nemuro Peninsula of Hokkaido, Japan (Utsu, 1972). Ellipses in Figure 7.25 denote the source areas of large earthquakes that have occurred over a certain period of time, and numerical values written in individual ellipses denote the year of earthquake occurrence and its magnitude. A major earthquake of $M7.4$ (called the Nemuro-Hanto-Oki earthquake) actually occurred on July 17, 1973 in region C. This was well known at that time as a case of successful earthquake prediction based on a seismic gap.

Such an appearance of a seismic gap in the seismic zone can be accounted for in terms of the earthquake generation cycle model (described in Section 7.1), in which large shallow-focus earthquakes having specific magnitudes and source areas occur one after another at quasi-periodic time intervals along a plate boundary. The entire seismic zone along a plate boundary is in general not broken down by a single earthquake rupture but is broken down locally and at different times one after another by a number of earthquake ruptures. Therefore, an area in the seismic zone necessarily remains unbroken for a certain period of time, but will be broken down by the occurrence of an earthquake some time in the future. Such an unbroken area can be recognized as a seismic gap, a potential earthquake source area. Thus, it is meaningful from the standpoint of long-term forecasting to identify such a seismic gap.

As noted in Section 7.1, recurrence time intervals for inter-plate earthquakes occurring on a plate boundary fault were of the order of 10^2 years; however, recurrence time intervals for intra-plate earthquakes occurring on an active fault in a region away from plate boundaries are of the order of 10^3 to 10^4 years or longer, and are widely distributed around their average value T_{av}. Therefore, it is particularly important to make a probabilistic forecast about the occurrence of such intra-plate earthquakes. Thus, the other primary role of the long-term forecast is in narrowing down which faults have a high probability of generating large earthquakes within a time scale of the order of 10 to 100 years, and then in evaluating the probability of large earthquake occurrence to make an earthquake hazard assessment (Ohnaka and Matsu'ura, 2002). The long-term forecasting in this sense has already been put into practical use in Japan.

If the location of a fault, recurrence time intervals of previous earthquakes on the fault, and the time of the last earthquake occurrence on the fault are known, then the probability that an earthquake will occur on the fault in the future can be expressed as a function of the length of time that has elapsed since the last earthquake occurrence (see Eq. (7.4)). The probability of a future earthquake occurrence can be calculated from Eq. (7.4), given the length of time that has elapsed since the last earthquake occurrence, and by imposing the condition that $\beta > 1$ (see Section 7.1), since we focus attention on large earthquakes that occur intermittently at quasi-periodic intervals. The probability thus calculated may contain errors arising from uncertain estimates of recurrence time intervals and of the time of the last earthquake occurrence. We have to keep this in mind when the individual values of recurrence time intervals and/or of last earthquake occurrence time estimated from paleoseismological or geological data bear uncertainties.

7.3.3 Intermediate-term forecasting

If a certain fault segment zone is specifically identified by a long-term forecast as having a high probability that a large earthquake will occur in the future, a network of observing/monitoring instruments of various types can be densely deployed in and around the particular fault segment zone, and in the neighboring region. For example, such a network of instrumentation has been deployed in the Tokai district in central Japan as part of a national project to forecast a large earthquake (to be called the Tokai earthquake) likely to occur along the Nankai Trough in the vicinity of Suruga Bay and the Enshunada region. In the time frame of an intermediate-term forecast, by use of the data observed and/or monitored with such a network of instrumentation, a forecast would need to be made in regard to the time of earthquake occurrence with a margin of error of less than one year, and of a magnitude with a margin of error of less than ± 1 (Ohnaka and Matsu'ura, 2002).

Let us address physical modeling for intermediate-term forecasting. When a certain fault segment zone specifically identified by long-term forecast as having a high probability for a typical, large earthquake is in a seismogenic region where seismicity becomes progressively active with time during the crustal deformation process, the time-to-failure function model (e.g., Bufe and Varnes, 1993; Bufe *et al.*, 1994; Jaume and Sykes, 1999) presented in Section 7.2 may be useful for an intermediate-term forecast. In the time-to-failure function model, not only the occurrence time t_f but also the magnitude M of the ensuing event expected at $t = t_f$ on the master fault can be estimated when $n < 1$ (see Eq. (7.10)). In this case, the occurrence time t_f and the magnitude M of a typical earthquake may be forecast, if both t_f and M are evaluated by the best-fit regression analysis using Eqs. (7.10), (7.11), and (7.12).

According to Bufe and Varnes (1993), ω_i in Eq. (7.11) in general represents any of several measurable quantities, such as the Benioff strain release, event count and seismic moment, for each event, and is estimated from earthquake magnitude M_i using an empirical formula of the form

$$\log \omega_i = cM_i + d, \tag{7.26}$$

where c is a coefficient representing the degree of magnitude dependence.

In the time to failure function model, the quantification of seismic activity in a region is made by using Eqs. (7.11) and (7.26). For example, the simplest way of quantifying seismic activity in a region is to define Ω ($=\sum \omega_i$, see Eq. (7.11)) as representing the cumulative number of earthquakes in the region. In this quantification, Ω does not depend on earthquake magnitude (because $c = 0$); in other words, the weight given for seismicity to small earthquakes is equal to that given to larger earthquakes in the region. In contrast, if ω_i is defined as representing the seismic energy E_{si} or the seismic moment M_{0i} of the ith event in the sequence of N earthquakes that have occurred in a certain region, the coefficient c has a value of 1.5. This coefficient value is derived from the Gutenberg and Richter energy–magnitude relationship (Gutenberg and Richter, 1956) or the Kanamori moment–magnitude relationship (Kanamori, 1977; Hanks and Kanamori, 1979). In this quantification, ω_i depends on earthquake magnitude M_i, and larger-magnitude earthquakes in the region are more weighted in specifically quantifying Ω (because $c = 1.5 > 1$). If, however, ω_i is defined as the Benioff strain release, the coefficient c is 0.75 (see Eq. (7.12)), which is a value between 0 and 1.5. In this quantification, therefore, though ω_i depends on earthquake magnitude M_i, the magnitude dependence is moderately weighted in quantifying Ω.

Whether or not the coefficient value given for seismicity quantification is zero or relatively large or moderate, may influence the estimate of the occurrence time and magnitude expected for the event on the master fault. If this is the case, the possibility for successfully estimating both the occurrence time and the magnitude of the ensuing event depends on how to quantify Ω. Thus, it would be important to know how much of the magnitude dependence of ω_i is best suited for the quantification of seismicity.

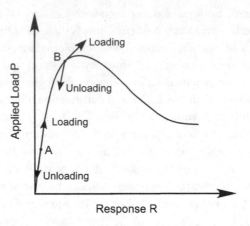

Fig. 7.26 Constitutive relation between load P applied to an inhomogeneous body and its response R to the applied load (Yin *et al.*, 2000). Loading and unloading responses are reversible in the elastic regime, whereas loading and unloading responses are irreversible in the non-elastic regime.

Specifically, both the occurrence time and the magnitude of the expected event were successfully estimated from the best-fit regression analysis using a sequence of northern California earthquakes prior to the 1989 Loma Prieta earthquake, only when ω_i was defined as representing the Benioff strain release (Bufe and Varnes, 1993). In other words, the time-to-failure function model is capable of forecasting both t_f and M, if $\Omega(t)$ is defined as the cumulative Benioff strain release (Bufe and Varnes, 1993). This may be partly because the magnitude dependence of ω_i was moderate ($c = 0.75$) when ω_i was defined as the Benioff strain release. In addition, we have to keep in mind that the best-fit solution for t_f, n, and Ω_f in Eq. (7.10) may be affected by the specific conditions under which the seismicity data are selected, such as spatial window, time span, and lower-limit magnitude. These issues remain to be investigated in detail and to be fully elucidated.

Another model that may also be useful for an intermediate-term forecast is the load–unload response ratio (LURR) model proposed by Yin *et al.* (e.g., 1995, 2000, 2002). At an early stage of phase II (see Figure 7.4 in Section 7.1), the crust behaves elastically. At later stages of phase II, however, it progressively behaves in-elastically, as the tectonic stress approaches a critical strength level of the fault. The load–unload response ratio model can be applied to seismic phenomena observed at later stages of phase II. If a measure of the proximity to the critical state in a region is suitably defined, a stage in the process of crustal deformation leading up to a large earthquake with tectonic loading can be identified in terms of the measure of the proximity to the critical state, and thereby an ensuing earthquake rupture in the region may be forecast.

In general, the relation between load P applied to an inhomogeneous body and its response R to the applied load can be depicted as shown in Figure 7.26. The body deforms in response to the applied load, and the relation between the applied load and the resulting deformation (displacement, or strain) is referred to as the constitutive relation. Let ΔP and ΔR be increments of load P and of response R, respectively, and let us define the response rate X by (Yin et al., 2000)

$$X = \lim_{\Delta P \to 0} \frac{\Delta R}{\Delta P}, \tag{7.27}$$

and the load–unload response ratio Y by (Yin et al., 2000)

$$Y = \frac{X_+}{X_-}, \tag{7.28}$$

where X_+ denotes the response rate during loading, and X_- denotes the response rate during unloading. In the perfectly elastic regime, the deformation during the loading process is completely recovered during the unloading process. Thus, $X_+ = X_-$, and consequently $Y = 1$. Non-elastic deformation proceeds beyond the elastic limit, however. In the regime of non-elastic deformation, $X_+ > X_-$ and therefore $Y > 1$. In particular, $Y \gg 1$ at the stage where the stress is in close proximity to the critical level at which an earthquake occurs.

The change in Coulomb failure stress ΔCFS is defined as

$$\Delta \text{CFS} = \Delta \tau_s - \mu \Delta \sigma_n^{\text{eff}} \cong \Delta \tau_s - \mu' \Delta \sigma_n, \tag{7.29}$$

where $\Delta \tau_s$ is the change in shear stress acting in the slip direction on a fault, $\Delta \sigma_n$ is the change in normal stress σ_n acting across the fault plane, $\Delta \sigma_n^{\text{eff}}$ is the change in the effective normal stress defined by $\sigma_n^{\text{eff}} = \sigma_n - P$ (P, pore pressure), μ is the coefficient of internal friction, μ' is the apparent coefficient of friction into which a change in pore pressure P is incorporated, and compressive stress and pressure have been set as positive.

The change in Coulomb failure stress, ΔCFS, provides an indication of whether or not a fault plane is prone to slip failure. Specifically, the fault plane is brought closer to slip failure if $\Delta \text{CFS} > 0$, and is moved away from slip failure if $\Delta \text{CFS} < 0$. When $\Delta \text{CFS} > 0$, it is referred to as loading, and when $\Delta \text{CFS} < 0$, it is referred to as unloading (Yin et al., 2000). At stages where the tectonic stress is in close proximity to the critical level, a slight change in stress induced by the Earth's tides can trigger the occurrence of an earthquake. In the load–unload response ratio model, ΔCFS is calculated to examine whether ΔCFS is positive or negative for each event in the sequence of N earthquakes that have occurred in a certain region. Of the N earthquakes, let us assume that L events have had positive ΔCFS, and that M events have had negative ΔCFS ($L + M \leq N$). In this case, the load–unload response ratio Y defined as the ratio of the cumulative Benioff strain release during loading to that during unloading is expressed as follows (Yin et al., 2000):

$$Y = \frac{\left(\sum_{i=1}^{L} \sqrt{E_{si}} \right)_+}{\left(\sum_{i=1}^{M} \sqrt{E_{si}} \right)_-}, \tag{7.30}$$

where E_{si} represents the seismic energy for each event in the sequence of N earthquakes, and signs "+" and "–" mean "loading" and "unloading," respectively. E_{si} is commonly calculated from the Gutenberg–Richter energy–magnitude relationship. The load–unload

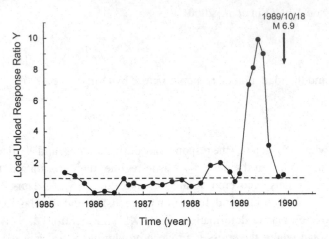

Fig. 7.27 A specific example of the variation of load–unload response ratio Y with time before a major earthquake ($M6.9$) that occurred in California (Yin *et al.*, 2000).

response ratio Y thus defined has been practically used, and determined by calculating Earth-tide-induced perturbations in the Coulomb failure stress on optimally oriented faults (Yin *et al.*, 2000, 2002).

The load–unload response ratio Y, defined as Eq. (7.30), provides a measure of the proximity to the critical level, and high values (> 1) of the load–unload response ratio indicate that the region is in close proximity to the critical state that has the potential to cause a major earthquake, as exemplified in Figure 7.27. This figure shows a plot of the load–unload response ratio Y defined by Eq. (7.30) against time during the crustal deformation process leading to a major earthquake ($M6.9$) that occurred on October 18, 1989 in southern California, USA (Yin *et al.*, 2000). In like manner, values for the load–unload response ratio have been calculated for various regions of different tectonic regimes (Yin *et al.*, 2000, 2002), and the results suggest that the model provides a useful means for identifying the stage in the process of crustal deformation leading up to a major or great earthquake with tectonic loading.

The scaling relation between the critical region size and the magnitude of the final event (Bowman *et al.*, 1998), as shown in Figure 7.10, estimated from the time-to-failure function model, agrees with that estimated from the load–unload response ratio model (Yin *et al.*, 2002). This agreement is not by coincidence. The time-to-failure function model and the load–unload response ratio model are both based on the common underlying physical process during which non-elastic crustal deformation progresses with tectonic loading at later stages of the recurrence time interval, and during which activity in moderate-size earthquakes progressively increases with tectonic loading at the later stages until a major or great earthquake (final event) occurs, as modeled in Figure 7.5, in a relatively wide region surrounding a master fault on which the major or great earthquake is expected to occur (for details, see Sections 7.1 and 7.2), even though the two models have been formulated in different terms from different perspectives.

Other precursory phenomena that may be helpful for an intermediate-term forecast

In general, a variety of physical and chemical phenomena can be activated at high stresses. For example, when water is present in interstitial pores existing in crustal rocks, the chemical reaction between silicate rock-forming minerals and water molecules operating as a corrosive agent, is more likely to occur at higher stresses. Consequently, some silicate rocks can be fractured by this stress-aided corrosion (called delayed fracture), even though the applied stress level is lower than the rock strength (see subsection 7.2.1). If interstitial pore water present in a fault zone migrates into pores newly created by dilatancy, changes in telluric current and in electrical resistance are likely to occur in the zone. The occurrence of dilatancy is also capable of causing anomalous ground uplift and groundwater level change. Anomalous ground uplift due to dilatancy necessarily causes a change in gravity on the ground.

Laboratory experiments show that electromagnetic waves are radiated by the piezoelectric effect of quartz, a typical rock-forming mineral, and by the electrokinetic effect resulting from pore water migration (e.g., Yoshida et al., 1994, 1998). This suggests the possibility that electromagnetic waves are radiated by stress changes and/or pore water migration accompanying local fractures that occur when tectonic stress attains a high level in proximity to the critical fault-strength level, in the vicinity of the source area of an anticipated large earthquake.

If radon, a radioactive element with a 3.8 day half-life, is released into pore water in the course of crack formation in rock at the stage where tectonic stress is as high as in proximity to the critical fault-strength level, a change in radon concentration in groundwater can also be observed prior to an anticipated earthquake. In fact, the change in radon concentration was found in well water prior to the 1966 strong Tashkent earthquake (Sadovsky et al., 1972), and this precursory abnormality in radon concentration was well known at that time, and recognized as an effective tool for forecasting seismic events. Since then, indeed, many observations of radon concentrations in well water and soil gas have been made (e.g., Hirotaka et al., 1988; Igarashi et al., 1995; Richon et al., 2003; Choubey et al., 2009). It has also been confirmed in laboratory experiments that radon is actually released in the course of rock fracture (e.g., Holub and Brady, 1981).

When tectonic stress is in proximity to the critical level, there is a possibility that a variety of anomalous phenomena, such as the ones mentioned above, will be observed in a region in which a large earthquake is anticipated on a master fault. However, as noted in subsection 7.3.1, observable precursory phenomena may differ according to specific properties (geological, tectonic structure, fault-surface geometric irregularity, thermal, and electro-magnetic) of individual seismogenic fault zone and its surrounding region. Thus, the forecast scenario (or model) must be modified according to regional differences in the properties and/or according to the changes in some specific properties with the passage of time. With this in mind, we need to build specific and scientifically testable forecast scenarios (or models), in order to examine to what extent such anomalous phenomena can be useful for the intermediate-term forecast of individual earthquakes in a specific seismogenic zone. This will require further detailed research and will remain a scientific challenge.

7.3.4 Short-term forecasting

Tectonic deformation eventually concentrates locally in the vicinity of a master fault, and an earthquake rupture begins to nucleate at a location of lowest resistance to rupture growth on the fault. In other words, the nucleation process leads to an unstable, dynamic high-speed rupture that propagates along the master fault (mainshock). This earthquake nucleation process is the transition process from a stable and quasi-static state to an unstable and dynamic state of high-speed rupture propagation, and the nucleation process is an integral part of the mainshock rupture. The critical size of earthquake nucleation and the critical time length of its duration differ according to a range of environmental conditions in seismogenic fault zones. For example, the critical size of rupture nucleation and the critical time length of its duration depend on the characteristic length representing the geometric irregularity of fault surfaces. This has been demonstrated by high-resolution laboratory experiments (see Chapters 5 and 6). In addition, the larger size of a nucleation zone results in a larger earthquake; in other words, the nucleation zone size of a larger earthquake is greater (see Figure 5.22). Specifically, the critical size of the nucleation zone is about 10 km for a typical earthquake of magnitude 8, and roughly 3 km for a typical earthquake of magnitude 7 (Ohnaka and Matsu'ura, 2002). Thus, continuous monitoring of the ongoing nucleation process will be appropriate and justifiable as a means of making a short-term forecast of a typical, large earthquake, if the ongoing nucleation process itself can be identified and monitored by any observational means.

The shear rupture nucleation process is necessarily accompanied by changes in slip displacement and in shear stress. In the nucleation zone, slip (pre-slip) proceeds and shear stress changes with time, initially both at very slow rates, and gradually at accelerating rates. However, both local slip amounts and local shear stress (or shear strain) changes during the nucleation process are very small in comparison to the slip mount and the shear stress change during the dynamic mainshock breakdown process, as presented in Chapter 5. In addition, the nucleation zone is locally limited, and therefore its zone size is small if compared with the entire fault size of the mainshock. Such small amounts of slip and stress (or strain) changes in a locally limited zone may pose a challenge to monitoring them in the field. Nevertheless, if that challenge is overcome, monitoring both local pre-slip and local stress (or strain) changes is the most reliable means for tracking the ongoing nucleation process. If there exists any observable quantity necessarily accompanying slip displacement and stress change during the nucleation process, monitoring the quantity would be one way of achieving the tracking of the nucleation process.

As a rupture nucleation grows with time locally on a fault, the degree of coupling strength between the fault surfaces degrades locally, and a certain amount of the strain stored in the surrounding medium (rock body) is released by the elastic rebound of the deformed medium on either side of the fault. As a result, the rate of elastic strain accumulation is lowered (Figure 7.28). The rupture nucleation develops into an unstable, dynamic rupture in due course, and a great amount of the stored strain is released rapidly at the time of the unstable, dynamic rupture occurrence. This sequence of changes may cause changes in the spatially patterned crustal strain on the ground surface adjacent to the fault. If this is the case, casting

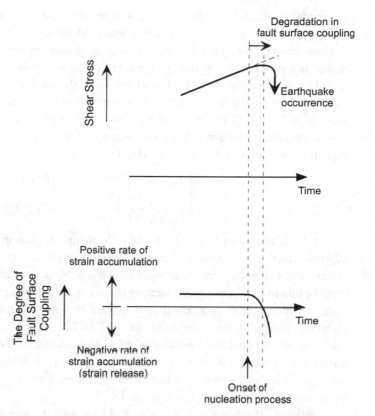

Fig. 7.28 Diagram of the degradation of fault surface coupling and its strength during the nucleation process (Ohnaka and Matsu'ura, 2002).

a spotlight on changes in the crustal strain pattern created by both the slip displacement progress and the stress change accompanying the growth of the rupture nucleation zone may also be one way to track the ongoing nucleation process of a shallow-focus earthquake in the field.

Although it is most reasonable to make a short-term forecast of a large earthquake by monitoring the ongoing nucleation process, practical methods of how to make such a short-term forecast have not been established, as noted above. Further detailed research will be required for this. Here let us make an attempt to build a physical model for a short-term forecast of a typical large earthquake, based on the results of high-resolution laboratory experiments on the slip failure (or shear rupture) nucleation process.

As noted previously, the power law of Eq. (7.9) or Eq. (7.10) used for an intermediate-term forecast possibly has a singularity at $t = t_f$, depending on a value evaluated for the exponent n. The specific reason for the possible singularity at $t = t_f$ is that Eq. (7.9) has been derived without considering the physical process immediately before the imminent event, although the transition process of crustal deformation from a stable state to an unstable state in a relatively wide region on a time scale of a few years to a few decades is considered for the modeling. Equation (7.9) or (7.10) is therefore not suitable for a short-term forecast, although it may be certainly useful for an intermediate-term forecast.

For short-term forecast modeling, the physical transition process leading to a dynamic mainshock rupture must be incorporated. The physical transition process leading up to an unstable, dynamic rupture is what is referred to as the nucleation process, as noted above. It is therefore crucial to incorporate the nucleation process into the physical model for a short-term forecast. It has been derived in previous chapters that the rupture growth length $L(t)$ during the nucleation process increases with time t, in accordance with the following equation (see Chapters 5 and 6; in particular, Eq. (6.3)):

$$L(t) = \begin{cases} L_0 + V_{st}t & (0 < t < t_{sc}) \\ L_c \left(\dfrac{t_a - t_c}{t_a - t} \right)^{1/(n-1)} & (t_{sc} \le t \le t_c), \end{cases}$$

where L_0 is the initial length, L_c is the critical rupture growth length, V_{st} is the steady rupture growth rate controlled by an applied loading rate, t_{sc} is the critical time at which the behavior of the rupture growth changes from a steady, quasi-static phase controlled by the loading rate to a self-driven, accelerating phase controlled by inertia, t_c is the other critical time at which $L = L_c$ is attained, t_a is defined by Eq. (5.16), and n is a dimensionless constant. The value for n has been obtained as 7.31 in laboratory experiments on the stick-slip rupture nucleation process on precut rock interfaces of different roughnesses (see Chapter 5). It has also been confirmed that the rupture growth length $L(t)$ expressed by the above equation agrees well with experimental data on the stick-slip rupture nucleation on precut rock interfaces (see Figures 6.6 and 6.7).

Equation (5.16) indicates that $t_c < t_a$, and therefore, the above equation of a power law type does not have a singularity at $t = t_c$. The seismic moment M_0 of a typical earthquake is directly related to the critical length L_c of its nucleation zone by the following equation (Eq. (5.33)):

$$M_0 = 1.0 \times 10^9 (2L_c)^3.$$

Hence, the occurrence time t_c and the seismic moment M_0 for a mainshock can be forecast, if both t_c and L_c are evaluated by the best-fit regression analysis from time-sequence data observed over a certain period of time t within $0 < t < t_c$, using the following equation of a power law type (Ohnaka and Matsu'ura, 2002):

$$L(t) = L_c \left(\frac{t_a - t_c}{t_a - t} \right)^A, \tag{7.31}$$

where A is a constant. In order to use Eq. (7.31) for a short-term forecast, it is necessary to track the rupture growth length $L(t)$ with time t during the nucleation process in the neighborhood of the fault.

Let us consider a short-term forecast by using time-sequence data on slip displacement on a fault, observed at a fixed position within the nucleation zone on the fault plane. The slip displacement, $D(t)$, observed at a fixed position in the nucleation zone, proceeds with time t, initially at a very slow and steady rate, and gradually at accelerating rates, as shown in Figure 7.29. It has been demonstrated empirically that $D(t)$ measured in laboratory

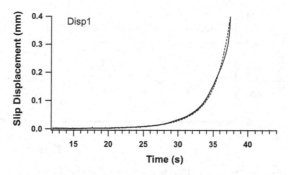

Fig. 7.29   A specific example of slip displacement $D(t)$ versus time t records (solid curved-line) measured at a fixed position within the nucleation zone of a stick-slip rupture on a precut fault in the laboratory (Ohnaka and Matsu'ura, 2002). The broken curved-line in the figure graphically represents Eq. (7.32), with $D_{c0} = 0.4117$ mm, $B = 5.5, t_b = 50$ s, and $t_c = 37.56$ s.

experiments can be expressed by the formula of the same power law type as Eq. (7.31); that is (Ohnaka and Matsu'ura, 2002),

$$D(t) = D_{c0} \left(\frac{t_b - t_c}{t_b - t} \right)^B , \tag{7.32}$$

where D_{c0} is the slip displacement at $t = t_c$, t_b is a constant meeting the condition of $t_c < t_b$, and B is a constant.

Figure 7.29 is a specific example of slip displacement versus time records (solid curved-line) measured at a fixed position within the nucleation zone of a stick-slip rupture on a pre-cut fault observed in the laboratory. The broken curved-line in Figure 7.29 graphically represents formula (7.32) indicating the relationship between $D(t)$ and t where $D_{c0} = 0.4117$ mm, $B = 5.5, t_b = 50$ s, and $t_c = 37.56$ s have been substituted (Ohnaka and Matsu'ura, 2002). From Figure 7.29, we can see that the power law type formula (7.32) can well represent actual data measured. If slip displacement can be measured at a fixed location in the nucleation zone on a master fault in the field, and if the parameters in Eq. (7.32) can be evaluated by using time-sequence data on slip displacement measured during a time period up to time $t(< t_c)$, then formula (7.32) may be useful for a short-term forecast.

A physical model for the preparation process leading to an imminent earthquake has been presented, based on the stick-slip rupture nucleation process observed in high-resolution laboratory experiments. We have seen that the preparation process leading to an unstable, dynamic rupture obeys power laws. The power law type formula proposed for a short-term forecast does not have any singularity at a critical point in time, and hence not only the occurrence time but also the magnitude can be forecast for an anticipated earthquake, at least theoretically. Hereafter, the short-term forecast model presented in this subsection needs to be verified by data on real earthquakes observed in the field.

The short-term forecast model presented in this subsection has been created for typical earthquakes. Note, therefore, that it may not be suitable to apply the model to silent earthquakes or aseismic slips occurring in the field, without considering the underlying physics.

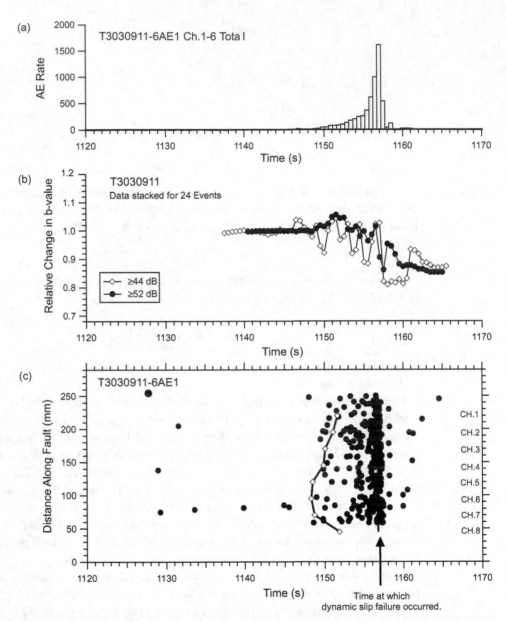

Fig. 7.30 (a) Time series variation of acoustic emission activities induced during the nucleation process, prior to the dynamic instability of a stick-slip rupture on a precut fault having surfaces with fractal roughness. (b) Relative change in *b*-value with time during the nucleation process. (c) Spatiotemporal distribution of acoustic emission sources (black dots). The white squares connected with solid lines indicate the bi-directional growth of the nucleation zone with time. Unpublished data from Ohnaka *et al.* (1994) are plotted.

Whether an earthquake rupture nucleation process on a fault is accompanied by fore-shocks or not depends upon the fault zone property and the conditions in the fault zone. Therefore, foreshocks are not always induced during the nucleation process, as noted previously. However, in the case of a particular fault zone in which foreshocks are fully activated in the course of the nucleation process, foreshock activity data may be used beneficially for a short-term forecast. Tracking how the zone of immediate foreshock activity grows with time would be helpful for judging whether or not the fault behavior is in the ongoing nucleation process leading to a mainshock rupture. Tracking the b-value for immediate foreshock activity would also be helpful for judging whether or not a mainshock rupture will occur shortly. In fact, a decrease in the b-value was observed for immediate foreshock activity preceding the L'Aquila earthquake ($M_w6.3$) that occurred on April 6, 2009 in Italy (Papadopoulos *et al.*, 2010). This is consistent with the result, observed in high-resolution laboratory experiments, that b-values for acoustic emission activity decrease with time during the nucleation process leading to the dynamic instability of a stick-slip rupture on a precut rock interface, as shown in Figure 7.30.

Some of the precursory phenomena described in subsection 7.3.3 may also be observed immediately before a mainshock. For example, electromagnetic waves may be radiated during the nucleation process depending on the circumstances of the fault zone. Changes in telluric current and in electrical resistance may occur during the nucleation process. In the future, it may become possible to track a change in the internal state of the fault zone occurring during the nucleation process by active means of radiated microwaves and/or cosmic ray elementary particles. Possible short-term forecast scenarios based on these phenomena will be challenges for the future. In order to develop practical methods for forecasting major or great earthquakes that cause serious damage to human society, we need to further pursue a thorough investigation into earthquake forecasting technologies.

Illustration credits

Permission for reproduction of previously published figures was sought in accordance with the various policies of the rights holders and publishers. The sources of all reproduced figures are appropriately cited in this book. I would like to express my appreciation to all copyright holders of the figures reproduced in this book.

Copyright by the American Geophysical Union

Reproduced from *Journal of Geophysical Research*:

- Figure 2.4, Dieterich, **77**, 3690–3697, 1972a.
- Figures 3.5, 3.6, 3.7, 3.15, 3.16, 3.23, 3.24, 3.25, 3.26, 3.28, 5.28, and 6.3, Ohnaka, **108** (B2), 2080, doi: 10.1029/2000JB000123, 2003.
- Figures 3.8, 3.9, 3.11, 3.12, 3.13, 3.14, 5.1, 5.2, 5.3, 5.4, 5.5, 5.6, 5.7, 5.8, 5.9, 5.10, 5.11, 5.12, 5.13, 5.14, 5.15, 5.16, and 6.5, Ohnaka and Shen, **104**, 817–844, 1999.
- Figures 4.6, 4.7, 4.8, 4.9, 4.10, 4.11, 4.12, 4.13, 4.14, 4.15, 4.16, 4.17, 4.18, Tables 4.1 and 4.2, Ohnaka and Kato, **112**, B07201, doi: 10.1029/2006JB004260, 2007.
- Figures 5.30, 5.31, 5.32, and 5.33, Ohnaka and Yamashita, **94**, 4089–4104, 1989.
- Figure 7.9, Bufe and Varnes, **98**, 9871–9883, 1993.
- Figure 7.10, Bowman *et al.*, **103**, 24 359–24 372, 1998.

Reproduced from *Earthquake Source Mechanics* (edited by S. Das, J. Boatwright, and C. H. Scholz), *Geophysical Monograph 37*:

- Figures 5.35, and 5.36, Ohnaka *et al.*, pp. 13–24, 1986.

Copyright by the American Association for the Advancement of Science

Reproduced from *Science*:

- Figure 7.17, US Geological Survey Staff, **247**, 286–293, 1990.

Copyright by the Seismological Society of America

Reproduced from *Bulletin of the Seismological Society of America*:

- Figure 5.29, Hanks, **72**, 1867–1879, 1982.

Copyright by Birkhauser Verlag

Reproduced from *Pure and Applied Geophysics*:

- Figure 4.2(c), Dieterich, **116**, 790–806, 1978a.
- Figures 5.17, 5.20, 5.21, 5.22, and 5.23, Ohnaka, **157**, 2259–2282, 2000.
- Figure 5.24, Maeda, **155**, 381–394, 1999.
- Figures 7.26 and 7.27, Yin *et al.*, **157**, 2365–2383, 2000.
- Figures 7.6, 7.7, 7.12, 7.14, 7.15, and 7.16, Ohnaka, **122**, 848–862, 1984/85.

Copyright by Elsevier Science Publishers

Reproduced from *Tectonophysics*:

- Figures 3.1, 3.2, and 3.3, Ohnaka *et al.*, **277**, 1–27, 1997.
- Figures 3.17, Ohnaka and Kuwahara, **175**, 197–220, 1990.
- Figures 7.1 and 7.2, Senatorski, **344**, 37–60, 2002.
- Figures 7.18, 7.20, and 7.21, Ohnaka, **211**, 149–178, 1992.

Copyright by Polish Scientific Publishers PWN

Reproduced from *Theory of Earthquake Premonitory and Fracture Processes* (edited by R. Teisseyre):

- Figure 7.19, Ohnaka, Part I, pp. 45–76, 1995b.

Copyright by TERRAPUB

Reproduced from *Earth Planets Space*:

- Figures 6.7, 6.8, and 7.4, Ohnaka, **56**, 773–793, 2004a.

Copyright by the University of Tokyo Press

Reproduced from *The Physics of Earthquake Generation* (co-authored by M. Ohnaka and M. Matsu'ura):

- Figures 3.37, 5.26, 5.27, 7.3, 7.5, 7.8, 7.11, 7.28, and 7.29, pp. 1–378, 2002 (in Japanese).

References

Abe, K., Reliable estimation of the seismic moment of large earthquakes, *Journal of Physics of the Earth*, 23, 381–390, 1975.

Abercrombie, R. E., D. C. Agnew, and F. K. Wyatt, Testing a model of earthquake nucleation, *Bulletin of the Seismological Society of America*, 85, 1873–1878, 1995.

Aki, K., Generation and propagation of G waves from the Niigata earthquake of June 16, 1964: part 2. Estimation of earthquake moment, released energy and stress drop from the G wave spectra, *Bulletin of Earthquake Research Institute, The University of Tokyo*, 44, 73–88, 1966.

Aki, K., Characterization of barriers on an earthquake fault, *Journal of Geophysical Research*, 84, 6140–6148, 1979.

Aki, K., A probabilistic synthesis of precursor phenomena. In *Earthquake Prediction: An International Review* (edited by D. W. Simpson and P. G. Richards), pp. 566–574. Washington, DC: American Geophysical Union, 1981.

Aki, K., Asperities, barriers, characteristic earthquakes and strong motion prediction, *Journal of Geophysical Research*, 89, 5867–5872, 1984.

Aki, K., Scale dependence in earthquake phenomena and its relevance to earthquake prediction, *Proceedings of the National Academy of Sciences of the United States of America*, 93, 3740–3747, 1996.

Aki, K. and P. G. Richards, *Quantitative Seismology: Theory and Methods*, W. H. Freeman, San Francisco, California, 1980.

Aki, K. and P. G. Richards, *Quantitative Seismology*, 2nd edition, University Science Books, Sausalito, California, 2002.

Anderson, T. L., *Fracture Mechanics – Fundamentals and Applications*, 2nd edition. Boca Raton, CA: CRC Press, 1995.

Ando, M., Seismo-tectonics of the 1923 Kanto earthquake, *Journal of Physics of the Earth*, 22, 263–277, 1974.

Andrews, D. J., Rupture propagation with finite stress in antiplane strain, *Journal of Geophysical Research*, 81, 3575–3582, 1976a.

Andrews, D. J., Rupture velocity of plane strain shear cracks, *Journal of Geophysical Research*, 81, 5679–5687, 1976b.

Andrews, D. J., Dynamic plane-strain shear rupture with a slip-weakening friction law calculated by a boundary integral method, *Bulletin of the Seismological Society of America*, 75, 1–21, 1985.

Aochi, H. and M. Matsu'ura, Slip- and time-dependent fault constitutive law and its significance in earthquake generation cycles, *Pure and Applied Geophysics*, 159, 2029–2044, 2002.

Aochi, H. and R. Madariaga, The 1999 Izmit, Turkey, earthquake: nonplanar fault structure, dynamic rupture process, and strong ground motion, *Bulletin of the Seismological Society of America*, 93, 1249–1266, 2003.

Atkinson, B. K., A fracture mechanics study of subcritical tensile cracking of quartz in wet environments, *Pure and Applied Geophysics*, 117, 1011–1024, 1979.

Atkinson, B. K., Subcritical crack propagation in rocks: theory, experimental results and applications, *Journal of Structural Geology*, 4, 41–56, 1982.

Atkinson, B. K., Subcritical crack growth in geological materials, *Journal of Geophysical Research*, 89, 4077–4114, 1984.

Atkinson, B. K. (editor), *Fracture Mechanics of Rock*. London: Academic Press, 1987.

Atkinson, B. K. and P. G. Meredith, The theory of subcritical crack growth with applications to minerals and rocks. In *Fracture Mechanics of Rock* (edited by B. K. Atkinson), pp. 111–166. London: Academic Press, 1987.

Aviles, C. A., C. H. Scholz, and J. Boatwright, Fractal analysis applied to characteristic segments of the San Andreas fault, *Journal of Geophysical Research*, 92, 331–344, 1987.

Barenblatt, G. I., The formation of equilibrium cracks during brittle fracture. General ideas and hypotheses. Axially–symmetric cracks, *Journal of Applied Mathematics and Mechanics (PMM)*, 23, 622–636, 1959.

Beeler, N. M., T. E. Tullis, and J. D. Weeks, The roles of time and displacement in the evolution effect in rock friction, *Geophysical Research Letters*, 21, 1987–1990, 1994.

Ben-Zion, Y. and J. R. Rice, Slip patterns and earthquake populations along different classes of faults in elastic solids, *Journal of Geophysical Research*, 100, 12959–12983, 1995.

Beroza, G. C. and T. Mikumo, Short slip duration in dynamic rupture in the presence of heterogeneous fault properties, *Journal of Geophysical Research*, 101, 22449–22460, 1996.

Beroza, G. C. and W. L. Ellsworth, Properties of the seismic nucleation phase, *Tectonophysics*, 261, 209–227, 1996.

Bizzarri, A., On the deterministic description of earthquakes, *Reviews of Geophysics*, 49, RG3002, doi:10.1029/2011/2011RG000356, 2011.

Bizzarri, A. and M. Cocco, Slip-weakening behavior during the propagation of dynamic ruptures obeying rate- and state-dependent friction laws, *Journal of Geophysical Research*, 108, 2373, doi: 10.1029/2002JB002198, 2003.

Bizzarri, A., M. Cocco, D. J. Andrews, and E. Boschi, Solving the dynamic rupture problem with different numerical approaches and constitutive laws, *Geophysical Journal International*, 144, 656–678, 2001.

Blanpied, M. L., T. E. Tullis, and J. D. Weeks, Frictional behavior of granite at low and high sliding velocities, *Geophysical Research Letters*, 14, 554–557, 1987.

Bouchon, M., The state of stress on some faults of the San Andreas system as inferred from near-field strong motion data, *Journal of Geophysical Research*, 102, 11731–11744, 1997.

Bouchon, M., H. Karabulut, M. Aktar, S. Ozalaybey, Jean Schmittbuhl, and M.-P. Bouin, Extended nucleation of the 1999 M_w 7.6 Izmit earthquake, *Science*, 331, 877–880, 2011.

Bowden, F. B. and D. Tabor, *The Friction and Lubrication of Solids*. Oxford: Clarendon Press, 1954.

Bowman, D. D., G. Ouillon, C. G. Sammis, A. Sornette, and D. Sornette, An observational test of the critical earthquake concept, *Journal of Geophysical Research*, 103, 24359–24372, 1998.

Brace, W. F., Laboratory studies of stick-slip and their application to earthquakes, *Tectonophysics*, 14, 189–200, 1972.

Brace, W. F. and J. D. Byerlee, Stick-slip as a mechanism for earthquakes, *Science*, 153, 990–992, 1966.

Brace, W. F. and J. D. Byerlee, California earthquakes: why only shallow focus? *Science*, 168, 1573–1575, 1970.

Brace, W. F. and R. J. Martin, III, A test of the law of effective stress for crystalline rocks of low porosity, *International Journal of Rock Mechanics and Mining Sciences*, 5, 415–426, 1968.

Brace, W. F. and D. L. Kohlstedt, Limits on lithospheric stress imposed by laboratory experiments, *Journal of Geophysical Research*, 85, 6248–6252, 1980.

Brehm, D. J. and L. W. Braile, Intermediate-term earthquake prediction using precursory events in the New Madrid seismic zone, *Bulletin of the Seismological Society of America*, 88, 564–580, 1998.

Brune, J. N., Tectonic stress and the spectra of seismic shear waves from earthquakes, *Journal of Geophysical Research*, 75, 4997–5009, 1970.

Bufe, C. G. and D. J. Varnes, Predictive modeling of the seismic cycle of the greater San Francisco Bay region, *Journal of Geophysical Research*, 98, 9871–9883, 1993.

Bufe, C. G., S. P. Nishenko, and D. J. Varnes, Seismicity trends and potential for large earthquakes in the Alaska-Aleutian region, *Pure and Applied Geophysics*, 142, 83–99, 1994.

Byerlee, J. D., Theory of friction based on brittle fracture, *Journal of Applied Physics*, 38, 2928–2934, 1967.

Byerlee, J. D. and W. F. Brace, Stick-slip, stable sliding, and earthquakes – effect of rock type, pressure, strain rate, and stiffness, *Journal of Geophysical Research*, 73, 6031–6037, 1968.

Campillo, M. and I. R. Ionescu, Initiation of antiplane shear instability under slip dependent friction, *Journal of Geophysical Research*, 102, 20363–20371, 1977.

Campillo, M., P. Favreau, I. R. Ionescu, and C. Voisin, On the effective friction law of a heterogeneous fault, *Journal of Geophysical Research*, 106, 16307–16322, 2001.

Carlson, J. M. and J. S. Langer, Mechanical model of an earthquake fault, *Physical Review A*, 40, 6470–6484, 1989.

Carlson, J. M., J. S. Langer, B. E. Shaw, and C. Tang, Intrinsic properties of a Burridge–Knopoff model of an earthquake fault, *Physical Review A*, 44, 884–897, 1991.

Carter, N. L., Steady state flow of rocks, *Reviews of Geophysics and Space Physics*, 14, 301–360, 1976.

Carter, N. L. and H. G. Ave'Lallemant, High temperature flow of dunite and peridotite, *Geological Society of America Bulletin*, 81, 2181–2202, 1970.

Charles, R. J., Dynamic fatigue of glass, *Journal of Applied Physics*, 29, 1657–1662, 1958.

Chopra, P. N. and M. S. Paterson, The experimental deformation of dunite, *Tectonophysics*, 78, 453–473, 1981.

Choubey, V. M., N. Kumar, and B. R. Arora, Precursory signatures in the radon and geohydrological borehole data for M 4.9 Kharsali earthquake of Garhwal Himalaya, *Science of the Total Environment*, 407, 5877–5883, 2009.

Coastal Movements Data Center (CMDC), *Tables and Graphs of Annual Mean Sea Level along the Japanese Coast 1894–1975*, published by CMDC, Geospatial Information Authority of Japan, pp. 1–37, 1976.

Cocco, M. and A. Bizzarri, On the slip-weakening behavior of rate- and state dependent constitutive laws, *Geophysical Research Letters*, 29(11), doi: 10.1029/2001GL013999, 2002.

Day, S. M., Three-dimensional simulation of spontaneous rupture: the effect of nonuniform prestress, *Bulletin of the Seismological Society of America*, 72, 1881–1902, 1982.

Dieterich, J. H., Time–dependent friction in rocks, *Journal of Geophysical Research*, 77, 3690–3697, 1972a.

Dieterich, J. H., Time–dependent friction as a possible mechanism for aftershocks, *Journal of Geophysical Research*, 77, 3771–3781, 1972b.

Dieterich, J. H., Time-dependent friction and the mechanics of stick-slip, *Pure and Applied Geophysics*, 116, 790–806, 1978a.

Dieterich, J. H., Preseismic fault slip and earthquake prediction, *Journal of Geophysical Research*, 83, 3940–3948, 1978b.

Dieterich, J. H., Modeling of rock friction, 1. Experimental results and constitutive equations, *Journal of Geophysical Research*, 84, 2161–2168, 1979.

Dieterich, J. H., Constitutive properties of faults with simulated gouge. In *Mechanical Behavior of Crystal Rocks* (edited by N. L. Cater, M. Friedman, J. M. Logan, and D. W. Stearns), *Geophysical Monograph 24*, pp. 103–120. Washington, DC: American Geophysical Union, 1981.

Dieterich, J. H., A model for the nucleation of earthquake slip. In *Earthquake Source Mechanics* (edited by S. Das, J. Boatwright, and C. H. Scholz), *Geophysical Monograph 37*, pp. 37–47. Washington, DC: American Geophysical Union, 1986.

Dieterich, J. H. and B. Kilgore, Implications of fault constitutive properties for earthquake prediction, *Proceedings of the National Academy of Sciences of the United States of America*, 93, 3787–3794, 1996.

Dieterich, J. H., D. W. Barber, G. Conrad, and Q. A. Gorton, Preseismic slip in a large scale friction experiment, *Proceedings of the 19th US Rock Mechanics Symposium*, pp. 110–117, 1978.

Dmowska, R. and V. C. Li, A mechanical model of precursory source processes for some large earthquakes, *Geophysical Research Letters*, 9, 393–396, 1982.

Dodge, D. A., G. C. Beroza, and W. L. Ellsworth, Evolution of the 1992 Landers, California, foreshock sequence and its implications for earthquake nucleation, *Journal of Geophysical Research*, 100, 9865–9880, 1995.

Dodge, D. A., G. C. Beroza, and W. L. Ellsworth, Detailed observations of California foreshock sequence: implications for the earthquake initiation process, *Journal of Geophysical Research*, 101, 22371–22392, 1996.

Ellsworth, W. L. and G. C. Beroza, Seismic evidence for an earthquake nucleation phase, *Science*, 268, 851–855, 1995.

Engdahl, E. R. and A. Villasenor, Global seismicity: 1900–1999. In *International Handbook of Earthquake and Engineering Seismology, Part A* (edited by W. H. K. Lee, H. Kanamori, P. C. Jennings, and C. Kisslinger), pp. 665–690. New York: Academic Press, 2002.

Fedotov, S. A., Regularities of the distribution of strong earthquakes in Kamchatka, the Kurile Islands and northeastern Japan, *Academy of Sciences of the USSR, Trudy Institute of Physics of the Earth*, 36, 66–93, 1965.

Fukuyama, E. and R. Madariaga, Dynamic propagation and interaction of a rupture front on a planer fault, *Pure and Applied Geophysics*, 157, 1959–1979, 2000.

Fukuyama, E. and K. B. Olsen, A condition for super-shear rupture propagation in a heterogeneous stress field, *Pure and Applied Geophysics*, 159, 2047–2056, 2002.

Fukuyama, E., C. Hashimoto, and M. Matsu'ura, Simulation of the transition of earthquake rupture from quasi-static growth to dynamic propagation, *Pure and Applied Geophysics*, 159, 2057–2066, 2002.

Gilman, J. J., Cleavage, ductility and tenacity in crystals. In *Fracture* (edited by B. L. Averbach, D. K. Felbeck, G. T. Hahn, and D. A. Thomas), pp. 193–221. New York: Wiley, 1959.

Gilman, J. J., Strength of ceramic crystals, *Journal of the American Ceramic Society*, April, 79–102, 1962.

Griffith, A. A., The phenomena of rupture and flow in solids, *Philosophical Transactions of the Royal Society of London*, Series A, 221, 163–198, 1920.

Griggs, D. T., F. J. Turner, and H. C. Heard, Deformation of rocks at 500° to 800 °C. In *Rock Deformation* (edited by D. Griggs and J. Handin), *The Geological Society of America Memoir* 79, 39–104, 1960.

Gu, J.-C., J. R. Rice, A. L. Ruina, and S. T. Tse, Slip motion and stability of a single degree of freedom elastic system with rate and state dependent friction, *Journal of the Mechanics and Physics of Solids*, 32, 167–196, 1984.

Guatteri, M., P. Spudich, and G. C. Beroza, Inferring rate and state friction parameters from a rupture model of the 1995 Hyogo-ken Nanbu (Kobe) Japan earthquake, *Journal of Geophysical Research*, 106, 26511–26521, 2001.

Gutenberg, B. and C. F. Richter, Magnitude and energy of earthquakes, *Annali di Geofisica*, 9, 1–15, 1956.

Hamada, N., *Re-examination of Seismicity Associated with Destructive Inland Earthquakes of Japan and Its Seismological Significance*, Doctor Thesis, The University of Tokyo, 1986.

Hamada, N., Re-examination of seismicity associated with destructive inland earthquakes of Japan and its seismological significance, *Papers in Meteorology and Geophysics*, 38, 77–156, 1987.

Hanks, T. C., f_{max}, *Bulletin of the Seismological Society of America*, 72, 1867–1879, 1982.

Hanks, T. C. and H. Kanamori, A moment magnitude scale, *Journal of Geophysical Research*, 84, 2348–2350, 1979.

Hansen, F. D. and N. L. Carter, Creep of selected crustal rocks at 1000 MPa, *EOS*, Transactions, American Geophysical Union, 63, p. 437, 1982.

Hashimoto, C. and M. Matsu'ura, 3-D physical modeling of stress accumulation processes at transcurrent plate boundaries, *Pure and Applied Geophysics*, 157, 2125–2147, 2000.

Hashimoto, C. and M. Matsu'ura, 3-D simulation of earthquake generation cycles and evolution of fault constitutive properties, *Pure and Applied Geophysics*, 159, 2175–2199, 2002.

Hawking, S., *A Brief History of Time*. London: Bantam Press, Transworld Publishers, 1988.

Hirotaka, U. I., H. Moriuchi, Y. Takemura, H. Tsuchida, I. Fujii, and M. Nakamura, Anomalously high radon discharge from Atotsugawa fault prior to the western Nagano Prefecture earthquake (M 6.8) of September 14, 1984, *Tectonophysics*, 152, 147–152, 1988.

Holub, R. F. and B. T. Brady, The effect of stress on radon emanation from rock, *Journal of Geophysical Research*, 86, 1776–1784, 1981.

Ida, Y., Cohesive force across the tip of a longitudinal–shear crack and Griffith's specific surface energy, *Journal of Geophysical Research*, 77, 3796–3805, 1972.

Ida, Y., The maximum acceleration of seismic ground motion, *Bulletin of the Seismological Society of America*, 63, 959–968, 1973.

Ide, S. and M. Takeo, Determination of constitutive relations of fault slip based on seismic wave analysis, *Journal of Geophysical Research*, 102, 27379–27391, 1997.

Igarashi, G., S. Saeki, N. Takahata, and K. Sumikawa, Ground-water radon anomaly before the Kobe earthquake in Japan, *Science*, 269, 60–61, 1995.

Imamura, A., On the seismic activity of central Japan, *Japanese Journal of Astronomy and Geophysics, Transactions*, National Research Council of Japan, 6, 119–137, 1928/29.

Irwin, G. R., Onset of fast crack propagation in high strength steel and aluminum alloys, *Sagamore Research Conference Proceedings*, 2, 289–305, 1956.

Irwin, G. R., Analysis of stresses and strains near the end of a crack traversing a plate, *Journal of Applied Mechanics*, 24, 361–364, 1957.

Ishibashi, K. and K. Satake, Problems on forecasting great earthquakes in the subduction zones around Japan by means of paleoseismology, *Journal of the Seismological Society of Japan* (*Zisin*), Second Series, 50 (Supplement), 1–21, 1998 (in Japanese).

Japan Meteorological Agency (JMA), Catalogue of relocated major earthquakes in and near Japan (1926–1960), *Seismological Bulletin of the JMA, Supplementary Volume No. 6*, 1–109, 1982.

Jaume, S. C. and L. R. Sykes, Evolving towards a critical point: a review of accelerating seismic moment/energy release prior to large and great earthquakes, *Pure and Applied Geophysics*, 155, 279–306, 1999.

Jones, L. M. and P. Molner, Some characteristics of foreshocks and their possible relationship to earthquake prediction and premonitory slip on faults, *Journal of Geophysical Research*, 84, 3596–3608, 1979.

Kanamori, H., The energy release in great earthquakes, *Journal of Geophysical Research*, 82, 2981–2987, 1977.

Kanamori, H. and M. Ando, Fault parameters of the great Kanto earthquake of 1923, *Publications for the 50th Anniversary of the Great Kanto Earthquake, 1923*, pp. 89–101, 1973 (in Japanese).

Kanamori, H. and G. S. Stewart, Seismological aspects of the Guatemala earthquake of February 4, 1976, *Journal of Geophysical Research*, 83, 3427–3434, 1978.

Kato, A., M. Ohnaka, and H. Mochizuki, Constitutive properties for the shear failure of intact granite in seismogenic environments, *Journal of Geophysical Research*, 108(B1), 2060, doi: 10.1029/2001JB000791, 2003a.

Kato, A., M. Ohnaka, S. Yoshida, and H. Mochizuki, Effect of strain rate on constitutive properties for the shear failure of intact granite in seismogenic environments, *Geophysical Research Letters*, 30(21), 2108, doi: 10.1029/2003GL018372, 2003b.

Kato, A., S. Yoshida, M. Ohnaka, and H. Mochizuki, The dependence of constitutive properties on temperature and effective normal stress in seismogenic environments, *Pure and Applied Geophysics*, 161, 1895–1913, 2004.

Kato, A., K. Obara, T. Igarashi, H. Tsuruoka, S. Nakagawa, and N. Hirata, Propagation of slow slip leading up to the 2011 M_w9.0 Tohoku-oki earthquake, *Science*, 335, 705–708, 2012.

Kikuchi, M. and K. Sudo, Inversion of teleseismic P waves of Izu-Oshima, Japan earthquake of January 14, 1978, *Journal of Physics of the Earth*, 32, 161–171, 1984.

Kilgore, B. D., M. L. Blanpied, and J. H. Dieterich, Velocity-dependent friction of granite over a wide range of conditions, *Geophysical Research Letters*, 20, 903–906, 1993.

Kirby, S. H., Tectonic stresses in the lithosphere: Constraints provided by the experimental deformation of rocks, *Journal of Geophysical Research*, 85, 6353–6363, 1980.

Kirby, S. H., Rheology of the lithosphere, *Reviews of Geophysics*, 21, 1458–1487, 1983.

Kronenberg, A. and J. Tullis, Flow strengths of quartz aggregates: grain size and pressure effects due to hydrolytic weakening, *Journal of Geophysical Research*, 89, 4281–4297, 1984.

Kumamoto, T., Long-term conditional seismic hazard of Quaternary active faults in Japan, *Journal of the Seismological Society of Japan (Zisin)*, Second Series, 50 (Supplement), 53–71, 1998 (in Japanese).

Langer, J. S., J. M. Carlson, C. R. Myers, and B. E. Shaw, Slip complexity in dynamic models of earthquake faults, *Proceedings of the National Academy of Sciences of the United States of America*, 93, 3825–3829, 1996.

Lawn, B. R. and T. R. Wilshaw, *Fracture of Brittle Solids*, Cambridge Solid State Science Series. Cambridge: Cambridge University Press, 1975.

Lay, T., H. Kanamori, and L. Ruff, The asperity model and the nature of large subduction zone earthquakes, *Earthquake Prediction Research*, 1, 3–71, 1982.

Li, V. C. and J. R. Rice, Preseismic rupture progression and great earthquake instabilities at plate boundaries, *Journal of Geophysical Research*, 88, 4231–4246, 1983a.

Li, V. C. and J. R. Rice, Precursory surface deformation in great plate boundary earthquake sequences, *Bulletin of the Seismological Society of America*, 73, 1415–1434, 1983b.

Linker, M. F. and J. H. Dieterich, Effects of variable normal stress on rock friction: observations and constitutive equations, *Journal of Geophysical Research*, 97, 4923–4940, 1992.

Lockner, D. A., J. D. Byerlee, V. Kuksenko, A. Ponomarev, and A. Sidorin, Quasi-static fault growth and shear fracture energy in granite, *Nature*, 350, 39–42, 1991.

Lockner, D. A., J. D. Byerlee, V. Kuksenko, A. Ponomarev, and A. Sidorin, Observations of quasistatic fault growth from acoustic emissions. In *Fault Mechanics and Transport*

Properties of Rocks (edited by B. Evans and T.-F. Wong), pp. 3–31, A Festschrift in Honor of W.F. Brace. London: Academic Press, 1992.

Madariaga, R., Dynamics of an expanding circular fault, *Bulletin of the Seismological Society of America*, 66, 639–666, 1976.

Madariaga, R., Implications of stress-drop models of earthquakes for the inversion of stress drop from seismic observations, *Pure and Applied Geophysics*, 115, 301–316, 1977.

Madariaga, R. and K. B. Olsen, Criticality of rupture dynamics in 3-D, *Pure and Applied Geophysics*, 157, 1981–2001, 2000.

Madariaga, R. and K. B. Olsen, Earthquake dynamics. In *International Handbook of Earthquake and Engineering Seismology, Part A* (edited by W. H. K. Lee, H. Kanamori, P. C. Jennings, and C. Kisslinger), pp. 175–194. New York: Academic Press, 2002.

Madariaga, R., K. B. Olsen, and R. J. Archuleta, Modeling dynamic rupture in a 3-D earthquake fault model, *Bulletin of the Seismological Society of America*, 88, 1182–1197, 1998.

Maeda, K., Time distribution of immediate foreshocks obtained by a stacking method, *Pure and Applied Geophysics*, 155, 381–394, 1999.

Marone, C., The effect of loading rate on static friction and the rate of fault healing during the earthquake cycle, *Nature*, 391, 69–72, 1998a.

Marone, C., Laboratory-derived friction laws and their application to seismic faulting, *Annual Review of Earth and Planetary Sciences*, 26, 643–696, 1998b.

Masuda, K., H. Mizutani, and I. Yamada, Experimental study of strain-rate dependence and pressure dependence of failure properties of granite, *Journal of Physics of the Earth*, 35, 37–66, 1987.

Masuda, K., H. Mizutani, I. Yamada, and Y. Imanishi, Effects of water on time-dependent behavior of granite, *Journal of Physics of the Earth*, 36, 291–313, 1988.

Matsuda, T., Present state of long-term prediction of earthquakes based on active fault data in Japan – an example for the Itoigawa–Shizuoka tectonic line active fault system, *Journal of the Seismological Society of Japan (Zisin)*, Second Series, 50 (Supplement), 23–33, 1998 (in Japanese).

Matsu'ura, M., H. Kataoka, and B. Shibazaki, Slip-dependent friction law and nucleation processes in earthquake rupture. In *Earthquake Source Physics and Earthquake Precursors* (edited by T. Mikumo, K. Aki, M. Ohnaka, L. J. Ruff, and P. K. P. Spudich), *Special Issue of Tectonophysics*, 211, 135–148, 1992.

Matsu'ura, S. R., I. Karakama, and K. Tsumura, *List of Earthquakes in the Kanto Area and Its Vicinity*, Earthquake Research Institute, The University of Tokyo, Tokyo, 1988.

Meredith, P. G. and B. K. Atkinson, Fracture toughness and subcritical crack growth during high-temperature tensile deformation of Westerly granite and Black gabbro, *Physics of the Earth and Planetary Interiors*, 39, 33–51, 1985.

Michalske, T. A. and S. W. Freiman, A molecular interpretation of stress corrosion in silica, *Nature*, 295, 511–512, 1982.

Mikumo, T. and Y. Yagi, Slip-weakening distance in dynamic rupture of in-slab normal-faulting earthquakes, *Geophysical Journal International*, 155, 443–455, 2003.

Mikumo, T., K. B. Olsen, E. Fukuyama, and Y. Yagi, Stress-breakdown time and slip-weakening distance inferred from slip-velocity functions on earthquake faults, *Bulletin of the Seismological Society of America*, 93, 264–282, 2003.

Miyake, H., T. Iwata, and K. Irikura, Source characterization for broadband ground-motion simulation: kinematic heterogeneous source model and strong motion generation area, *Bulletin of the Seismological Society of America*, 93, 2531–2545, 2003.

Mogi, K., Sequential occurrences of recent great earthquakes, *Journal of Physics of the Earth*, 16, 30–36, 1968.

Murrell, S. A. F. and I. A. H. Ismail, The effect of decomposition of hydrous minerals on the mechanical properties of rocks at high pressures and temperatures, *Tectonophysics*, 31, 207–258, 1976.

Nakanishi, H., Earthquake dynamics driven by a viscous fluid, *Physical Review A*, 46, 4689–4692, 1992.

Nielsen, S., J. Taddeucci, and S. Vinciguerra, Experimental observation of stick-slip instability fronts, *Geophysical Journal International*, 180, 697–702, 2010.

Odedra, A., *Laboratory studies on shear fracture of granite under simulated crustal conditions*, Ph.D. Thesis, pp. 1–269, University College London, London, 1998.

Odedra, A., M. Ohnaka, H. Mochizuki, and P. Sammonds, Temperature and pore pressure effects on the shear strength of granite in the brittle–plastic transition regime, *Geophysical Research Letters*, 28, 3011–3014, 2001.

Ohnaka, M., Experimental studies of stick-slip and their application to the earthquake source mechanism, *Journal of Physics of the Earth*, 21, 285–303, 1973.

Ohnaka, M., Frictional characteristics of typical rocks, *Journal of Physics of the Earth*, 23, 87–112, 1975.

Ohnaka, M., A physical basis for earthquakes based on the elastic rebound model, *Bulletin of the Seismological Society of America*, 66, 433–451, 1976.

Ohnaka, M., Acoustic emission during creep of brittle rock, *International Journal of Rock Mechanics and Mining Sciences & Geomechanics Abstracts*, 20, 121–134, 1983.

Ohnaka, M., A sequence of seismic activity in the Kanto area precursory to the 1923 Kanto earthquake, *Pure and Applied Geophysics*, 122, 848–862, 1984/85.

Ohnaka, M., Nonuniformity of crack-growth resistance and breakdown zone near the propagating tip of a shear crack in brittle rock: A model for earthquake nucleation to dynamic rupture, *Canadian Journal of Physics*, 68, 1071–1083, 1990.

Ohnaka, M., Earthquake source nucleation: a physical model for short-term precursors. In *Earthquake Source Physics and Earthquake Precursors* (edited by T. Mikumo, K. Aki, M. Ohnaka, L. J. Ruff, and P. K. P. Spudich), *Special Issue of Tectonophysics*, 211, 149–178, 1992.

Ohnaka, M., Critical size of the nucleation zone of earthquake rupture inferred from immediate foreshock activity, *Journal of Physics of the Earth*, 41, 45–56, 1993.

Ohnaka, M., A shear failure strength law of rock in the brittle–plastic transition regime, *Geophysical Research Letters*, 22, 25–28, 1995a.

Ohnaka, M., Earthquake source nucleation model and immediate seismic precursors. In *Theory of Earthquake Premonitory and Fracture Processes* (edited by R. Teisseyre), Part I, pp. 45–76. Warszawa: Polish Scientific Publishers PWN, 1995b.

Ohnaka, M., Constitutive equations for shear failure of rocks. In *Theory of Earthquake Pre-monitory and Fracture Processes* (edited by R. Teisseyre), Part I, pp. 26–44. Warszawa: Polish Scientific Publishers PWN, 1995c.

Ohnaka, M., Nonuniformity of the constitutive law parameters for shear rupture and quasistatic nucleation to dynamic rupture: a physical model of earthquake generation processes, *Proceedings of the National Academy of Sciences of the United States of America*, 93, 3795–3802, 1996.

Ohnaka, M., Earthquake generation processes and earthquake prediction: implications of the underlying physical law and seismogenic environments, *Journal of the Seismological Society of Japan (Zisin)*, Second Series, 50 (Supplement), 129–155, 1998 (in Japanese).

Ohnaka, M., A physical scaling relation between the size of an earthquake and its nucleation zone size, *Pure and Applied Geophysics*, 157, 2259–2282, 2000.

Ohnaka, M., A constitutive scaling law and a unified comprehension for frictional slip failure, shear fracture of intact rock, and earthquake rupture, *Journal of Geophysical Research*, 108 (B2), 2080, doi: 10.1029/2000JB000123, 2003.

Ohnaka, M., Earthquake cycles and physical modeling of the process leading up to a large earthquake, *Earth Planets Space*, 56, 773–793, 2004a.

Ohnaka, M., A constitutive scaling law for shear rupture that is inherently scale-dependent, and physical scaling of nucleation time to critical point, *Pure and Applied Geophysics*, 161, 1915–1929, 2004b.

Ohnaka, M. and T. Yamashita, A cohesive zone model for dynamic shear faulting based on experimentally inferred constitutive relation and strong motion source parameters, *Journal of Geophysical Research*, 94, 4089–4104, 1989.

Ohnaka, M. and Y. Kuwahara, Characteristic features of local breakdown near a crack-tip in the transition zone from nucleation to unstable rupture during stick-slip shear failure. In *Earthquake Source Processes* (edited by S. Das and M. Ohnaka), *Special Issue of Tectonophysics*, 175, 197–220, 1990.

Ohnaka, M. and L. F. Shen, Scaling of the shear rupture process from nucleation to dynamic propagation: implications of geometric irregularity of the rupturing surfaces, *Journal of Geophysical Research*, 104, 817–844, 1999.

Ohnaka, M. and M. Matsu'ura, *The Physics of Earthquake Generation*. Tokyo: University of Tokyo Press, 2002 (in Japanese).

Ohnaka, M. and A. Kato, Depth dependence of constitutive law parameters for shear failure of rock at local strong areas on faults in the seismogenic crust, *Journal of Geophysical Research*, 112, B07201, doi: 10.1029/2006JB004260, 2007.

Ohnaka, M., Y. Kuwahara, and K. Yamamoto, Constitutive relations between dynamic physical parameters near a tip of the propagating slip zone during stick-slip shear failure. In *Mechanics of Earthquake Faulting* (edited by R. L. Wesson), *Special Issue of Tectonophysics*, 144, 109–125, 1987a.

Ohnaka, M., Y. Kuwahara, and K. Yamamoto, Nucleation and propagation processes of stick-slip failure and normal stress dependence of the physical parameters of dynamic slip failure, *Natural Disaster Science*, 9, 1–21, 1987b.

Ohnaka, M., K. Yamamoto, Y. Kuwahara, and T. Hirasawa, Dynamic processes during slip of stick-slip as an earthquake fault model, *Journal of the Seismological Society of Japan (Zisin)*, Second Series, 36, 53–62, 1983 (in Japanese).

Ohnaka, M., Y. Kuwahara, K. Yamamoto, and T. Hirasawa, Dynamic breakdown processes and the generating mechanism for high-frequency elastic radiation during stick-slip instabilities. In *Earthquake Source Mechanics* (edited by S. Das, J. Boatwright, and C. H. Scholz), *Geophysical Monograph 37*, pp. 13–24. Washington, DC: American Geophysical Union, 1986.

Ohnaka, M., S. Yoshida, and L.-F. Shen, Slip failure nucleation process and induced micro-seismicity, paper presented at Workshop 5 on *Earthquake Source Modeling and Fault Mechanics*, International Association of Seismology and Physics of the Earth's Interior 27th General Assembly held in Wellington, New Zealand, January 1994.

Ohnaka, M., M. Akatsu, H. Mochizuki, A. Odedra, F. Tagashira, and Y. Yamamoto, A constitutive law for the shear failure of rock under lithospheric conditions. In *Earthquake Generation Processes: Environmental Aspects and Physical Modeling* (edited by M. Matsu'ura, C. J. Marone, S. R. McNutt, M. Takeo, I. G. Main, and J. B. Rundle), *Special Issue of Tectonophysics*, 277, 1–27, 1997.

Okubo, P. G., Dynamic rupture modeling with laboratory-derived constitutive relations, *Journal of Geophysical Research*, 94, 12321–12335, 1989.

Okubo, P. G. and J. H. Dieterich, State variable fault constitutive relations for dynamic slip. In *Earthquake Source Mechanics* (edited by S. Das, J. Boatwright, and C. H. Scholz), *Geophysical Monograph 37*, pp. 25–35. Washington, DC: American Geophysical Union, 1986.

Okubo, P. G. and K. Aki, Fractal geometry in the San Andreas fault system, *Journal of Geophysical Research*, 92, 345–355, 1987.

Orowan, E., Fracture and strength of solids, *Reports on Progress in Physics*, 12, 185–232, 1948.

Palmer, A. C. and J. R. Rice, The growth of slip surfaces in the progressive failure of over–consolidated clay, *Proceedings of the Royal Society of London*, A-**332**, 527–548, 1973.

Papadopoulos, G. A., M. Charalampakis, A. Fokaefs, and G. Minadakis, Strong foreshock signal preceding the L'Aquila (Italy) earthquake (M_w6.3) of 6 April 2009, *Natural Hazards and Earth System Sciences*, 10, 19–24, 2010.

Papageorgiou, A. S. and K. Aki, A specific barrier model for the quantitative description of inhomogeneous faulting and the prediction of strong ground motion. I. Description of the model, *Bulletin of the Seismological Society of America*, 73, 693–722, 1983a.

Papageorgiou, A. S. and K. Aki, A specific barrier model for the quantitative description of inhomogeneous faulting and the prediction of strong ground motion. Part II. Applications of the model, *Bulletin of the Seismological Society of America*, 73, 953–978, 1983b.

Paterson, M. S., *Experimental Rock Deformation – The Brittle Field*. Berlin: Springer-Verlag, 1978.

Perrin, G., J. R. Rice, and G. Zheng, Self-healing slip pulse on a frictional surface, *Journal of the Mechanics and Physics of Solids*, 43, 1461–1495, 1995.

Rabinowicz, E., *Friction and Wear of Materials*. New York: John Wiley and Sons, 1965.

Raleigh, C. B. and M. S. Paterson, Experimental deformation of serpentinite and its tectonic implications, *Journal of Geophysical Research*, 70, 3965–3985, 1965.

Raleigh, C. B. and W. H. K. Lee, Sea-floor spreading and island-arc tectonics, *Proceedings of the Andesite Conference* (edited by A. R. McBirney), Oregon, Department of Geology and Mineral Industries Bulletin, 65, 99–110, 1969.

Reid, H. F., The elastic-rebound theory of earthquakes, *University of California Publications, Bulletin of the Department of Geology*, 6, 413–444, 1911.

Rice, J. R., A path independent integral and the approximate analysis of strain concentration by notches and cracks, *Journal of Applied Mechanics*, 35, 379–386, 1968.

Rice, J. R., The mechanics of earthquake rupture. In *Physics of the Earth's Interior, Proceedings of the International School of Physics "Enrico Fermi"* (edited by A. M. Dziewonski and E. Boschi), pp. 555–649. Amsterdam: North-Holland, 1980.

Rice, J. R., Constitutive relations for fault slip and earthquake instabilities, *Pure and Applied Geophysics*, 121, 443–475, 1983.

Rice, J. R., Shear instability in relation to the constitutive description of fault slip, *Proceedings of the 1st International Congress on Rockbursts and Seismicity in Mines*, pp. 57–62, Johannesburg, 1984.

Rice, J. R., Spatio-temporal complexity of slip on a fault, *Journal of Geophysical Research*, 98, 9885–9907, 1993.

Rice, J. R. and A. L. Ruina, Stability of steady frictional slipping, *Journal of Applied Mechanics, Transactions of the American Society of Mechanical Engineers*, 50, 343–349, 1983.

Rice, J. R. and S. T. Tse, Dynamic motion of a single degree of freedom system following a rate and state dependent friction law, *Journal of Geophysical Research*, 91, 521–530, 1986.

Rice, J. R. and Y. Ben-Zion, Slip complexity in earthquake fault models, *Proceedings of the National Academy of Sciences of the United States of America*, 93, 3811–3818, 1996.

Richon, P., J.-C. Sabroux, M. Halbwachs, J. Vandemeulebrouck, N. Poussielgue, J. Tabbagh, and R. Punongbayan, Radon anomaly in the soil of Taal volcano, the Philippines: A likely precursor of the M 7.1 Mindoro earthquake (1994), *Geophysical Research Letters*, 30, No. 9, 1481, doi: 10.1029/2003GL016902, 2003.

Rudnicki, J. W., Fracture mechanics applied to the Earth's crust, *Annual Review of Earth and Planetary Sciences*, 8, 489–525, 1980.

Rudnicki, J. W., Physical models of earthquake instability and precursory processes, *Pure and Applied Geophysics*, 126, 531–554, 1988.

Ruina, A., Slip instability and state variable friction laws, *Journal of Geophysical Research*, 88, 10359–10370, 1983.

Ruina, A. L., Constitutive relations for frictional slip. In *Mechanics of Geomaterials* (edited by Z. Bazant), pp. 169–188. New York: John Wiley, 1985.

Ruiz, S. and R. Madariaga, Determination of the friction law parameters of the Mw 6.7 Michilla earthquake in northern Chile by dynamic inversion, *Geophysical Research Letters*, 38, L09317, doi:10.1029/2011GL047147, 2011.

Rummel, F., H. J. Alheid, and C. Frohn, Dilatancy and fracture-induced velocity changes in rock and their relation to frictional sliding, *Pure and Applied Geophysics*, 116, 743–764, 1978.

Rundle, J. B., W. Klein, D. L. Turcotte, and B. D. Malamud, Precursory seismic activation and critical-point phenomena, *Pure and Applied Geophysics*, 157, 2165–2182, 2000.

Sadovsky, M. A., I. L. Nersesov, S. K. Nigmatullaev, L. A. Latynina, A. A. Lukk, A. N. Semenov, I. G. Simbereva, and V. I. Ulomov, The processes preceding strong earthquakes in some regions of Middle Asia, *Tectonophysics*, 14, 195–307, 1972.

Saito, M., Forecasting time of slope failure by tertiary creep, *Proceedings of 7th International Conference on Soil Mechanics and Foundation Engineering*, 2, 677–683, 1969.

Sammonds, P. and M. Ohnaka, Evolution of microseismcity during frictional sliding, *Geophysical Research Letters*, 25, 699–702, 1998.

Sammonds, P. R., P. G. Meredith, and I. G. Main, Role of pore fluids in the generation of seismic precursors to shear fracture, *Nature*, 359, 228–230, 1992.

Savage, J. C. and R. O. Burford, Geodetic determination of relative plate motion in central California, *Journal of Geophysical Research*, 78, 832–845, 1973.

Scholz, C. H., Mechanism of creep in brittle rock, *Journal of Geophysical Research*, 73, 3295–3302, 1968.

Scholz, C. H., Static fatigue of quartz, *Journal of Geophysical Research*, 77, 2104–2114, 1972.

Scholz, C. H., Mechanisms of seismic quiescences, *Pure and Applied Geophysics*, 126, 701–718, 1988.

Scholz, C. H. and J. T. Engelder, The role of asperity indentation and ploughing in rock friction – I. Asperity creep and stick–slip, *International Journal of Rock Mechanics and Mining Sciences & Geomechanics Abstracts*, 13, 149–154, 1976.

Scholz, C., P. Molnar, and T. Johnson, Detailed studies of frictional sliding of granite and implications for the earthquake mechanism, *Journal of Geophysical Research*, 77, 6392–6406, 1972.

Schwartz, D. P. and K. J. Coppersmith, Fault behavior and characteristic earthquakes: examples from the Wasatch and San Andreas fault zones, *Journal of Geophysical Research*, 89, 5681–5698, 1984.

Segall, P. and D. D. Pollard, Mechanics of discontinuous faults, *Journal of Geophysical Research*, 85, 4337–4350, 1980.

Senatorski, P., Slip-weakening and interactive dynamics of an heterogeneous seismic source, *Tectonophysics*, 344, 37–60, 2002.

Shibazaki, B. and M. Matsu'ura, Foreshocks and pre-events associated with the nucleation of large earthquakes, *Geophysical Research Letters*, 22, 1305–1308, 1995.

Shibazaki, B. and M. Matsu'ura, Transition process from nucleation to high-speed rupture propagation: scaling from stick-slip experiments to natural earthquakes, *Geophysical Journal International*, 132, 14–30, 1998.

Shimazaki, K. and P. Somerville, Static and dynamic parameters of the Izu-Oshima, Japan earthquake of January 14, 1978, *Bulletin of the Seismological Society of America*, 69, 1343–1378, 1979.

Sibson, R. H., Fault rocks and fault mechanisms, *Journal of the Geological Society of London*, 133, 191–213, 1977.

Sibson, R. H., Rupture interaction with fault jogs. In *Earthquake Source Mechanics* (edited by S. Das, J. Boatwright and C. H. Scholz), *Geophysical Monograph 37, Maurice Ewing Volume 6*, pp. 157–167. Washington, DC: American Geophysical Union, 1986.

Sieh, K., Lateral offsets and revised dates of large prehistoric earthquakes at Pallett Creek, southern California, *Journal of Geophysical Research*, 89, 7641–7670, 1984.

Sieh, K., The repetition of large-earthquake ruptures, *Proceedings of the National Academy of Sciences of the United States of America*, 93, 3764–3771, 1996.

Somerville, P., K. Irikura, R. Graves, S. Sawada, D. Wald, N. Abrahamson, Y. Iwasaki, T. Kagawa, N. Smith, and A. Kowada, Characterizing crustal earthquake slip models for the prediction of strong ground motion, *Seismological Research Letters*, 70, 59–80, 1999.

Stesky, R. M., Mechanisms of high temperature frictional sliding in Westerly granite, *Canadian Journal of Earth Sciences*, 15, 361–375, 1978.

Stesky, R. M., W. F. Brace, D. K. Riley, and P.-Y. Bobin, Friction in faulted rock at high temperature and pressure, *Tectonophysics*, 23, 177–203, 1974.

Stuart, W. D., Strain softening prior to two-dimensional strike slip earthquakes, *Journal of Geophysical Research*, 84, 1063–1070, 1979a.

Stuart, W. D., Strain-softening instability model for the San Fernando earthquake, *Science*, 203, 907–910, 1979b.

Stuart, W. D. and G. M. Mavko, Earthquake instability on a strike-slip fault, *Journal of Geophysical Research*, 84, 2153–2160, 1979.

Stuart, W. D. and T. E. Tullis, Fault model for preseismic deformation at Parkfield, California, *Journal of Geophysical Research*, 100, 24079–24099, 1995.

Sykes, L. R. and S. P. Nishenko, Probabilities of occurrence of large plate rupturing earthquakes for the San Andreas, San Jacinto, and Imperial faults, California, 1983–2003, *Journal of Geophysical Research*, 89, 5905–5927, 1984.

Sykes, L. R. and S. C. Jaume, Seismic activity on neighboring faults as a long-term precursor to large earthquakes in the San Francisco Bay area, *Nature*, 348, 595–599, 1990.

Tanaka, S., M. Ohtake, and H. Sato, Evidence for tidal triggering of earthquakes as revealed from statistical analysis of global data, *Journal of Geophysical Research*, 107(B10), 2211, doi: 10.1029/2001JB001577, 2002.

Tse, S. T. and J. R. Rice, Crustal earthquake instability in relation to the depth variation of frictional slip properties, *Journal of Geophysical Research*, 91, 9452–9472, 1986.

Tsumura, K., I. Karakama, I. Ogino, and M. Takahashi, Seismic activities before and after the Izu-Oshima-Kinkai earthquake of 1978, *Bulletin of the Earthquake Research Institute, The University of Tokyo*, 53, 675–706, 1978 (in Japansese).

Tullis, T. E., Rock friction and its implications for earthquake prediction examined via models of Parkfield earthquakes, *Proceedings of the National Academy of Sciences of the United States of America*, 93, 3803–3810, 1996.

Tullis, T. E. and J. D. Weeks, Constitutive behavior and stability of frictional sliding of granite, *Pure and Applied Geophysics*, 124, 383–414, 1986.

Turcotte, D. L. and D. A. Spence, An analysis of strain accumulation on a strike slip fault, *Journal of Geophysical Research*, 79, 4407–4412, 1974.

US Geological Survey Staff, The Loma Prieta, California, earthquake: an anticipated event, *Science*, 247, 286–293, 1990.

Utsu, T., Large earthquakes near Hokkaido and the expectancy of the occurrence of a large earthquake off Nemuro, *Reports of the Coordinating Committee for Earthquake Prediction, Japan*, 7, 7–13, 1972 (in Japanese).

Utsu, T., Seismicity of Japan from 1885 through 1925 – A new catalog of earthquakes of $M \geq 6$ felt in Japan and smaller earthquakes which caused damage in Japan, *Bulletin of the Earthquake Research Institute, The University of Tokyo*, 54, 253–308, 1979 (in Japanese).

Utsu, T., Seismicity of central Japan from 1904 through 1925, *Bulletin of the Earthquake Research Institute, The University of Tokyo*, 56, 111–137, 1981 (in Japanese).

Utsu, T., Seismicity of Japan from 1885 through 1925 (Correction and Supplement), *Bulletin of the Earthquake Research Institute, The University of Tokyo*, 57, 111–117, 1982a (in Japanese).

Utsu, T., Catalog of large earthquakes in the region of Japan from 1885 through 1980, *Bulletin of the Earthquake Research Institute, The University of Tokyo*, 57, 401–463, 1982b (in Japanese).

Utsu, T., Seismicity patterns and long-term prediction of large earthquakes – seismic cycles, gaps, quiescence, precursory activities, migration, correlation, etc., *Journal of the Seismological Society of Japan (Zisin), Second Series*, 50 (Supplement), 73–82, 1998 (in Japanese).

Utsu, T., *Seismicity Studies: A Comprehensive Review*, University of Tokyo Press, Tokyo, pp. 1–876, 1999 (in Japanese).

Utsu, T., Statistical features of seismicity. In *International Handbook of Earthquake and Engineering Seismicity, Part A* (edited by W. H. K. Lee, H. Kanamori, P. C. Jennings, and C. Kisslinger), pp. 719–732. New York: Academic Press, 2002.

Varnes, D. J., Predicting earthquakes by analyzing accelerating precursory seismic activity, *Pure and Applied Geophysics*, 130, 661–686, 1989.

Wesnousky, S. G., Seismological and structural evolution strike-slip faults, *Nature*, 335, 340–343, 1988.

Wiederhorn, S. M., E. R. Fuller, Jr., and R. Thomson, Micromechanisms of crack growth in ceramics and glasses in corrosive environments, *Metal Science*, 14, 450–458, 1980.

Williams, M. L., On the stress distribution at the base of a stationary crack, *Journal of Applied Mechanics*, 24, 109–114, 1957.

Wong, T.-F., Shear fracture energy of Westerly granite from post-failure behavior, *Journal of Geophysical Research*, 87, 990–1000, 1982a.

Wong, T.-F., Effects of temperature and pressure on failure and post-failure behavior of Westerly granite, *Mechanics of Materials*, 1, 3–17, 1982b.

Wong, T.-F., On the normal stress dependence of the shear fracture energy. In *Earthquake Source Mechanics* (edited by S. Das, J. Boatwright, and C. H. Scholz), *Geophysical Monograph 37*, pp. 1–11. Washington, DC: American Geophysical Union, 1986.

Yamanaka, Y. and M. Kikuchi, Asperity map along the subduction zone in northeastern Japan inferred from regional seismic data, *Journal of Geophysical Research*, 109, B07307, doi: 10.1029/2003JB002683, 2004.

Yamashita, T. and M. Ohnaka, Precursory surface deformation expected from a strike-slip fault model into which rheological properties of the lithosphere are incorporated. In *Earthquake Source Physics and Earthquake Precursors* (edited by T. Mikumo, K. Aki, M. Ohnaka, L. J. Ruff, and P. K. P. Spudich), *Special Issue of Tectonophysics*, 211, 179–199, 1992.

Yin, X.-C., X.-Z. Chen, Z.-P. Song, and C. Yin, A new approach to earthquake prediction: the load/unload response ratio (LURR) theory, *Pure and Applied Geophysics*, 145, 701–715, 1995.

Yin, X.-C., Y.-C. Wang, K.-Y. Peng, Y.-L. Bai, H.-T. Wang, and X.-F. Yin, Development of a new approach to earthquake prediction: load/unload response ratio (LURR) theory, *Pure and Applied Geophysics*, 157, 2365–2383, 2000.

Yin, X.-C., P. Mora, K. Peng, Y. C. Wang, and D. Weatherley, Load–unload response ratio and accelerating moment/energy release critical region scaling and earthquake prediction, *Pure and Applied Geophysics*, 159, 2511–2523, 2002.

Yokobori, T., *Strength, Fracture and Fatigue of Materials*. Tokyo: Gihodo, 1955 (in Japanese).

York, D., Least-squares fitting of a straight line, *Canadian Journal of Physics*, 44, 1079–1086, 1966.

Yoshida, S., P. Manjgaladze, D. Zilpimiani, M. Ohnaka, and M. Nakatani, Electromagnetic emissions associated with frictional sliding of rock. In *Electromagnetic Phenomena Related to Earthquake Prediction* (edited by M. Hayakawa and Y. Fujinawa), pp. 307–322. Tokyo: Terra Scientific Publishing Company, 1994.

Yoshida, S., O. C. Clint, and P. R. Sammonds, Electric potential changes prior to shear fracture in dry and saturated rocks, *Geophysical Research Letters*, 25, 1577–1580, 1998.

Zhang, W., T. Iwata, K. Irikura, H. Sekiguchi, and M. Bouchon, Heterogeneous distribution of the dynamic source parameters of the 1999 Chi-Chi, Taiwan, earthquake, *Journal of Geophysical Research*, 108, B5, 2232, doi:10.1029/2002JB001889, 2003.

Index

Printed in the United States
By Bookmasters